LabVIEW® Programming, Data Acquisition and Analysis

ISBN 0-13-030367-4

VIRTUAL INSTRUMENTATION SERIES

Mahesh L. Chugani • Abhay R. Samant • Michael Cerna
- LabVIEW Signal Processing

Rahman Jamal • Herbert Pichlik
- LabVIEW Applications

Shahid Khalid
- LabWindows/CVI Programming For Beginners

Barry Paton
- Sensors, Transducers & LabVIEW

Jeffrey Travis
- Internet Applications in LabVIEW

Lisa K. Wells • Jeffrey Travis
- LabVIEW For Everyone

LabVIEW® Programming, Data Acquisition and Analysis

Jeffrey Y. Beyon

Prentice Hall PTR
Upper Saddle River, New Jersey 07458
www.phptr.com

Library of Congress Cataloging-in-Publication Data Available

Acquisitions Editor: *Bernard Goodwin*
Production Editor: *Rose Kernan*
Cover Design: *Nina Scuderi*
Cover Design Director: *Jerry Votta*
Manufacturing Manager: *Alexis R. Heydi*

Prentice-Hall, Inc.
Upper Saddle River, NJ 07458

Prentice Hall books are widely used by corporations and government agencies for training, marketing, and resale.

The publisher offers discounts on this book when ordered in bulk quantities.
For more information, contact Corporate Sales Department, phone: 800-382-3419;
fax: 201-236-7141; e-mail: corpsales@prenhall.com
or write: Prentice Hall PTR
Corporate Sales Department
One Lake Street
Upper Saddle River, NJ 07458

Printed in the United States of America
10 9 8 7 6 5 4 3 2 1

ISBN 0-13-030367-4

Prentice-Hall International (UK) Limited, **London**
Prentice-Hall of Australia Pty. Limited, **Sydney**
Prentice-Hall Canada Inc., **Toronto**
Prentice-Hall Hispanoamericana, S.A., **Mexico**
Prentice-Hall of India Private Limited, **New Delhi**
Prentice-Hall of Japan, Inc., **Tokyo**
Pearson Education Asia Pte. Ltd.
Editora Prentice-Hall do Brasil, Ltda., **Rio de Janeiro**

Contents

6 Arrays and Clusters 87

7 Prerequisite for Data Acquisition 107

8

Data Acquisition: Analog Input 129

9

Data Acquisition: Analog Output 153

10

Data Acquisition: Digital I/O and Counters 169

11 File Input and Output 187

12 String Manipulation 211

13 Instrument Control 227

14 Data Analysis 241

Preface

LabVIEW Programming, Data Acquisition and Analysis first edition, is intended as an introductory text for students who wish to learn LabVIEW as part of their programming repertoire. It has also been written to serve practicing engineers, scientists, and researchers, and even those who have been programming with LabVIEW, who wish to learn LabVIEW programming from the ground up or update their knowledge on different techniques of G-programming.

The book was started as a reference book for the biannual G-programming workshop that I offer to the local research community as well as a supplementary textbook for an introductory computer science class that I have taught. The ultimate goal of this book is to provide the most important aspects of LabVIEW programming to the research community and students, whether it is for scientific programming or for personal use and experience, in the most concise way so that readers can quickly start programming with LabVIEW.

Philosophy

There have already been many books published on LabVIEW programming, so why bother writing another one? The philosophy behind this book was not to write a huge volume with everything in it, nor to write a narrowly-focused technical manual. LabVIEW is a programming language, often referred to as G-language, and what beginning users need is to get the most essential techniques and understanding about where and how to start. Once the programming gets going, programmers can usually improve their skills on their own since they then know where to go with their questions and problems.

When the intention of a book is to deliver the most knowledge in the most concise way, caution must be exercised not to be so abstract that the book skips details about important topics, nor to be so verbose that the readers lose their focus. So the challenge in completing this book was to screen the topics that must be covered, and to present the material in layperson's terms so that anyone could understand them easily.

Programming should be fun; so should the learning process. If the learning part becomes wearisome, the fun in programming will never be accomplished. However, nothing can be fun at all times, so some material in this book may appear to be dull or dry. In these sections, every possible effort has been made to make them concise without losing the integrity of the information. Any topic that seemed to be useful only in rare cases was excluded.

If a book is about software, it can quickly become outdated as newer versions are introduced. In order to avoid such a case, all of the topics in this book are carefully selected and presented so that readers can comfortably apply the techniques to their applications regardless of their versions of LabVIEW. Such an effort can be seen clearly in many chapters, especially those about data acquisition. The result of such an endeavor is a book with key concepts in programming with LabVIEW. On the other hand, all of the examples and exercise problems are carefully selected so that they are not too general. The outcome of such an effort is template-like examples, especially in the latter part of this book, such as Chapters 11, 12, 14, and 15. The examples in the latter chapters naturally seem to be more practical and more directly related to real applications, since the early part of this book covers the basic tools of G-programming. Most of the examples are based on actual applications that I have written for private companies, so their usefulness should be apparent.

LabVIEW is a great programming language. It allows programmers to develop an application, either with or without data acquisition and instrument control, in the shortest time period with the greatest amount of functionality. When I first encountered LabVIEW, however, that was not my first impression since I was more used to text-based programming languages. It looked different, but did not seem to be as powerful. (Icons for programming?) However, it did not take long to upset such a prejudice. It is not too difficult to find consultants or LabVIEW programmers who try to do *everything* in their applications with LabVIEW. Therefore, I strongly believe that readers will soon agree with many experienced LabVIEW programmers and myself about the unlimited capability of LabVIEW.

During the process of completing this book, I kept in mind one thought: "Remember when I was first learning LabVIEW." Without putting yourself in other people's shoes, you never understand their problems. Interestingly enough, people tend to forget about their past problems once they solve them. However, I remembered my experiences with learning LabVIEW throughout the entire period of preparation for this book. Therefore, the book may sometimes seem to be too descriptive, especially to those with previous experience in LabVIEW, but this is intentional because most readers will not have had such experiences, and what seems to be obvious to one person may not be so to others.

As for the chapters about data acquisition and instrument control, all of the examples that come with LabVIEW are categorized by their functionality. I believe that it is often a waste of time to write a data acquisition or instrument control application from scratch. Instead, the examples in LabVIEW should be either directly implemented or used as a starting point. It is important to learn how to *modify* them for your application, not how to build them from the ground up. Such a strategy applies not only to data acquisition and instrument control cases, but to any application in general. Therefore, a categorized list follows immediately after the detailed discussion about each topic.

This book contains many examples as a template for your future use. If the examples were complete, they would most likely look so complicated that readers would hesitate even to study them in the first place. To make matters even worse, a complete example can hardly be useful for your application unless your application is exactly the same as what is shown in that particular example. Such complete examples can be used to show what can be done with LabVIEW, but they often become a mere showoff. Based on such a belief, all of the examples are carefully written so that they are simple enough to be understood, but their objective remains apparent. Then they can easily be expanded by readers for their own applications. Chapter 15, in particular, illustrates many examples that are extracted from actual applications written in LabVIEW and modified so that they show a unique functionality with a simple look. This is one of the most important principles behind this book: maintaining the simplicity for easy understanding and expandability while keeping each example practical and meaningful. (You are reading this because you want to learn something that can be used for your application, not to be impressed by LabVIEW.) Also, the content in this book is independent of *different versions of LabVIEW* and *different platforms* except for the locations of some VIs and functions. (VI stands for Virtual Instrumentation, which is the basic file unit in LabVIEW.) For instance, the **Analysis** subpalette in the **Functions** palette has been divided into two new subpalettes, **Signal Processing** and **Mathematics**, with some new VIs in LabVIEW 5.1. The information about such changes can be found in your LabVIEW package, and you can concentrate on learning the material in this book.

To summarize, this book is written to help anyone who intends to learn G-programming with LabVIEW, and start programming in the least amount of time while learning the most important aspects of LabVIEW programming. The topics are carefully selected so that readers can do almost anything with those covered in this book. The examples are extracted from the actual applications that I have written for companies and can be used as a template for your application. Data acquisition and instrument control chapters include lists of categorized LabVIEW examples, so that readers can easily decide which one to use where and when, after modifying them. The content in this book is independent of the version of LabVIEW, although version 3.0 or higher is recommended. The changes in the locations of some functions and VIs of LabVIEW can be found in your LabVIEW package.

Last but not least, LabVIEW is a powerful language, and it is worth spending some time on mastering it. I hope this book helps bring all readers the full power and the exultation of G-programming.

Key Features

The major features of this book are the exercise problems at the end of each chapter except Chapters 1 and 15, and the accompanying hands-on exercise manual, which comes with a set of complete solution VIs. Also, an evaluation copy of LabVIEW is included on the accompanying CD-ROM, and readers can test or create their VIs with limited functionality.

End-of-Chapter Problems

Whenever needed, a proper example is provided in the body of each chapter with step-by-step instructions. In addition to that, more problems are included at the end of each chapter, except for Chapters 1 and 15. If you intend to teach LabVIEW in a formal classroom environment, those exercise problems can be useful for weekly assignments or as stepping stones for term projects. Each problem set covers the material in each corresponding chapter, and the solutions to all of the problems are available to instructors upon their request. Any learning process cannot be complete without practice, and those problems will uphold such a principle.

LabVIEW Evaluation Copy on CD-ROM

The accompanying CD-ROM contains an evaluation copy of LabVIEW with limited functionality. Therefore, even if you do not own the full version, you may be able to complete and run most of the examples in this book with the evaluation copy. Some of the examples may require the full version of LabVIEW to run.

Accompanying Hands-on Exercise Manual

This is an extended version of the exercise manual that is being used for the biannual G-programming workshop that I have been offering to the research community in Virginia. Each problem has step-by-step instructions on how to create it, and the solutions to all of the problems are available. This manual is divided into six sections, and each one is designed to be finished each day, completing the entire sections within six days. The problems cover a broad range of important topics in G-programming: from the simple wiring technique to a complicated programming example, including data acquisition and instrument control. The problems for data acquisition are designed in such a way that any data acquisition device can be used.

Contents of the Book

Chapter 1 begins with a brief review of the history of computers and programming languages since they are closely related. Following the review, the result of a case study is presented, where a comparison test was performed between LabVIEW and C++ on a group of college students. The purpose of such a case study was to see if a graphical language such as LabVIEW would have any benefit in the programming language and concept learning process over the text-based programming languages. Then the conventions in this textbook are summarized.

Chapter 2 introduces the basic structure of LabVIEW source files that have the extension **.vi** (Virtual Instrumentation) and continues with the objects in the **Controls** palette, and the functions and the VIs in the **Functions** palette. The importance of using shortcuts is discussed along with a list of useful ones.

Chapter 3 explains the concept behind sub VIs and how to create them. The various user option settings of sub VIs are presented and studied in detail. The effect of each of the settings and where it can be used are also discussed.

Chapter 4 explores the loops and conditional statements of LabVIEW. The unique features, such as autoindexing and shift registers, and the differences from those of text-based languages are highlighted. Also, **Global Variable**s and **Local Variable**s of LabVIEW are presented and compared to those of text-based languages. Since LabVIEW represents programming graphically in a two-dimensional diagram window, the execution order can sometimes be opaque and nebulous. Such a problem can be resolved by using the **Sequence** structure, and a detailed discussion about its usage is presented in this chapter.

Chapter 5 examines different ways of displaying data and explains different situations where each type is preferred to the others. It also describes how to prepare the data properly so that user specifications, such as display scales of each axis can be reflected correctly on the graphs. Such user specification can be controlled programmatically by Attribute Nodes, and their properties are explored in this chapter, too.

Chapter 6 introduces the two commonly used data types in LabVIEW: arrays and clusters. Each type has its advantages and disadvantages, and they are discussed in detail. Due to the similarity between the two types, LabVIEW functions for the two types are compared to distinguish them clearly. The correlation between memory usage and arrays and clusters is also studied.

Chapter 7 presents the prerequisite material for data acquisition, including sampling theorem, a brief preview of data acquisition VIs, different types of input signal modes, and different types of data acquisition, such as analog input, analog output, digital input and output, and counter operation. The sampling theorem is formally defined, and the effects of ideal sampling using the impulse train and of practical sampling using a square pulse train are mathematically driven for interested readers. This chapter also discusses what to look for when selecting a data acquisition device.

Chapter 8 provides detailed discussion of data acquisition. The first category of data acquisition, analog input (AI), is studied in detail. As mentioned earlier, all of the examples of LabVIEW pertaining to AI are categorized based on their functionality. Such an effort is made in order to help readers choose which one to use easily, since the size of the example pool is so large that it can be somewhat frustrating to find the proper one. The selection of the LabVIEW examples covers almost all of the possible data acquisition methods; therefore, learning *how* to use or modify *which* example should be your top priority, instead of building a new VI from scratch. Of course, a detailed discussion about each AI VI is presented for those who prefer to start a completely new VI on their own. However, it is worth emphasizing again that adapting prewritten LabVIEW examples to your application should be the first attempt.

Chapter 9 deals with the second category of data acquisition, analog output (AO), and comprises the complete discussion of every aspect of AO. All of the prewritten LabVIEW AO examples are studied and categorized based on their functionality. Due to the number of examples available, the sorted list will assist readers in deciding which one to select. In addition to the list of categorized AO examples, a detailed discussion of AO VIs is also presented for those who prefer to build their own AO application.

Chapter 10 covers the last two categories of data acquisition: digital input and output and counters. Different types, such as handshaking and pattern generation, of digital input and output acquisition methods are defined and studied. Buffered and non-buffered digital acquisition techniques are also illustrated using LabVIEW examples. A variety of tasks that can be performed using counters on a data acquisition board are described individually in detail. Some examples include frequency measurements of high and low frequency Transistor Transistor Logic (TTL) signals, event counting where counting the number of high states is the goal of the operation, and continuous or single burst TTL signal generation. All of the prewritten LabVIEW examples are again categorized in addition to the detailed discussion about the VIs to help readers create their own application.

Chapter 11 discusses one of the most important tasks in data acquisition: saving or reading data to or from a file. Without knowing how to save the data, all of the painstaking hours of the data acquisition process can turn out to be nothing but a waste of time. This chapter illustrates all of the possible ways of saving and reading data with template-like examples, and they can be implemented directly without any modification. Since all of the templates are using basic LabVIEW functions and VIs, readers can easily add more functionality to them at their will. Such intention also coincides with one of the most important principles of this book: being simple, but practical and meaningful.

Chapter 12 can be considered as a continuation of Chapter 11 since file input and output are closely related to string manipulation. This chapter also plays an important role in instrument control, which is the topic of the following chapter. In order to write the data in a meaningful way, the data need to be prepared properly, either before being saved or after being read. This chapter covers all of the techniques necessary to complete such preparation stages.

Chapter 13 introduces instrument control with LabVIEW. The functions and the VIs for General Purpose Interface Bus (GPIB), serial communication, and Virtual Instrument Software Architecture (VISA) are closely studied. Some simple troubleshooting procedures for serial communication are presented in this chapter.

Chapter 14 covers a variety of data analysis techniques. In the early part of the chapter, some important data analysis theories are reviewed for interested readers. The categories are linear and nonlinear system analysis, stochastic and deterministic data analysis, and time and frequency data analysis. Also, some useful digital signal processing techniques, such as decimation and interpolation, are reviewed. Then matrix and vector representation of data is discussed to prepare readers for linear algebra VIs, followed by the study of analysis VIs in LabVIEW. Some of the topics may seem too advanced for some readers, but they can safely be skipped without any deficiency in learning G-programming techniques, and readers can proceed to the analysis VI section.

Chapter 15, the last chapter of the book, covers many issues in creating application and stand-alone executables and concludes with an overall review and many additional important issues of G-programming, which are not covered in the previous chapters. It also presents several practical examples that are actually being used in many real-world applications. Readers will find those examples unique and directly applicable to their applications with a great amount of flexibility. Examples include dynamic sub VI load and unload, creating a professional startup screen, toggle switch design, and many more. Chapter 15 also presents an in-depth discussion about maximizing the efficiency of applications. How to cope with the overhead incurred by the large size of VIs is also one of the unique topics in this chapter. Some possible workarounds are suggested along with a test result. The last topic pertains to Application Builder, which allows LabVIEW users to create a stand-alone executable from a VI. Detailed discussion of its feature is included to assist readers in using it properly.

This book includes two appendices, which contain the material that could not be included in the main body of a chapter. Appendix A contains step-by-step instructions on how to create a Code Interface Node using Code Warrior Professional Release on Power Macintosh platforms, and sample C++ codes for 1-D array and 2-D array cases are attached at the end. Appendix B consists of the front panels and the diagram windows of each sub VI that are used for the error handler design example presented in Chapter 15. Those sub VIs can be used as a template to create your own error handler VIs.

Flexibility in Coverage

Chapters 7 through 10 cover data acquisition, and Chapter 13 discusses instrument control. Therefore, if you are reading this book only for general programming with LabVIEW, you can skip those chapters. In addition, you can skip Chapter 14 if data analysis is not the purpose of your application. However, you should cover Chapter 15 regardless of your application type, since it covers many important topics that every LabVIEW programmer should know.

If you are using this book to teach yourself LabVIEW, you can first review each chapter while duplicating examples introduced in each chapter. Then you can use the exercise problems in the accompanying hands-on exercise manual to complete the learning process. The two books can be covered in parallel, too, since the order of topics in them is relatively independent.

If you intend to teach LabVIEW in a semester-long class, either as supplementary material or as a main topic, you can teach each chapter and assign the end-of-chapter problems as weekly homework. The solutions to these problems are available to instructors upon their request by email to jbeyon@pcs.cnu.edu. All of the topics except for those in Chapter 14 can be covered regardless of the level of students. Some of the material in Chapter 14 may not be suitable for those without any system theory or stochastic data analysis background; therefore, the coverage of Chapter 14 can be optional. The examples in each chapter, especially in Chapter 15, can be used as a stepping stone for a project assignment. For example, students may be assigned to write a professional database application as a term project, utilizing all of the topics and techniques covered in Chapters 2 through 15, excluding Chapters 7–10, and 13–14.

The hands-on exercise manual can also be used in a week-long short course where both data acquisition and data analysis techniques are covered. In such a case, this book can be used as a reference, and each section in the hands-on exercise manual can be covered on a daily basis.

Acknowledgments

A large number of people contributed to the completion of this book. I am grateful for the support of Professors George Webb, Randy Caton, and H. Lee Beach, Jr. of Christopher Newport University in making the biannual G-programming workshop successful, where the idea for this book originated. I also thank George D. Allison of NASA at Langley AFB, Virginia, who also contributed to the success of the workshop.

I would like to thank the many students from my previous classes, who tried the early stage of this book and gave me valuable feedback. My special thanks also go to Jesse Madden, who spent her valuable time proofreading the manuscripts.

I would also like to thank many individuals at National Instruments, including, but not limited to, Dr. Ravi Marawar, for their ongoing support for this project, and Abel Wakrim, for many of his general comments on the organization of the contents in this book.

My special gratitude goes to Bernard M. Goodwin, Rose Kernan, and all of the editors at Prentice Hall for their assistance in moving this project from an idea to a complete text.

Lastly, my greatest thanks are owed to my family, Mina, and Jeffrey, Jr. Without their support, encouragement, and perseverance, this book could never have been completed.

1 Graphical Language

1.1 History of Computers and Languages
1.2 Graphical Programming Language LabVIEW
1.3 Results of Case Study
1.4 Conventions

1.1 History of Computers and Languages

The history of programming languages accompanies that of calculators, which are the early form of computers. Some examples of the first kind are Pascal's mechanical calculator, designed by French mathematician B. Pascal in 1642, and the Adding Machine by J. Napier, who was a mathematician in Scotland. During 1820–1871, the Difference Engine followed by the Analytic Engine were developed by C. Babbage, and they already had the fundamental structure of the computers of these days, such as memory and the arithmetic logic unit (ALU). They used punch cards for their input and executed a sequence of programming commands. Then calculators evolved into machines closer to "computers." For example, J. V. Atanasoff and C. E. Berry at Iowa State College (now Iowa State University) designed a calculator using about 300 vacuum tubes during 1938–1942, and H. Aiken and IBM designed MARK I using 3000 relays. Then came MARK II, III, IV, and Electronic Numerical Integrator And Computer (ENIAC), Electronic Discrete Variable Automatic Computer (EDVAC), Electronic Delay Storage Automatic Calculator (EDSAC), ORDinance Variable Automatic Computer (ORDVAC), and ILLinois Institute for Advanced Study (ILLIAC).

As the calculators developed into computers, computer languages (or compilers) were also created and improved for (specific) computational purposes. FORmular TRANslation (FORTRAN) was developed at IBM in 1957, becoming FORTRAN II in 1958, then FORTRAN IV in 1962, and it has continued to be revised. FORTRAN is still a popular programming language for scientific computations. ALGOrithmic Language (ALGOL) was also for scientific computations and was developed during 1958–1960. COmmon Business Oriented Language (COBOL) was designed in 1960 for business programming purposes. Programming Language One (PL/1) was announced at IBM in 1963, combining FORTRAN, ALGOL, and COBOL. Beginner's All-purpose Symbolic Instruction Code (BASIC) was announced in 1965 and became very popular among all levels of programmers. Pascal was developed by N. Wirth in Zürich, Switzerland in 1971 and is also one of the popular choices among various programming languages. Some other examples are DYNAmic MOdels (DYNAMO); A Programming Language (APL), General Purpose Simulation System (GPSS), and Report Program Generator (RPG) at International Business Machines Corp. (IBM); Automatically Programmed Tools (APT) and LISt Processing (LISP) at Massachusetts Institute of Technology (MIT); and StriNg Oriented symBOlic Language (SNOBOL) at Bell Laboratory. Lastly, C was developed by D. M. Ritchie of AT&T Bell

Laboratories in 1972, and C++ by B. Stroustrup of AT&T Bell Laboratories in the early 1980s. The latest paradigms in text-based programming language are object-oriented programming languages, including C++, and some popular compilers are Visual Basic and Visual C++ by Microsoft.

As programming languages have evolved, the power of the languages has also grown, and the ease of writing and compiling the codes has improved significantly as well. A revolutionary concept in programming was introduced by National Instruments in 1986 with LabVIEW 1. It was a big program with very limited programming functionality at first and was written especially for data acquisition purposes. Ever since the advent of LabVIEW 1, its functionality and programming power have grown, and LabVIEW has now become one of the most popular data acquisition programs among engineers, researchers, and scientists. Unlike the text-based programming languages, where you have to learn the syntax and master all of the details, the concept of programming in LabVIEW is more visually intuitive and user friendly due to its graphical representation instead of lines of codes. Such a new concept has created a new expression, *G-language*, which many programmers accept and understand.

1.2 Graphical Programming Language LabVIEW

When you write a set of codes and link and compile them in a conventional compiler program, you really do not know if there are any errors until the compilation process. LabVIEW, however, indicates the status of error in programming as you develop the code. Conventional programming languages use a regular text file as their source codes, whereas LabVIEW uses a file called VI (Virtual Instrumentation) as its source code.

Conventional languages usually have three levels of codes: (1) high level: the source codes that programmers create by typing in the line commands; (2) middle level: the object codes containing assembly codes; and (3) low-level machine codes. On the other hand, LabVIEW has only two levels: high and low. The high-level codes are the VIs that the programmers create, and the low-level codes are what LabVIEW compiles the VIs into, and they are not visible to programmers at all. Since LabVIEW represents programming in terms of wires and icons, you might think that LabVIEW could have more overhead than text-based programming languages. This is somewhat true, but LabVIEW does a pretty good job of optimizing the VIs when it compiles them, so the result of the overhead is negligible for applications of a reasonable size. When the size of the VIs becomes large, special attention must be paid while developing the VIs. The following chapters (especially Chapter 15) discuss this thoroughly.

The ultimate goal of G-programming with LabVIEW is usually understood as data acquisition and instrument control, which is not quite true due to the variety and power of LabVIEW. LabVIEW can be used not only for data acquisition or instrument control applications but for general-purpose applications, such as database, data analysis program, network communication program, or even a simple game program. LabVIEW also has a variety of add-on toolkits for any user-specific application, such as image processing, Digital Signal Processing (DSP) data analysis, Internet application, and many others. As a result, the size as well as the applicability of LabVIEW has grown far beyond what it was in 1986, and LabVIEW surely has its place as a new programming language G in the realm of data acquisition and analysis, as well as other everyday applications.

1.3 Results of Case Study

One of the unique features of LabVIEW is the fast learning curve. A group of about 40 college students in an introductory computer science class were monitored over three semesters while being taught programming concepts using both LabVIEW and C++. At the end of each semester, a survey, as well as a programming contest using LabVIEW, was conducted. The results were interesting: (1) The majority of the students favored LabVIEW and claimed that they could understand the programming concepts better with LabVIEW than with C++, and (2) students with no previous programming experience were already showing techniques that could not have been learned in the same amount of time with C++ in the VIs that they turned in for the contest. Of course, there were students who did not favor the G-programming concept, but this percentile was very low (less than 5%). Based on such a survey, it seems true that the concept of G-programming does raise the slope of the learning curve compared to that of the conventional text-based programming languages, and engineers and other programmers should see the same results as the university students studied.

Here is a list of some of the results of the case studies and unique features of G-programming with LabVIEW:

1. The learning time seems to be far less than for the text-based programming languages.
2. The programming concept can be taught and understood efficiently due to the graphical representation, at different levels of students.
3. The development time is much less than it is with text-based programming languages.
4. The programs for data acquisition and instrument controls can be developed easily in a minimal amount of time.
5. LabVIEW has superior graphical user interface (GUI) tools that are ready to be used.
6. A broad range of applications can be written with LabVIEW, from data acquisition and instrument control to everyday applications. For example, with some of its toy products LEGO® has had a set of VIs available for children to control the robotic movement of the parts.

1.4 Conventions

The following summarizes the conventions used in this book:

Convention	Definition
bold	directories; folders; file names; object labels; file extensions; functions and item names of LabVIEW; menu selections; option selections; vectors; matrices; random variables; random processes e.g., Save **test.vi** in the folder **My Project**. e.g., Wire a numeric control **input** to a numeric indicator **output**. e.g., LabVIEW has its own directory with extension **.llb**. e.g., The function **Build Array** can build an array. e.g., The **Case** structure is in the subpalette **Structures**.

	e.g., The while loop in C++ is different from the **While Loop** in LabVIEW.
	e.g., Choose **Save** from the pull-down menu **File**.
	e.g., Choose **Latch When Released** as the mechanical action.
	e.g., The vector **x** is the first column of the matrix **X**.
	e.g., **X** is a Random Variable (RV) if you can write $\mathbf{X}(\omega)$, where $\omega \in \Omega$, and a random process if you can write $\mathbf{X}(\omega, t)$.
italic	emphasized text; functions, arguments, and constants in mathematical expressions; reference title
	e.g., You *must* include the folder **Shared Libraries** in the same folder where the executable is saved.
	e.g., One of the unique features of LabVIEW is *autoindexing* and *shift registers* with loops.
	e.g., $y(t) = a \sin(t) + b$ is a function of the argument t, where a and b are constants.
	e.g., R. A. DeCarlo, *Linear Systems*, Prentice Hall, 1989.
`Courier`	The text you enter from the keyboard.
	e.g., Enter `Action List` in **Item Name**.
	Note that if the text you entered becomes an item, it will be in **boldface**. For example, if you enter a text `Option 1` in a **Text Ring**, then it will become an item **Option 1**.
Geneva	Text-based programming (pseudo) code such as C/C++.
	e.g., The expression int x[10]; declares an array of ten integer elements in C or C++.
>>	Path of files and LabVIEW functions on all platforms.
	e.g., Find **Functions** >> **Numeric** >> **Add**.
	e.g., Find the example, LabVIEW directory >> **examples** >> **general** >> **arrays.llb** >> **Building Arrays.vi**.

Ctrl-H

Key combination of Control key and H. The Control key can be replaced by the Command key for Macintosh platforms.

e.g., Press **Ctrl-H**, which is equivalent to **Command-H** on Macintosh platforms, for the online help.

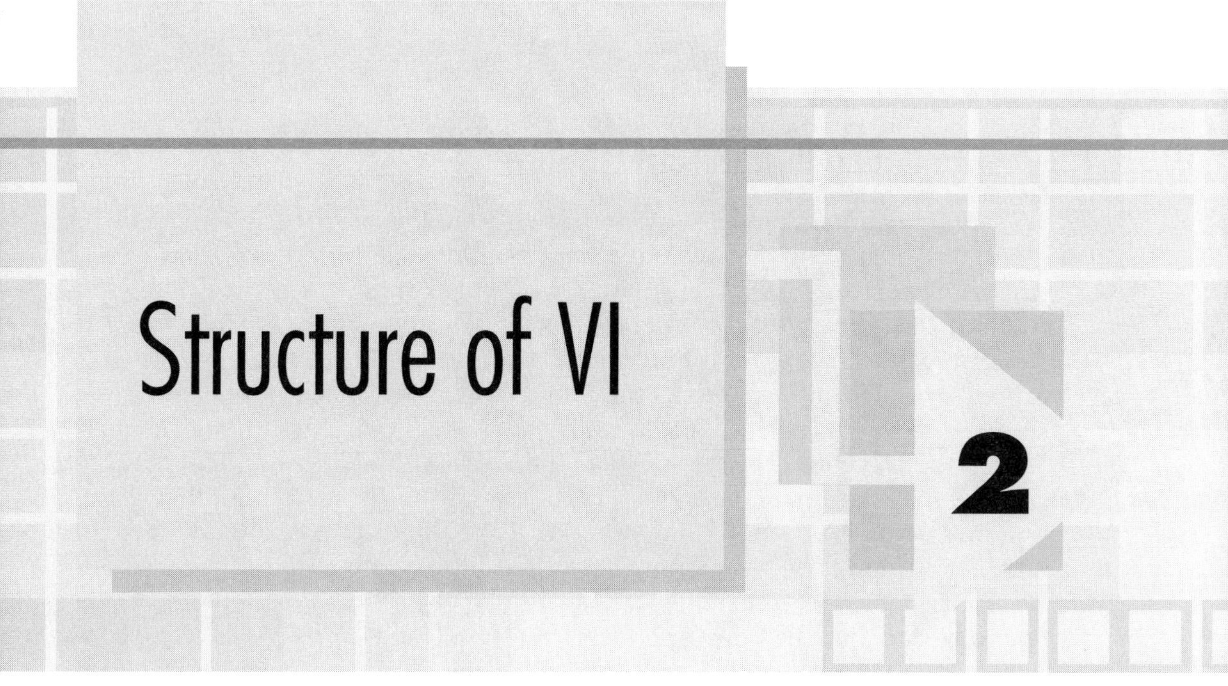

Structure of VI

2

2.1 Front Panel and Diagram Window

When you launch LabVIEW by double clicking the icon of LabVIEW, it will come up with two windows: one with a gray background, which is called the front panel, and the other with a white background, which is called the diagram window. Figures 2.1 and 2.2 show them, respectively. LabVIEW is a Graphical (G)-programming language for virtual instrumentation (VI), and these two windows reflect the characteristics of G-language.

The front panel resembles the front panel of an instrument: It has gauges, switches, knobs, and buttons. All of the wires and each individual component of the instrument are hidden in the chassis of the instrument that corresponds to the diagram window in LabVIEW. The files in LabVIEW have **.vi** extensions under PC environment, but not necessarily under Macintosh environment. (There is no extension for files on Macintosh computers.) Since the diagram window corresponds to the wires and connections in real instruments, you should expect to perform some wiring tasks in the diagram window. Wiring technique is very important and is addressed separately in a later section.

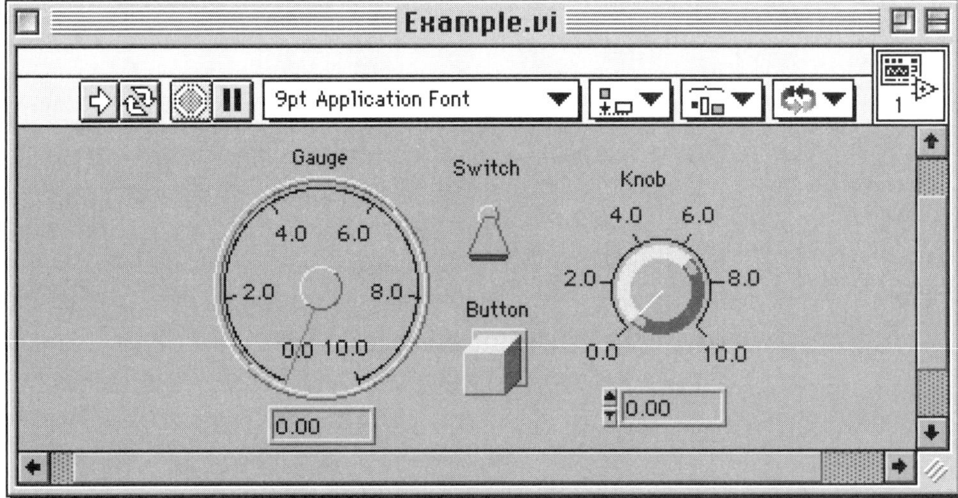

Figure 2.1
Front panel of a VI. (All of the screen shots of VIs in this book are used with the permission of National Instruments.)

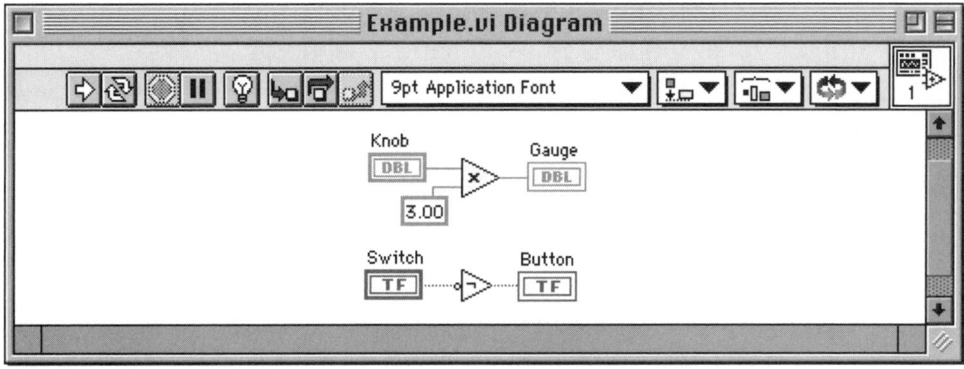

Figure 2.2
Diagram window of a VI.

First, examine the buttons in those two windows.

Run When this button is pressed, LabVIEW will execute the VI that is currently selected.

Run Continuously This button will run the VI continuously until the button **Abort Execution** is pressed. It is strongly recommended to avoid the use of this button until you become familiar with LabVIEW.

Abort Execution This stops the execution of the VI.

Pause This will pause the execution. To resume it, press the button again.

Highlight Execution The bulb will be lit if you click this button. If it is on, LabVIEW will display the signal flow while running the VI. A ball running along the wire while showing the current value at each node of the wire indicates the signal flow. This is very useful when debugging a VI since you can view the value as well as the execution flow; however, the speed of execution will be much slower than normal execution.

Step Into This is a different type of **Run** button that provides debugging capability. If you click this button instead of the **Run** button, LabVIEW will step through each node of wires and each element. The difference from the **Highlight Execution** button is that **Step**

Into will run one node and wait until you hit the button again, whereas **Highlight Execution** runs (slowly) without stopping.

Step Over While running a VI via **Step Into**, this button allows you to skip any node or element.

Step Out This exits the **Step Into** run mode.

9pt Application Font

Text Settings This allows you to change the feature of fonts, such as color, size, etc., with this menu ring.

Align Objects With this, you can vertically or horizontally align objects in the windows. First, select the objects you want to align, then select one of the modes from the pull-down menu.

Distribute Objects With this, you can control the space between objects. This is done by first selecting the objects and then choosing a mode from the pull-down menu.

Reorder With this feature, you can bring objects to the front, or to the back, or move them to the top or under the other objects. This is useful when multiple objects are overlapped and you are trying to change their visibility on the screen.

2.2 Objects in VI: Controls and Indicators

In the previous section, the word *objects* was used in relation to VIs, and it will be defined in this section. As discussed earlier, LabVIEW launches with the front panel and the diagram window. Consider the spectrum analyzer as an example to explain objects in the front panel and the diagram window. The analyzer may have switches to select different time bases or different sampling rates on the panel with a display of the spectrum. If you open the case, you will see wires, fast Fourier transform (FFT) chips, proper display units, and so on. In LabVIEW, these are referred to as *objects*, and they are the fundamental elements that constitute a VI.

In LabVIEW, there are two types of objects: controls and indicators. Controls are objects such as knobs, levers, or switches that allow you to enter or set values. Indicators are for display purposes only and include graphs, gauges, or meters. A simple VI that simulates a voltage source and a voltmeter is shown in Figures 2.3 and 2.4. There are two objects in both the front panel and the diagram window: **Input Voltage** and **Voltmeter**. The front panel in Figure 2.3 shows that the **Input Voltage** is a slide input switch, and

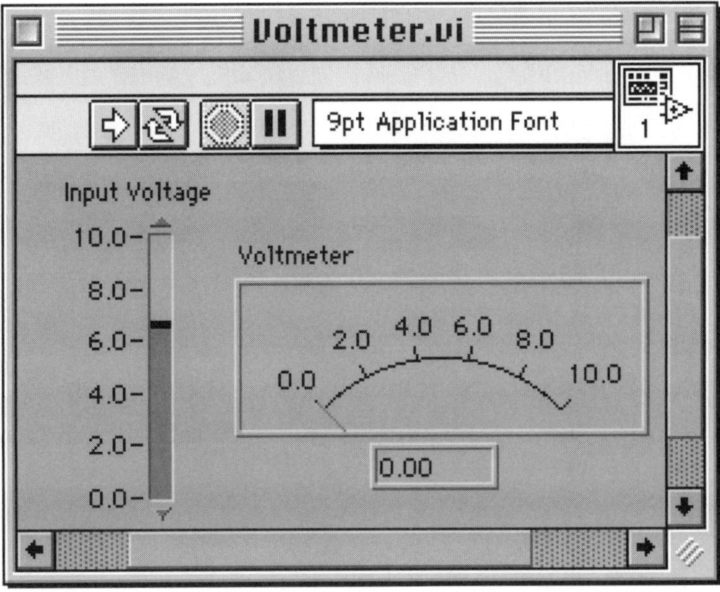

Figure 2.3
Voltmeter.vi (front panel).

the **Voltmeter** is a gauge that displays the input voltage. Figure 2.4 shows the wire connection of these two objects. In this case, the **Input Voltage** is the control and the **Voltmeter** is the indicator since you can *set* values to the **Input Voltage** and *display* them with the **Voltmeter**. However, just looking at these two objects in the front panel will not tell you which is the control

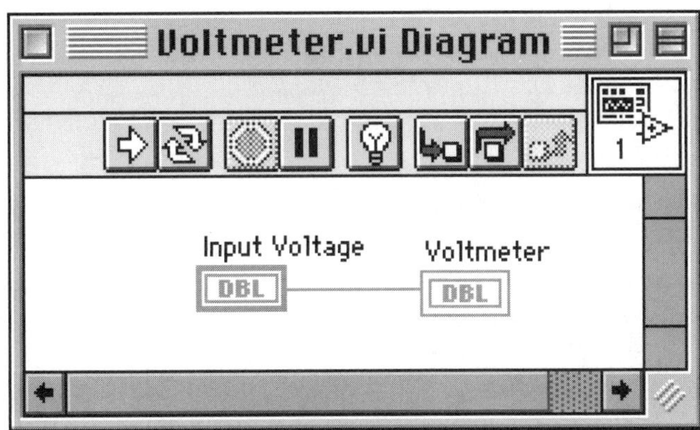

Figure 2.4 **Voltmeter.vi** (diagram window).

and which is the indicator. When trying to distinguish controls from indicators, remember the following statement:

If you are in doubt about any object, right click on it to bring up the pop-up menu. On Macintosh platforms, right clicking the mouse corresponds to the ***Command-click****.*

If you right click on either the **Input Voltage** or the **Voltmeter** in the front panel, the first option in the pop-up menu will be either **Change to Indicator** or **Change to Control**. If the menu shows **Change to Indicator**, it means that the object is a control; otherwise, it is an indicator. However, in the diagram window in Figure 2.4, you will see an easier way of telling them apart. The **Input Voltage** and the **Voltmeter** have DBL in the box in the diagram window and they are connected through a wire. The letter DBL implies that the data type of those two objects is double precision, which you can change to any other types, such as extended precision, single precision, or integers. (Can you guess how to change it?) The thickness of the box tells you the type of the object: Thicker lines indicate controls and thinner lines, indicators. You may ask the following question:

Why is it important to know the type of objects?

Recalling that VIs represent virtual instruments, what happens if you have two displays connected together? Assuming that displays can only receive data and *display* them on the screen, connecting two displays would be meaningless. The same logic applies to a situation where you have two knobs with which you set values. (Again, you are assuming that switches cannot receive data even though some real switches can receive data and display them.) If you try to wire two (or more) controls together, LabVIEW will return an error and will not execute the VI. This also applies to indicators. Surprisingly enough, one of the major errors that beginning LabVIEW programmers make results from this type of conflict!

One major error made by beginning LabVIEW programmers results from the following wiring conflict: Controls or indicators are wired to their own kind.

In debugging with text-based languages such as C or C++, simple mistakes such as spelling errors are often discovered after days of painful effort. Spelling errors are simple, but their simplicity causes programmers to be careless and therefore make such mistakes. Should you have an error in LabVIEW, the **Run** button will show up as a Broken Run Arrow. If you click on the Broken Run Arrow, it will display a list of errors. To locate the origin, double clicking on each error will bring you to the location. Make sure that you are not wiring controls to controls or indicators to indicators.

List Errors (Broken Run Arrow) This indicates that an error or errors exist in the VI. To find out what errors have been made, click the button and it will bring up a list of errors. If you double click on a specific error listing, LabVIEW will take you to the location where the error occurred.

2.2.1 Building an Example

In this section, you will build a simple example step by step. If you have not yet done so, invoke LabVIEW by double clicking its icon. It will come up with the startup window as shown in Figure 2.5. Note that LabVIEW tutorial is only available on PC platforms; therefore, it will be grayed out on Macintosh platforms, as shown in Figure 2.5. Click on **New VI**, and this will bring up two similar windows to those shown in Figures 2.1 and 2.2. Go to the front panel by selecting the window with the gray background. Right click on any place in the gray window to bring up the **Controls** palette. You will see many icons (subpalettes) that will be studied later. When the **Controls** palette comes up, fix it to the desktop by pointing the mouse arrow to the thumb pin located in the upper left corner of the palette. Once you fix it, point the mouse arrow to the first subpalette **Numeric**. (The label will show up as you point the mouse to the subpalettes.) Click the subpalette to open it, and fix it using the thumb pin in the manner just described.

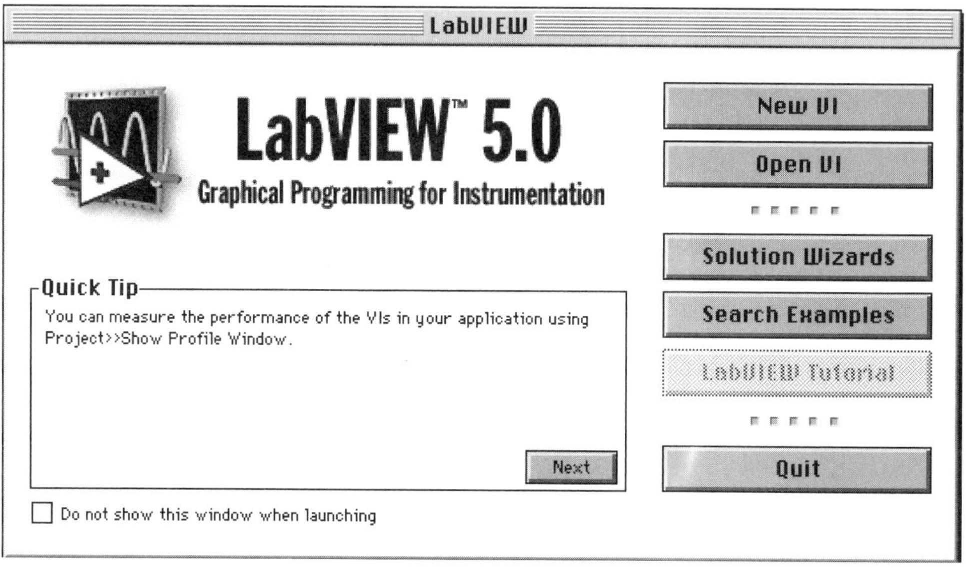

Figure 2.5
Startup window of LabVIEW 5.0 (Macintosh platforms). This book uses the names and VI paths of version 5.0. However, all of the topics and examples in this text are carefully chosen to be *independent* of any specific version, either lower or higher than 5.0. The locations of the VIs or functions in your LabVIEW may be different from those in this book. If so, refer to the information about the differences among different versions, which is provided in your LabVIEW package.

In the **Numeric** subpalette, you will see many different types of slides, gauges, and knobs. Place a **Vertical Fill Slide** in the front panel by clicking it once and dragging it to the front panel. When you put it in the front panel, it will be ready to take the label from the user. (The label is already selected.) Type `Input Voltage` and hit return. To hide the digital display, right click on the **Vertical Fill Slide** and uncheck **Show** >> **Digital Display**. If you clicked outside the label without typing in `Input Voltage`, you will realize that the label has disappeared. You can show it by right clicking on it and selecting **Show** >> **Label**. Once it shows up, you can then enter the label. Similarly, place a **Meter** from the **Numeric** subpalette next to the **Vertical Fill Slide** and label it **Voltmeter**.

Now, go to the diagram window by pressing **Ctrl-E**. This shortcut is very useful when navigating between the front panel and the diagram window.

However, you can do the same thing by selecting either **Show Diagram** or **Show Panel** under the **Window** pull-down menu. The importance of this shortcut is further emphasized in a later section. Until then, try to get used to the shortcuts that will be introduced in this section without questions.

Use shortcuts! ***Ctrl-E*** *toggles between the front panel and the diagram window.*

When you get to the diagram window, you will see two rectangular boxes (objects) with DBL in them. DBL indicates that the items (in this case, the **Input Voltage** and the **Voltmeter**) are double precision data type. If the labels are not shown, right click on the objects and select **Show** >> **Label** to make them appear in the diagram window or in the front panel. If you followed the instructions correctly until now, your mouse pointer on the screen should be the Hand Tool. Regardless of the tool that is currently selected, go to the pull-down menu **Windows** and select **Show Tools Palette**. Now select the Wire Tool from the **Tools** palette, and connect the **Input Voltage** to the **Voltmeter** using the Wire Tool.

 **Operate Value** (Hand Tool)  **Connect Wire** (Wire Tool)

Note that wiring technique is another factor that causes most of the errors that the beginning LabVIEW programmers make; therefore, the correct wiring method will be discussed in the following section separately.

2.3 Wiring Technique

Beginning LabVIEW users demonstrate a tendency to try wiring items together by holding the mouse button after clicking the first item and releasing it on the second one. If you follow this method, LabVIEW is just waiting to cause you problems. Fortunately, the correct wiring technique is simpler than this instinctive process. The following subsection discusses the correct way of wiring objects in the diagram window.

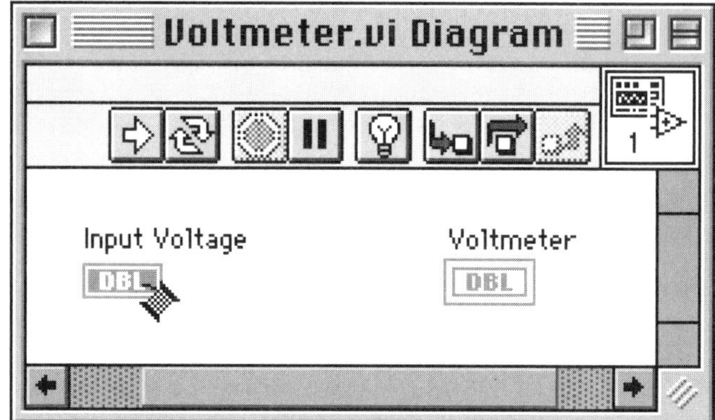

Figure 2.6 Wiring Step 1.

2.3.1 Wiring Steps

There are three steps in wiring: Click the origin, move to the target, and click the target. Note that the first step says *click* on the origin, not *click and hold* the mouse button on the origin. Now, if you have not yet done so, select the Wire Tool from the **Tools** palette and refer to Figure 2.6. (The **Tools** palette can be displayed by selecting **Windows >> Show Tools Palette**.) This illustrates the first step where you click the origin. When you place the wiring tool on the origin, **Input Voltage**, which is a control, the icon will start blinking to indicate that it has been selected. Once it starts blinking, click once and release the mouse button.

After completing the first step, drag the mouse. You will now see a dotted line attached to the Wire Tool. If your target is located in a place where you

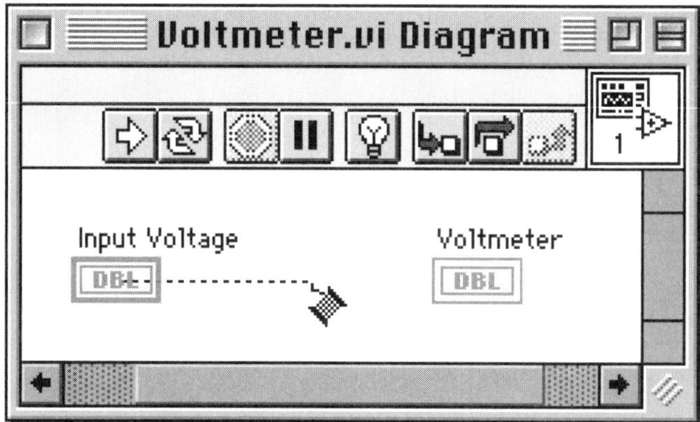

Figure 2.7 Wiring Step 2.

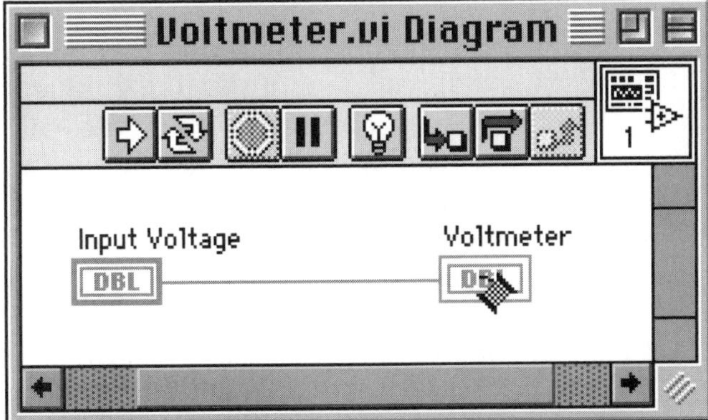

Figure 2.8 Wiring Step 3.

cannot reach it with a straight wire, you can set the break point by clicking once at a point where you want to make a turn. For now, drag the Wire Tool onto the target **Voltmeter**. This second step is shown in Figure 2.7.

The final step of completing the wiring is shown in Figure 2.8. When you place the Wire Tool on the target **Voltmeter**, it will start to blink, indicating that the wiring tool is right on the target. Once you confirm the blinking, click the mouse button once and release it; this will complete the wiring.

Good wiring habits will help LabVIEW programmers write well-organized G-programs and avoid a great deal of trouble. If nothing seems to be wrong with your VI but it still gives you the Broken Run Arrow, check if you wired controls to controls or indicators to indicators. Next, check for bad wires and make sure they are wired to the correct targets. These two troubleshooting methods should eliminate most of the errors typically made by beginning LabVIEW programmers.

If nothing seems to be wrong with your VI but it returns the Broken Run Arrow, first check for an object type conflict and then check for bad wires.

Once you finish wiring, go to the front panel and switch the Wire Tool to the Hand Tool from the **Tools** palette. Now assign a value to the **Input Voltage** and click the **Run** button to see if the value is displayed on the gauge.

2.3.2 Shortcuts

In this subsection, you will be introduced to more useful shortcuts in LabVIEW and understand why they are important. Some shortcuts will seem inevitable to be used and, therefore, you are strongly recommended to use them. First, we provide an explanation of **Ctrl-E** and its significance.

Assume that you have multiple VIs with their front panels and diagram windows both on screen. As a programmer, there may be situations where you assign similar names to VIs, such as **Test1.vi**, **Test2.vi**, and so on. If you simply jump between the front panels and the diagram windows by mouse, you may end up modifying the wrong diagram window by mistake. Of course, you can use the **Windows** pull-down menu to jump back and forth, but this will slow down programming and even become annoying as you develop into an experienced user. However, if you use **Ctrl-E**, you will quickly jump to the correct diagram window or front panel and thereby enhance your programming efficiency. The following is a list of useful shortcuts in LabVIEW:

Ctrl-E: Toggles between the front panel and the diagram window. (**Command-E** on Macintosh platforms.)

Ctrl-B: This will remove all of the bad wires appearing as dotted lines. (**Command-B** on Macintosh platforms.)

Ctrl-H: While pointing at an object, use this shortcut to bring up the online help for the object. (**Command-H** on Macintosh platforms.)

Shift-right click: This will pop up the **Tools** palette. (**Command-Shift**-click on Macintosh platforms.)

Tab key: This will shift between Hand Tool, Arrow Tool, Text Editor, and Wire Tool. See Section 2.4 for more details about those tools.

Space bar: This will toggle between the Arrow Tool and the Wire Tool. See Section 2.4 for more details about those tools.

Ctrl-Z: Undo shortcut. You can change the maximum undo steps per VI under **Edit** >> **Preferences** >> **Block Diagram**. LabVIEW allows a maximum of 99 steps of undo. This is a new feature of LabVIEW 5.0 or higher. (**Command-Z** on Macintosh platforms.)

Ctrl-drag: This creates a duplicate of objects. Using the Arrow Tool (**Position/Size/Select** in the **Tools** palette), select an object or multiple objects in the front panel or in the diagram window and

Figure 2.9 **Position/Size/Select** (Arrow Tool). This can be found in the **Tools** palette. It is used to select, position, or (re)size objects in the front panel or in the diagram window.

drag the selected object(s) while holding the **Ctrl** key. This will create a duplicate. (**Option**-drag on Macintosh platforms.)

To find out which objects in the diagram window correspond to those in the front panel, right click on the object and select **Find Terminal**. If you want to find the corresponding object in the front panel, double click on the object in the diagram window, and LabVIEW will take you to its counterpart in the front panel.

2.3.3 Creating and Deleting Objects

To create or place an object either in the front panel or in the diagram window from either the **Controls** or the **Functions** panel, first right click in the place where you want to place the object (the front panel or the diagram window). This will pop up either the **Controls** or the **Functions** panel. Then select an object and drop it in the proper window.

To delete an object, you must go to the place where the object was originally created. Once you are in the correct place, select an object or multiple objects using the Arrow Tool and hit the **Delete** key. This will remove the selected object(s). For example, if you created a control or an indicator in the *front* panel, you must go to the *front* panel to delete it; it cannot be deleted from the diagram window. Similarly, if you placed a function, such as **Add** from **Functions** >> **Numeric**, you must go to the *diagram* window to delete it. To select more than one object using the Arrow Tool, use the key combination **Shift**-click. The Arrow Tool can be found in the **Tools** palette, which is discussed in the next section and shown in Figure 2.9. The function **Add** is shown in Figure 2.10.

Figure 2.10 **Add** from **Functions** >> **Numeric**. To delete objects, you must go to the place (either the front panel or the diagram window) where they were originally created.

2.4 Tools, Controls, and Functions Palettes

You have covered the basic skills that will get you started writing simple VIs. However, it could be overwhelming to witness the wide variety of items in the palettes, and you may wonder what to use for your need. Therefore, each item in the **Tools** palette, **Controls** palette, and **Functions** palette will be examined in order to increase your confidence with using the many features of LabVIEW. The purpose of this preview is not intended to cover every detail about each subpalette but to provide a global view of the functionality that LabVIEW provides. Start with the **Tools** palette.

2.4.1 Tools Palette

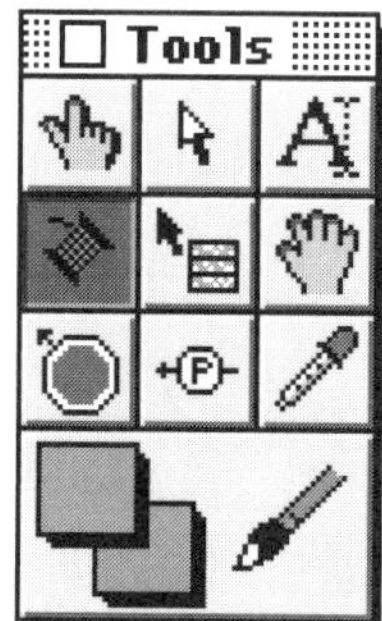

Operate Value (Hand Tool) With this, you can set values in the controls by clicking up or down arrows or by directly entering the value in the display.

Position/Size/Select (Arrow Tool) With this tool, you can position, (re)size, or select objects both in the front panel and in the diagram window. After selecting objects, **Ctrl**-drag will create duplicates of them. (**Option**-drag on Macintosh platforms.)

Edit Text (Text Editor) This tool allows you to edit text contents. Use this to enter or change the labels of objects, or enter comments in the front panel or in the diagram window.

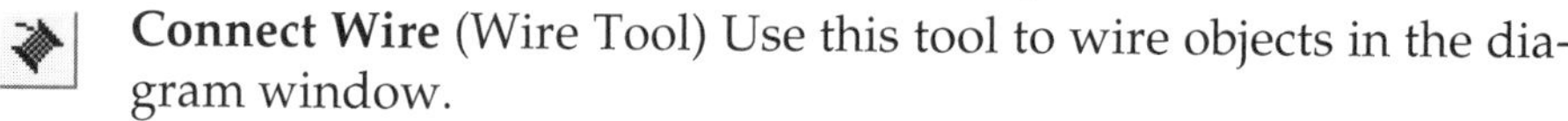

Connect Wire (Wire Tool) Use this tool to wire objects in the diagram window.

Object Pop-Up Clicking on objects while this tool is selected will show their pop-up menus without right mouse click.

Scroll Window This will grab the screen image and move it.

Set/Clear Breakpoint This will allow you to set or clear breakpoints while debugging VIs.

Probe Data With this option selected, you can create a probe on a wire by clicking on it. The probe allows you to view the value passed through the wire when the VI runs.

Get Color This samples the color from where the mouse pointer contacts.

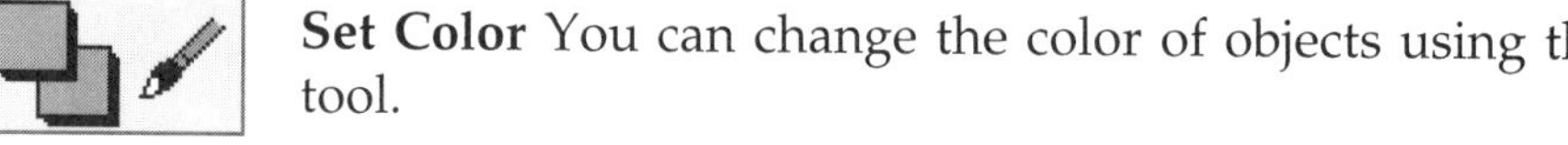

Set Color You can change the color of objects using this tool.

2.4.2 Controls Palette

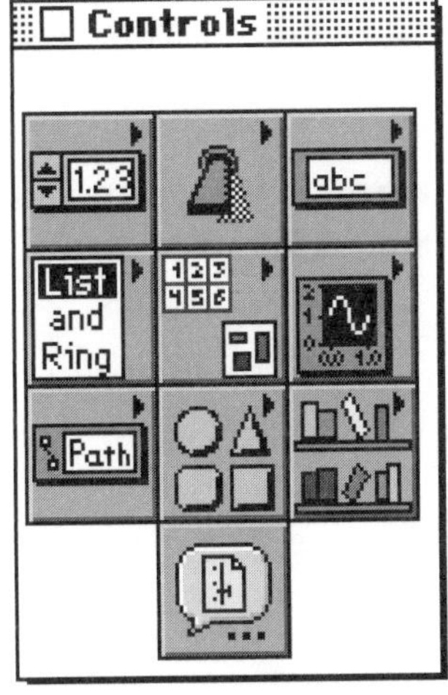

Numeric This contains all of the controls and indicators that can set or display numeric values.

Boolean This contains all of the Boolean controls and indicators, such as on/off switches, light emitting diodes (LEDs), etc.

String & Table This contains all of the string and table controls and indicators. String controls and indicators are important objects, especially in serial, GPIB, or VISA applications. See Chapters 12 and 13 for more details.

List & Ring Ring is a collection of string variables, an example of which could be the menu bar. This subpalette contains all of the list-and-ring controls and indicators.

Array & Cluster This contains all of the array and cluster controls and indicators. An array is a data structure that has multiple elements of the *same* data type, such as an array of integer or Boolean. If you need a data structure that contains *different* data types, LabVIEW provides one called *cluster* that can contain any type of data elements. For example, clusters could be used when you design a database containing different types of elements, such as age (integer), name (string), marital status (Boolean), etc.

Graph This contains two types of display objects: graphs and charts. The general rule of thumb is that charts are used for real-time display and graphs are used for offline display. More details are discussed in Chapter 5. Note that LabVIEW allows graphs and charts to be control type, but can you think of an example where graphs or charts need to be used to set values?

Path & Refnum This contains all of the controls and indicators for file path and task ID. Task ID is an integer that identifies tasks that you are currently involved. Tasks can be differentiated by identifying their task IDs.

Decorations This contains objects for decorations.

User Controls This is the place where you can save customized controls and objects for a personal use.

Select a Control... This allows you to select a customized control or indicator.

2.4.3 Functions Palette

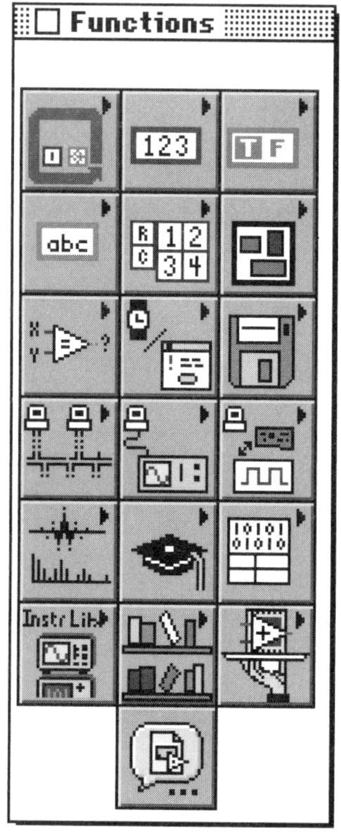

Structures This contains **While Loop**, **For Loop**, **Case** structure, and others related to loops. (Refer to Chapter 4.)

Numeric This contains all of the tools for processing numeric data.

Boolean This contains all of the tools for processing Boolean data.

String This contains all of the tools for processing string data. (Refer to Chapter 12.)

Array This contains all of the tools for processing array data. (Refer to Chapter 6.)

Cluster This contains all of the tools for processing cluster data. (Refer to Chapter 6.)

Comparison This contains all of the tools for comparing data.

Time & Dialog This contains all of the tools for timing and displaying dialog boxes on the screen. However, you can also design your own error dialog message. (Refer to Chapter 15.)

File I/O This contains all of the tools for saving data to a file or reading data from a file. This is an important tool, especially for data acquisition where you want to save data continuously or while offline. (Refer to Chapter 11.)

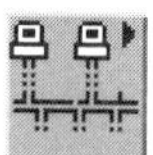

Communication This contains all of the tools for communication between computers, such as Transmission Control Protocol/Internet Protocol (TCP/IP), Apple Event for Macintosh, etc. The Internet Toolkit could be a powerful add-on to LabVIEW if you are interested in building communication applications.

Instrument I/O This contains all of the tools for GPIB, VISA, GPIB 488.2 protocols, and serial communications. (Refer to Chapter 13.)

Data Acquisition This contains all of the tools for analog input, analog output, digital input/output, and counter operations. (Refer to Chapters 8 through 10.)

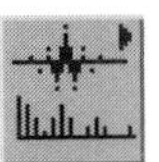

Analysis This provides all of the tools for the analysis of acquired data, such as filtering, power spectrum analysis, curve fitting, stochastic data analysis, etc. (Refer to Chapter 14.)

Tutorial This provides a tutorial of LabVIEW. This is only available on PC platforms and cannot be used on Macintosh or UNIX.

Advanced This provides more advanced tools, such as low-level data manipulation, calling Dynamic Link Libraries (DLLs) from LabVIEW, embedding C code in LabVIEW to expedite processing using Code Interface Node (CIN), or programming with sound files on both PC and Macintosh platforms. Step-by-step instructions on how to create a CIN are presented in Chapter 15.

Instrument Drivers This contains prewritten GPIB instrument driver VIs for a variety of GPIB and VISA instruments.

User Libraries This is where you can store your customized VIs and use them like other LabVIEW VIs. Procedures for customizing VIs are presented in Chapter 15.

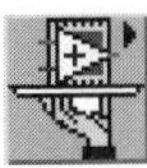

Application Control This contains all of the VIs that allow you to control third-party applications and VIs from different VIs and to customize the menu bar in LabVIEW. (Refer to Chapter 15.)

Select a VI . . . This is similar to **Select a Control** in the **Controls** palette. This allows you to select a VI and place it in the diagram window. Placing customized VIs in the diagram window can also be done by first including the customized VIs in the **User Libraries** palette and selecting them from there.

PROBLEMS

2.1 Create a VI whose front panel and diagram window are shown in Figure P2.1, and save it as **P02_01.vi**. Explain what this VI does. Note that the digital display of each object is hidden using its pop-up menu. Run the VI continuously to see the effect of using the button **Run Continuously** in the front panel. The objects (also known as VIs and functions) used in this problem are

Controls >> **Numeric** >> **Horizontal Fill Slide**
Functions >> **Numeric** >> **Subtract**

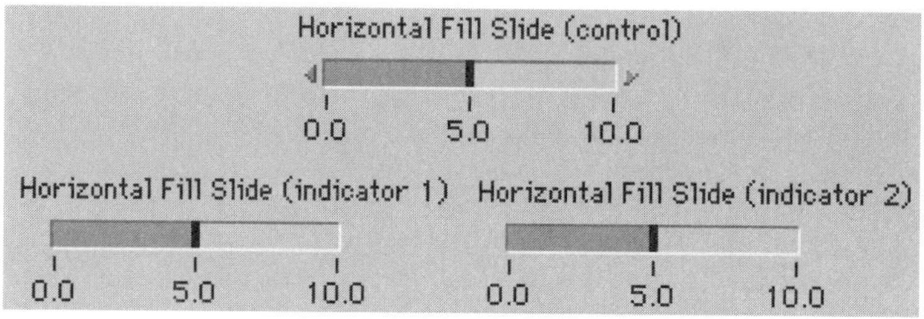

Figure P2.1

2.2 Repeat Problem 2.1 with a maximum value 1.0 instead of 10.0. In order to have the same effect as Problem 2.1, what change(s) should be made in the diagram window? Save the VI as **P02_02.vi**.

2.3 In Figure P2.1, there are three objects: one control and two indicators. How do you distinguish controls from indicators in the diagram window just by looking at their icons?

2.4 Create a VI in Figure P2.4 and save it as **P02_04.vi**. This VI shows the AND operation of two Boolean inputs **x** and **y**. Run the VI and see if the output is true only if both **x** and **y** are true. The VIs used in this problem are

Controls >> **Boolean** >> **Square Push Button 2** and **Square LED**
Functions >> **Boolean** >> **And**

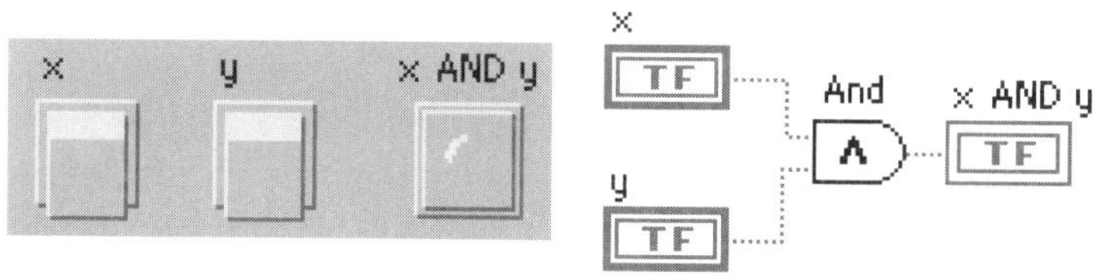

Figure P2.4

2.5 Repeat Problem 2.4 using OR gate and save it as **P02_05.vi**. Run the VI to see if the output **x OR y** is true if either **x** or **y** or both **x** and **y** are true.

2.6 Repeat Problem 2.4 using NAND (Not AND) and save it as **P02_06.vi**. Write the result **x NAND y** for each different state of **x** and **y** in a table.

2.7 Repeat Problem 2.4 using NOR (Not OR) and save it as **P02_07.vi**. Write the result **x NOR y** for each different state of **x** and **y** in a table.

2.8 Repeat Problem 2.4 using XOR (Exclusive OR) and save it as **P02_08.vi**. Write the result **x XOR y** for each different state of **x** and **y** in a table.

2.9 There are five objects in the front panel in Figure P2.9. Identify each one, and provide its path where it can be found.

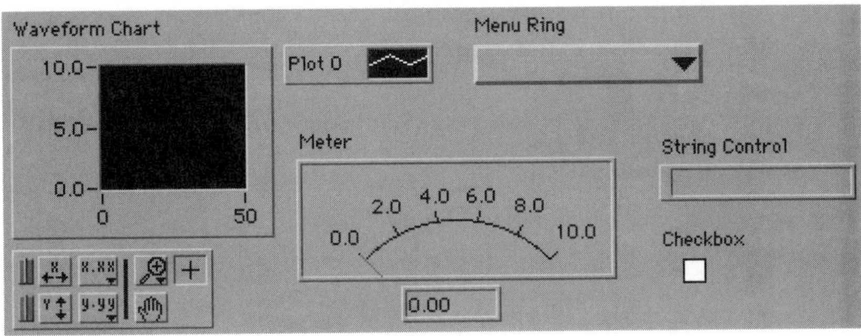

Figure P2.9

2.10 There are eight objects in the diagram window in Figure P2.10. Identify each one, and provide its path where it can be found. Also, briefly explain what each object does by reviewing online help. The online help can be accessed through the **Ctrl-H** shortcut on PC platforms or the **Command-H** shortcut on Macintosh platforms.

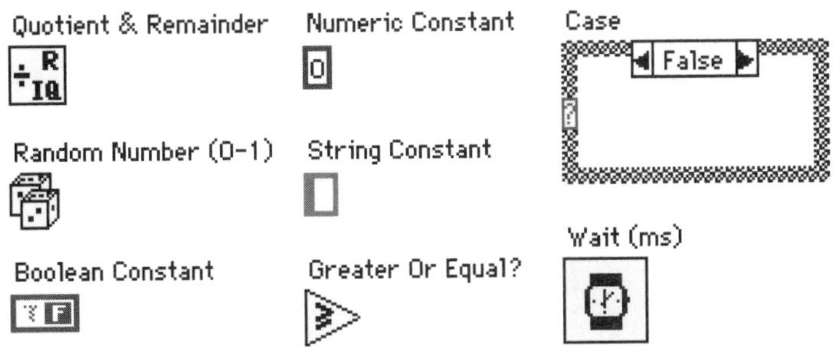

Figure P2.10

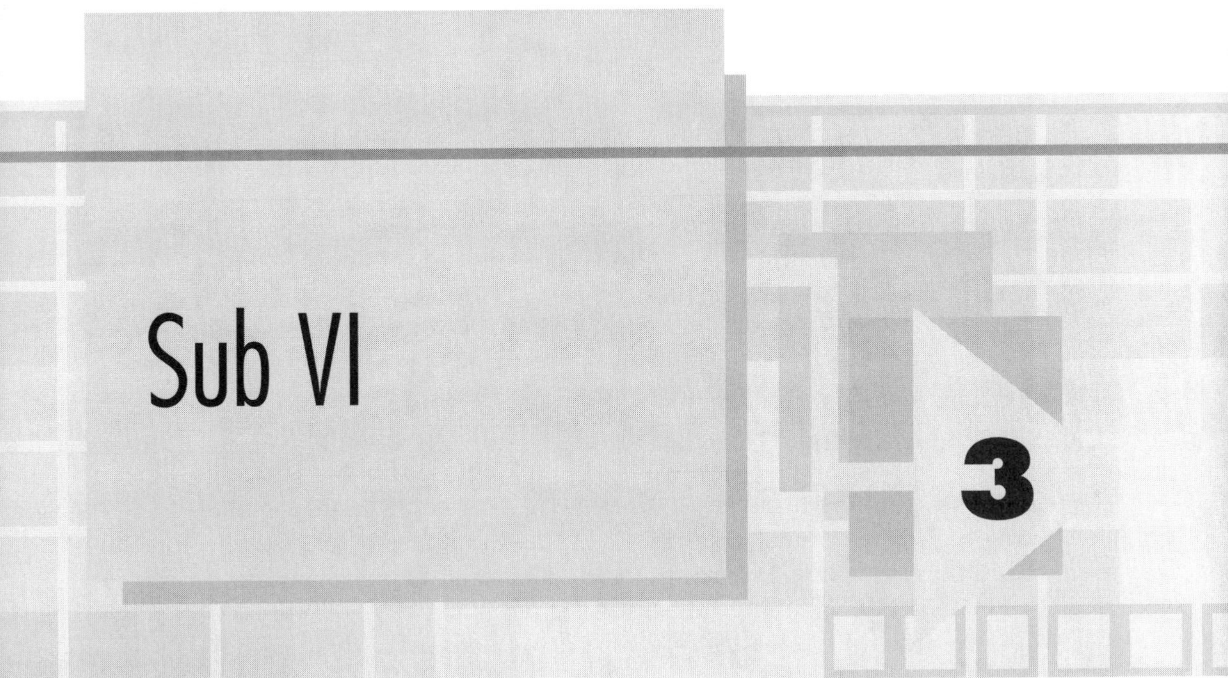

Sub VI

3

3.1 Definition of Sub VI

In the regular text-based programming languages such as FORTRAN, PASCAL, C, and others, the concept of subroutine is so essential that one may not be able to complete his or her code without using subroutines. Similarly, LabVIEW has something which corresponds to the subroutines and is called sub VI. As subroutines do, a LabVIEW sub VI can take parameters, perform a task, and return resulting values. Of course, a sub VI can have no input or no output parameters. Since LabVIEW is a Graphical (G)-language one should expect different ways of invoking the sub VIs. Since the VIs represent virtual instruments, you have wires for connection and objects for individual components. Therefore, you pass the parameters using wires, and call subroutines by including objects (sub VIs) in the diagram window. In other words, controlling the sub VIs is done *graphically*. In this chapter, you will see how to create sub VIs and control them using their settings.

3.2 Creating a Sub VI

Invoke LabVIEW if it is not running already. Create a new VI by pressing the button **New VI** in the startup window of LabVIEW or by choosing **New** under the **File** pull-down menu. Go to the front panel and place objects as shown in Figure 3.1 and save the front panel as **First Sub VI.vi**. There are two digital controls and one digital indicator that enable this sub VI to calculate the square of the sum of two variables x and y. When you finish this sub VI, you will be able to call on it whenever $(x + y)^2$ needs to be calculated. You can find the digital control and the digital indicator in the **Numeric** subpalette within the **Controls** palette. (See Figure 3.2.) Right clicking the mouse button in the front panel window gives access to these palettes.

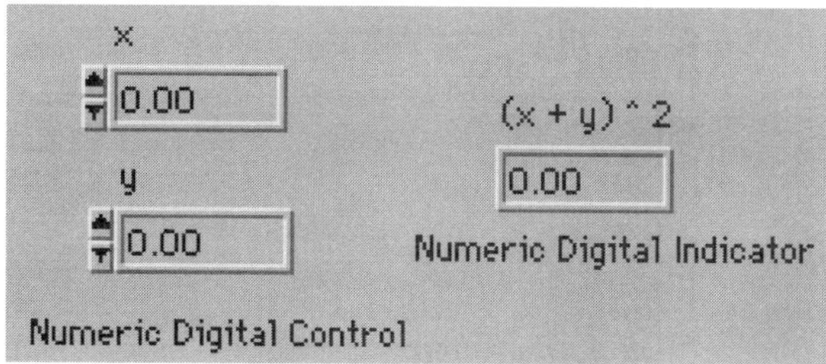

Figure 3.1 Front panel of **First Sub VI.vi**.

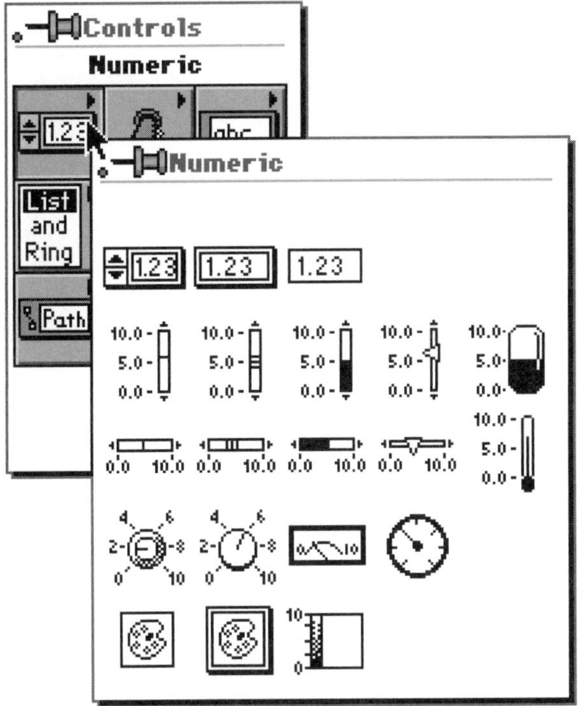

Figure 3.2 You can find the digital control and the digital indicator from **Controls** >> **Numeric**.

To distinguish each object, make sure to label them as **x**, **y**, and **(x + y)^2**. You can display the labels if they are not shown by right clicking on the object and selecting **Show** >> **Label** from the pop-up menu.

When in doubt, right click on the object of interest. This will display a pop-up menu from which you may choose an action. Online help is also available and can be accessed with the ***Ctrl-H*** *shortcut while the cursor rests on the object of interest.*

Now, go to the diagram window and complete it as shown in Figure 3.3. To do so, right click in the diagram window and access the **Functions** palette. Select **Numeric** to locate the **Add** function; it appears as a triangularly shaped icon with a plus sign in the center. To find **Power Of X** that calculates b^c for given b and c, pop up the **Functions** palette and go to **Numeric** >> **Logarithmic**. Once you have placed these objects, press either the tab key or space bar to change the cursor to the Wire Tool. Now wire them as shown in Figure 3.3.

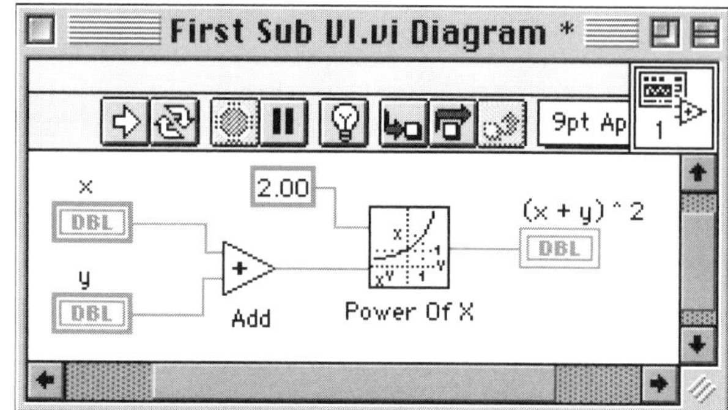

Figure 3.3 Diagram window of **First Sub VI.vi**.

 **Connect Wire** (Wire Tool)

Since wiring technique is important, here are the steps again.

Step 1: Place the Wire Tool on top of **x** (digital control: how do you tell if it is a control or an indicator?). It will start blinking.

Step 2: While the icon blinks, click once; that is, press the left mouse button once and release it.

Step 3: Drag the Wire Tool onto the top input of **Add**. Note that you just move the mouse without holding the button.

Step 4: When the Wire Tool is on top of the first input of **Add**, it (the top input) will also start blinking. Note that the label of the first input **x** is also shown. While the icon is blinking, click on the left mouse button, and it will complete the wire between **x** and the top input of **Add**.

Step 5: To wire the rest, follow the same procedure.

Now, you need to create terminals for the inputs and the output of this sub VI. Go to the front panel by using the shortcut **Ctrl-E** that jumps between the front panel and the diagram window. Locate the square icon in the upper right corner of the window, as shown in Figure 3.4. Right click on

Figure 3.4 Select **Show Connector** from the pop-up menu to display the terminals.

the icon and select **Show Connector**. This will turn the icon into a picture with three terminals as shown in Figure 3.4. Note that LabVIEW automatically knows how many terminals you need based on the number of controls and indicators in your VI. In the **First Sub VI.vi**, you have two controls, **x** and **y**, and one indicator, **(x + y)^2**; therefore, two input terminals and one output terminal are chosen by LabVIEW. However, a different pattern of terminals can be chosen manually.

To select a different terminal pattern, right click on the icon in the upper right corner of the front panel while the terminals are displayed. Go to **Patterns** and you will see all of the possible patterns of terminals. Ask yourself the following question about selecting terminal patterns: What happens if you need 50 input terminals and 50 output terminals? You immediately see a problem because no such pattern is available for selection. Furthermore, it will be difficult to identify each terminal so wiring would be challenging if not impossible. In order to avoid such a situation, you could use a different data structure, such as Array or Cluster (see Chapter 10), and bundle multiple terminals into a single data type.

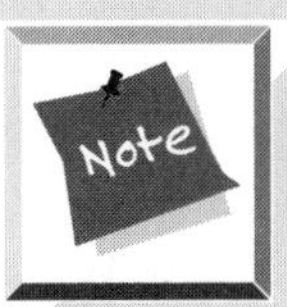

Try to minimize the number of terminals as much as possible. One way to do so is to use a data structure, such as Array or Cluster, that can have more than one element for a single object.

Keeping this in mind, finish your first sub VI example. When you select **Show Connector** from the pop-up menu after right clicking on the icon in the upper right corner of the front panel, LabVIEW automatically switches the mouse pointer to the Wire Tool. If not, you can select it from the **Tools** palette. After selecting the Wire Tool, click one terminal and click an object that you want to assign the terminal to. Continue to assign the terminals to the remaining controls and indicators. Note that you do not need an equal number of terminals and objects; therefore, it would be a good idea to have a few extra terminals available when you initially create them in case you need to modify subroutines (sub VIs). This will prevent you from having to reassign terminals and rewire other VIs to which the sub VIs were connected. Also, there is no requirement for input terminals to be placed on the left and output terminals on the right; however, this seems to be conventional for most LabVIEW programmers. When you finish assigning all of the terminals to each object, save the VI. By the way, you may have recognized the asterisk (*) next to the title of the VI in Figure 3.3. This indicates whether changes have been made since the last save. Therefore, it will disappear after saving the changes in the VI.

Now build another VI where you will call your first sub VI. Close **First Sub VI.vi** completely, create a new VI, and save it as **Main.vi**. (Saving changes on a regular basis, such as every 10 minutes, is a good programming habit.) Place objects as shown in Figure 3.5. Label each object as shown to distinguish between them easily. Both the **Knob** and the **Gauge** can be found under the **Controls** >> **Numeric** subpalette. By default, the maximum value of the knob is 10.0. Change the value to 1.0 by entering 1.0 over the 10.0. You will notice that the knob scale adjusts automatically. Explore the feature of the gauge for a moment. Notice that it can be rotated, resized, and the values can be changed. You can also show or hide the digital display from the pop-up menu of each item. When you have finished the front panel, switch to the diagram window by using **Ctrl-E**. You should have three objects in the diagram window: two controls and one indicator. (How do you tell if an object is a control or an indicator in the diagram window?) In text-based programming languages, you invoke subroutines by calling their names.

Similarly, you call sub VIs in LabVIEW by placing sub VIs in the diagram window. To do so, bring up the **Functions** panel by right clicking in the diagram window and locate the **Select a VI . . .** subpalette. Selecting it will bring up a dialog box asking for a VI to open. In the dialog window similar to Figure 3.6, locate and select **First Sub VI.vi** and place it in the diagram window. It will have the default LabVIEW icon, but it can be modified with

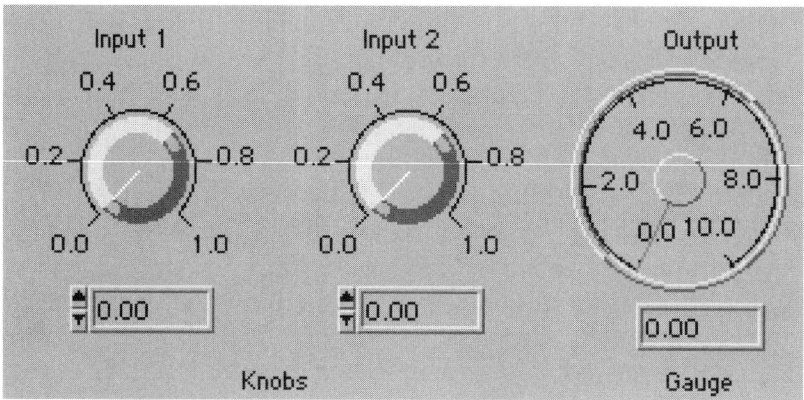

Figure 3.5
Front panel of **Main.vi**.

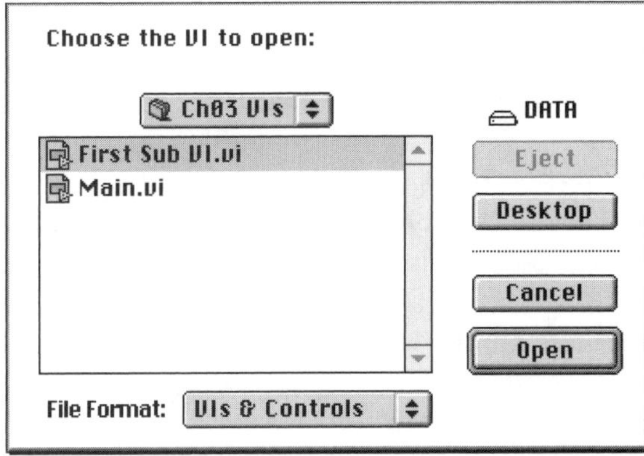

Figure 3.6 Dialog box to choose a VI (Macintosh platform).

the option **Edit Icon**. This option can be found in the pop-up menu of the icon in the upper right corner of **First Sub VI.vi**. (See Figure 3.4.)

Having put **First Sub VI.vi** in the diagram window of **Main.vi**, wire the two controls **Input 1** and **Input 2** to the input terminals (**x** and **y**) and the indicator **Output** to the output (**(x + y)^2**). (What is the shortcut to toggle between the Arrow Tool and the Wire Tool?) Right click on **First Sub VI.vi** to bring up its pop-up menu and select **Show** >> **Label** to display its name **First Sub VI.vi**. Note that when you place the Wire Tool on the first input terminal of the sub VI, LabVIEW displays **x** and similarly **y** on the second input terminal. In other words, LabVIEW displays the original labels that you assigned to each input and output terminals in your sub VI. This will help you identify terminals.

If you have followed the instructions correctly so far, the diagram window of **Main.vi** should look like Figure 3.7. Go to the front panel using the shortcut and select the Hand Tool either from the pop-up menu (shortcut: **Shift**-right click on PC platforms and **Command-Shift**-click on Macintosh platforms) or by tapping the tab key. Using the Hand Tool, set some values to the knobs (**Input 1** and **Input 2**) and run the VI by clicking on the **Run** button. Examine the VI further by playing with each object and different buttons. If you have the Broken Run Arrow indicating that an error or errors

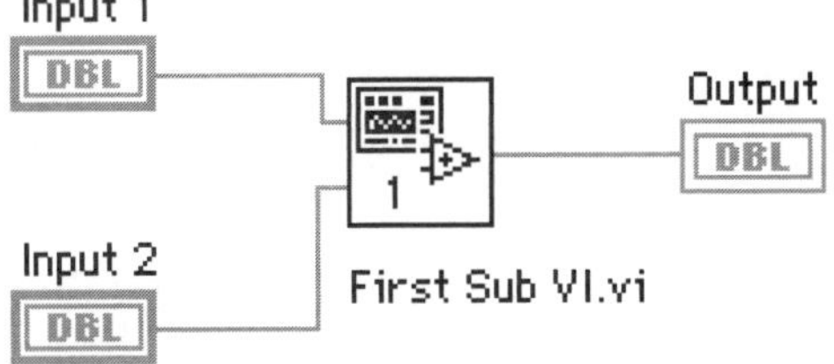

Figure 3.7 Diagram window of **Main.vi**.

have occurred, debug the VI using tools such as **Highlight Execution**, **Step Into**, **Step Out**, or **Probe Data**. (See Chapter 2 for their roles.)

By completing the preceding steps, you have just learned how to create a sub VI. Sub VIs are important stepping stones in LabVIEW programming. You can now call (invoke) **First Sub VI.vi** when you need to compute the square of the sum of two variables in other VIs just by placing the sub VI in their diagram windows. If the values you have set for all of the input parameters in the sub VI are values you will always start with, select **Make Current Values Default** from the **Operate** menu and save the VI, or you can save an individual control's default values by popping up the menu on the control, selecting **Data Operations**, then selecting **Make Current Value Default**, and saving the VI.

3.3 Creating Online Help Text for a Sub VI

LabVIEW provides online help for almost all of the objects, and the shortcut **Ctrl-H** will bring up the help text while pointing at objects of interest. You can also create an online help text for your own sub VIs that can be instantly accessed through the **Ctrl-H** shortcut. This will be illustrated by using the sub VI example you just created (**First Sub VI.vi**).

Open the example **First Sub VI.vi** by either double clicking the icon in **Main.vi** or opening it manually. Select **Show VI Info . . .** from the **Windows** pull-down menu. This will bring up the window **VI Information** with a text box under **Description**. Then the comments you enter in the **Description** will show up as the online help when you use **Ctrl-H** while the LabVIEW cursor is on **First Sub VI.vi**.

If you want to keep a history of modification of a VI, select **Show History** from the **Windows** pull-down menu. This will bring up a window revealing the VI history where you can add a new list or reset the previous one. This is helpful when multiple users are working on the same VI and one user wants to see what other users have done.

3.4 Security Settings of a Sub VI

One of the new features of LabVIEW 5.0 or higher is the password protection capability of VIs. If you select **Show VI Info . . .** from the **Windows** pull-down menu, it will bring up the window **VI Information** with the password

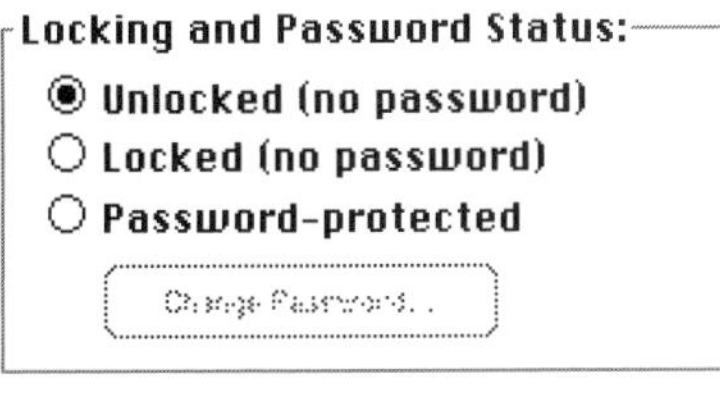

Figure 3.8 Password protection window of a VI.

setting window, as shown in Figure 3.8. Using the three setting buttons, you can lock or unlock the VI, or set, clear, or change its passwords. Locking a VI will allow you to open the diagram window but will not allow you to make any changes. In this mode, you only have limited debugging capability. (Debugging with light bulb will be disabled.) This is useful if you want to prevent any accidental modification during the debugging process.

When you set a password to a VI, you won't be able to open the diagram window without the correct password. The only way to recover the diagram window without the correct password is to *rewrite* the VI; therefore, passwords must be kept safe and used with caution. To activate this security feature, save the VI with your new password and quit LabVIEW. If you would like an immediate effect without quitting LabVIEW, select **Clear Password Cache** from the **Edit** pull-down menu followed by saving the VI. Next time you try to open the diagram window you will be prompted to enter the password.

3.5 Option Settings of a Sub VI

Knowing how to create sub VIs in LabVIEW would lead you to expect graphical control over them since LabVIEW is a G-language. What would you expect to happen if you call the *same* sub VI simultaneously with different input values? Can you somehow control the front panel of sub VIs, such as its location on screen? This section, on controlling sub VIs, will answer these questions and more. However, before you proceed, you should recognize that you can bring up the front panel of sub VIs by double clicking on their icons in the diagram window of the caller VI. Also, if you want to change the settings of a particular sub VI, you *must* change them *in* that particular sub VI. This implies that you need to bring up the front panel and its diagram window to modify it and save changes therein. Local options will be discussed in the later part of this chapter.

Go back to the **Main.vi** created in the previous section. Refer to the diagram window in Figure 3.7. In order to change the feature of **First Sub VI.vi**, you need to bring up its front panel by double clicking the icon. Proceed to do so to have its front panel on screen. If you need to change its diagram, you may use the shortcut **Ctrl-E** to go to the diagram window. For the moment, stay with its front panel. Right click on the icon in the upper right corner of the front panel of **First SubVI.vi**, as shown in Figure 3.4, to bring up the pop-up menu. Selecting **VI Setup . . .** from the menu will bring up a window, as shown in Figure 3.9. If you click on **Execution Options**, you will see the other two available options, but only the first two, **Execution Options** and **Window Options**, are discussed here. For those who are interested in the **Documentation** option, you may refer to the LabVIEW User Manual. Note that the menu type that contains **Execution Options**, **Window Options**, and **Documentation** is called *menu ring*.

The most common feature that LabVIEW programmers use is the second and the third option in the left column in Figure 3.9: **Show Front Panel When Called** and **Close Afterwards if Originally Closed**. The option **Close Afterwards if Originally Closed** becomes available once you select **Show Front Panel When Called**. To understand their usage, consider the following

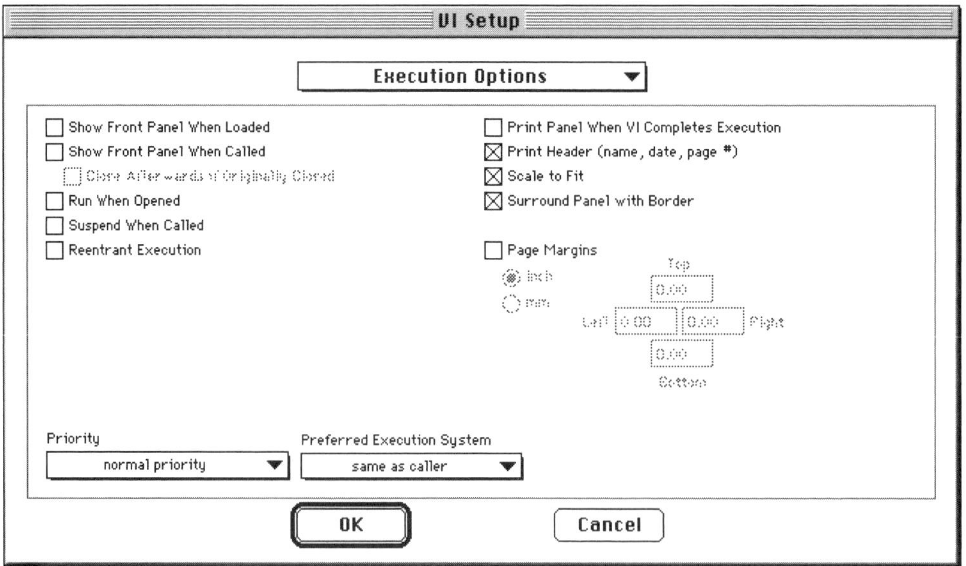

Figure 3.9
Execution Options in **VI Setup . . .** window.

example. Suppose you are writing a database program for your customer using LabVIEW. In the front panel, your customer wants to display multiple buttons for tasks such as Enter Password, Update Address, Display Check-In and Check-Out Time, and so forth. Your task is to write a VI that brings up the address page when the user presses the Update Address button. This could be done easily by using a sub VI containing the address information if you select **Show Front Panel When Called** and **Close Afterwards if Originally Closed**. (In order to have a button waiting for an action in this case, you will need to use a loop, which will be discussed in Chapter 4.) You can adjust the location of the window when it shows up on screen using the **Window Options**, which will be discussed in a moment. Another option you must be aware of whenever you create your own sub VI is the **Reentrant Execution** option.

In the beginning of this section, you were asked what would happen if you simultaneously called the same sub VI in multiple places in one VI. When you open a VI (main VI) that calls (contains) other VIs (sub VIs), LabVIEW will load the main VI as well as all of the sub VIs in the memory. If you call the same VI (sub VI) in different places, intending to run these simultaneously, LabVIEW may return wrong answers if you are not using the **Reentrant Execution** option. In fact, each copy of the sub VI will execute sequentially in random order without the option **Reentrant Execution** selected. Therefore, if the functionality of a sub VI is related to a timed simultaneous operation, it may bring an unexpected result. Also, if the sub VI has memory, the data from the previous call will be reused in the next call! (See Problems 4.2 and 4.3 for an example of the effect of **Reentrant Execution**.) To avoid such an incident, you must select **Reentrant Execution** so that LabVIEW will allocate separate memory blocks and treat each copy of the same VI as different ones. Having the option **Reentrant Execution** checked, identical copies of a sub VI will run independently of one another. However, the remaining choices in the **Execution Options** and the debugging buttons of a sub VI will become unavailable with the **Reentrant Execution** option selected.

Another option is **Window Options**, and its functionality is self-explanatory. With it, you may hide the toolbar of a VI, move or resize the windows to fit the entire screen, or perform many other tasks related to the windows of VIs. One thing to remember is that these options will take effect when you *execute* the VI. If a VI is a sub VI, then these options will apply when it gets *called* by the caller VI. **Window Options** is shown in Figure 3.10. Lastly, the option **Dialog Box** forces the front panel of the sub VI to be active on the top of the caller VI. This feature is useful if you want users not to lose the front panel of the sub VI. This can happen if users accidentally activate

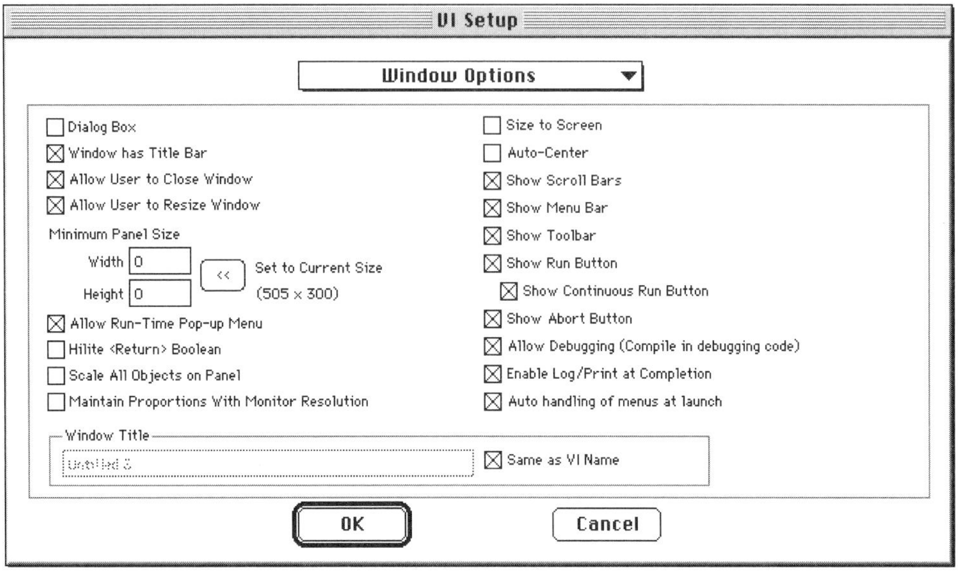

Figure 3.10
Window Options in **VI Setup . . .** window.

the caller window by clicking in it during the VI execution; then the front panel of the sub VI will be hidden behind that of the caller VI if the front panel of the sub VI is smaller than that of the caller VI. In order to avoid such an inconvenience—you have to move the front panel of the caller VI to retrieve the window of the sub VI—you can use the option **Dialog Box** selected for the sub VI.

3.6 SubVI Node Setup

The settings you make on a sub VI in its **VI Setup . . .** are global. In other words, if you pop up the front panel of a sub VI, bring up **VI Setup . . .**, and change any setting, it will apply whenever you call that sub VI. However, you may want to apply some settings to only a few of the copies of the same sub VI. You can do this by using **SubVI Node Setup**. Locate the sub VI to which you want to apply local settings. Right click on that sub VI to bring up the pop-up menu and select **SubVI Node Setup**; this will bring up a window as shown in Figure 3.11. You may realize that the settings in this win-

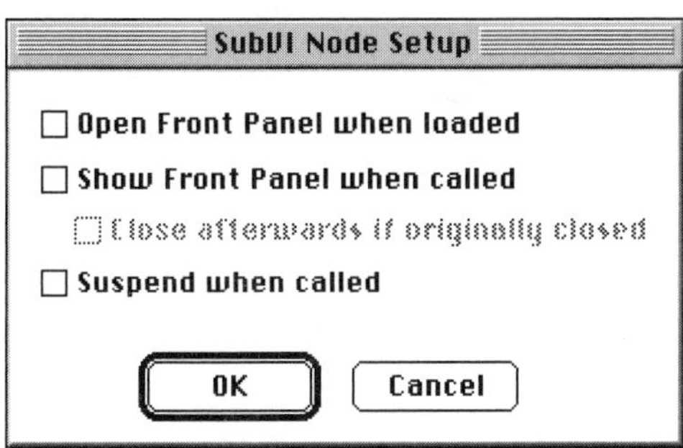

Figure 3.11 **SubVI Node Setup** window.

dow are four of the options available in the **Execution Option** in **VI Setup** that was shown in Figure 3.9. Therefore, you can only have a small set of options from **Execution Options** for sub VI local settings. These settings will not affect the other copies of the sub VI but will apply only to the one that has the settings invoked.

3.7 Alternative Way of Creating a Sub VI

There is an alternative way to create sub VIs, and it will be illustrated with the **First Sub VI.vi**, which computes **(x+y)^2** for given **x** and **y**. Suppose you have a VI whose diagram window is as shown in Figure 3.12.

Note that Figure 3.12 performs the same computation as **First Sub VI.vi**. In the previous section, you manually assigned each element to each terminal to complete **First Sub VI.vi**. However, LabVIEW provides an alternative way to create sub VIs instantly. The steps are as follows:

Step 1: Select the portion from which you would like to create a sub VI. In Figure 3.12, you will select the entire diagram.

Step 2: Select **Edit** >> **Create SubVI**. This will create a sub VI with three terminals **x**, **y**, and **(x + y) ^ 2**.

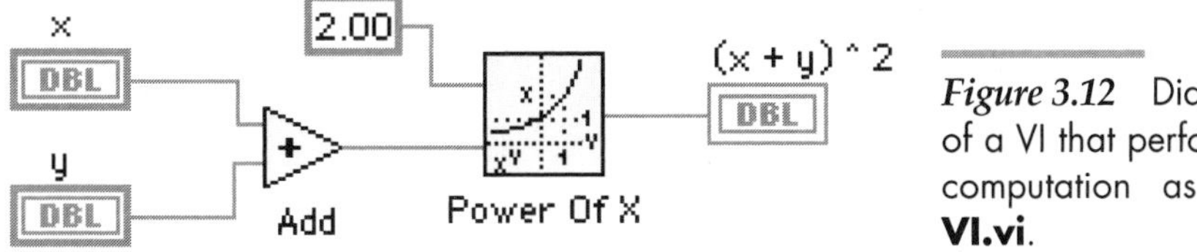

Figure 3.12 Diagram window of a VI that performs the same computation as **First Sub VI.vi**.

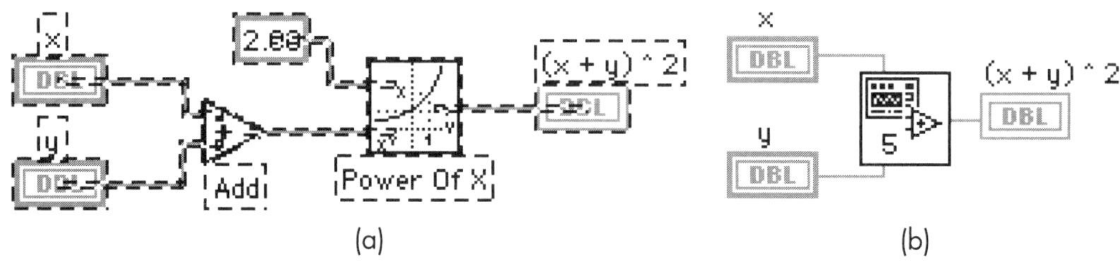

Figure 3.13
(a) A portion from which a sub VI is needed is selected in the diagram window. (b) Selecting **Edit** >> **Create SubVI** will create a VI from the portion selected.

Each step is shown in Figure 3.13. This method is very convenient since you do not have to assign each terminal to the objects in your sub VI; however, this can cause some confusion if the structure of input and output becomes complicated. Therefore, this method should not be abused in creating sub VIs unless the sub VI is as simple as the example in Figure 3.13.

In order to create sub VIs, always try to use the method described in Section 3.2, and minimize the use of the method using ***Edit >> Create Sub VI*** *unless the structure of input and output is extremely simple.*

PROBLEMS

3.1 In the following equation, identify control(s) and indicator(s). Then create a sub VI **P03_01.vi** that performs the equation, assuming that x is guaranteed to be nonzero. Make sure to create terminals for control(s) and indicator(s).

$$y = 20 \log_{10} (|x|)$$

3.2 Expand the sub VI you created in Problem 3.1 so it can now return the result of the following equation as well as y:

$$z = 3 \cos(x)$$

Save the modified sub VI as **P03_02.vi**. If you created a few extra terminals in Problem 3.1, expanding **P03_01.vi** could be trivial. Therefore, it is a good idea to have a few extra terminals available when you create sub VIs for the first time.

3.3 Create a sub VI **P03_03.vi** that shows its front panel when called to display a message HELLO in the front panel for 3 seconds. The front panel disappears afterwards. Test the sub VI to confirm that it behaves as described previously by calling **P03_03.vi** from another VI. Hint: You may use the following LabVIEW functions and options:

Controls >> String & Table >> Simple String
Functions >> Time & Dialog >> Wait (ms)
pull-down menu **Operate >> Make Current Values Default**
VI Setup . . . >> Execution Options >> Show Front Panel When Called, and **Close Afterwards if Originally Closed**

3.4 Modify the setting(s) of **P03_03.vi** so that the front panel shows up in the center of the screen when it is called and disappears afterwards. Save the VI as **P03_04.vi**.

3.5 Create a sub VI **P03_05.vi** that takes an angle in degree as input, converts it in radian, and returns the sine and cosine of the angle in radian as output. You may use the following LabVIEW functions:

Functions >> Numeric >> Additional Numeric Constants >> Pi
Functions >> Numeric >> Trigonometric >> Sine, and **Cosine**

Loops and Conditional Statements

4

An application cannot be complete without using for or while loops or conditional if and else statements. It may cause you to wonder how such statements can be achieved with a graphical language like LabVIEW. This chapter will show you how and illustrate a variety of interesting new features that cannot be encountered in text-based languages.

There are seven items on the **Structures** subpalette within the **Functions** palette that can be popped up by right clicking in the diagram window: **Sequence**, **Case**, **For Loop**, **While Loop**, **Formula Node**, **Global Variable**, and **Local Variable**. Start with the **For Loop**.

4.1 For Loop

This loop iterates the number of times specified; however, if the total number of iterations is zero or negative, it will not run. Note that the **For Loop** is available in the diagram window and not in the front panel. This is because the **For Loop** is related to a functionality of instruments and not to data display or input switches. You can put a **For Loop** in the diagram window by first selecting it from **Functions** >> **Structures**. When the **For Loop** is chosen, the cursor changes to N. Dragging it in the diagram will create a **For Loop**; then place commands that are meant to repeat in the loop. Consider the following pseudo code:

```
for i = 0 to 9
    display i;
    pause 1 second;
end;
```

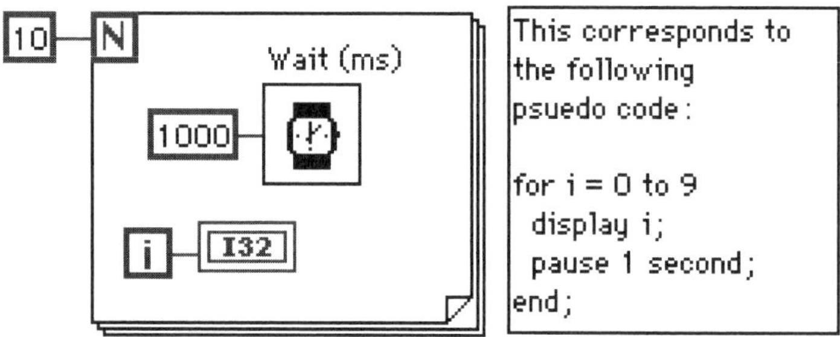

Figure 4.1
Conversion of conventional for loop into the **For Loop** of LabVIEW.

The corresponding **For Loop** in LabVIEW is shown in Figure 4.1. The *count terminal* [N] is for the total number of iterations. The value entered must be a positive number in order for the **For Loop** to iterate. The *iteration terminal* [i] returns the current iteration index. In Figure 4.1, you have a **Numeric Constant** 10 wired to the count terminal. This will cause the **For Loop** to repeat 10 times and the index will vary from 0 (not 1) to 9. Like the C programming language, the indexing in LabVIEW always starts at zero. Now you will be introduced to an important technique of creating input to LabVIEW VI or function terminals.

4.1.1 Creating Inputs to Terminals

The previous chapter discussed how to create sub VIs. It is true that almost everything in LabVIEW is a sub VI, which is equivalent to saying that everything in C is a function. Also, each sub VI can call other sub VIs like each function can call other functions in C. When you call functions in C, you pass arguments in parentheses, whereas you pass arguments through wires in LabVIEW. When you pass arguments in C, you should match the *type* of the arguments, such as float, integer, and so forth. If you want to force a certain data type to be a different type, you may use coercion or casting in C. LabVIEW also has coercion and casting features as well as data type conversion capability. In LabVIEW, you should create the correct type of variables or constants as input in order to avoid a type conflict. It is important to match the type because a type conflict causes an unnecessary use of memory. (Chapter 15 discusses such an issue in detail.)

When you created the diagram in Figure 4.1, you probably searched through subpalettes looking for the **Numeric Constant** to wire 10 to the count terminal, [N]. It expects an I32 long integer type. Therefore, if you wire an input of a different type to it, a type conflict will result and cause LabVIEW to allocate a new memory address for type coercion. Recall one of the statements made previously:

If you are in doubt, right click on the object.

If you right click on the iteration input terminal, you will see three options available in the pop-up menu: **Create Constant**, **Create Control**, and **Create**

Figure 4.2 Coercion dot.

Indicator. If you choose **Create Constant**, LabVIEW will automatically create a constant of correct type and even wire it for you! Of course, you can change the type of the constant manually. Right click on the constant 10 in Figure 4.1, and go to **Representation**. You will see the complete selection of available data types. Just for practice, change the data type to double precision DBL. After doing so, you will see that the color of the number 10 has changed from blue to orange. You will also see a gray dot on the iteration input terminal of the **For Loop**. This is called a *coercion dot* and it indicates the existence of a type conflict (Figure 4.2).

Now, correctly wire an I32 data type to the iteration terminal [i] with the right-click method. The **Wait (ms)** can be found under the **Time & Dialog** subpalette in the **Function** palette and its input type is an unsigned 32-bit integer (U32). To avoid a type conflict, right click on the input terminal of **Wait (ms)** and select **Create Constant** followed by entering 1000 for 1 second. This will provide us with enough time to see the increment of index. Go to the front panel and run the VI to view the increment of index from 0 to 9.

If you have more than one input to a VI and you want to create inputs automatically, select the Wire Tool and place it on the input of interest. (Its label will show up.) Now, bring up the pop-up menu by right clicking and select an option from **Create Constant**, **Create Control**, or **Create Indicator**. This is an essential addition to the wiring technique discussed previously.

An example of creating input terminals from the pop-up menu of the input terminals is shown in Figure 4.3. The VI used in the example is **Write To Spreadsheet File.vi**, which can be found in **Functions >> File I/O**. Currently, the Wire Tool is placed on the second input **2D data** so that its label is being displayed. If you select **Create Control** or **Create Constant** from the pop-up menu, LabVIEW will automatically create a control or a constant of correct data type. Note that if a wire in Figure 4.3 has a square dot at the be-

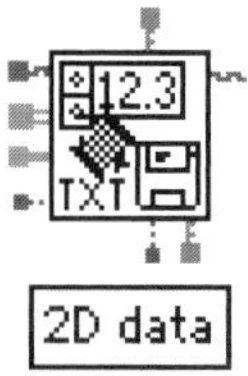

Figure 4.3 Creating inputs from the pop-up menu of the input terminals.

ginning of it, the corresponding terminal expects a control or a constant as its input. If a wire does not have such a square dot at its beginning, the corresponding terminal expects an indicator as its input. Therefore, in Figure 4.3, the input **2D data** expects either a control or a constant as its input due to the square dot at the beginning of the wire.

Note

To create and wire correct inputs to input terminals of a VI or a function, select the Wire Tool and locate the input of interest by monitoring the label of input. Bring up the pop-up menu and select either ***Create Constant, Create Control****, or* ***Create Indicator****. If a wire has a square dot, as shown in Figure 4.3, the corresponding terminal expects a control or a constant as its input. If not, the corresponding terminal expects an indicator as its input. If you want to change the input data type manually, right click on the input, go to* ***Representation****, and select the preferred type.*

4.1.2 Color of Different Data Types

You have just seen that blue is an integer data type and orange is a float type. This color differentiation allows you to identify easily the data type in much the same way that you identify controls and indicators by the thickness of the borderline. The following is the complete list of colors and their corresponding data types. (Clusters are discussed in detail in Chapter 6, where arrays are studied.)

blue	integer
orange	float
green	Boolean
pink	string
brown/pink	cluster (blend of different types)

4.2 While Loop

A **While Loop** is a structure that repeats its body until a test condition fails. The **While Loop** can be placed in a diagram window in the same manner as the **For Loop**. Consider the following pseudo code, which will be converted to LabVIEW G-language code:

```
x = 0;
While (x < 10)
     display x;
     pause 1 second;
     x = x + 1;
end;
```

In this example, the condition (x < 10) is tested *before* the loop starts. So if *x* was initialized with 10, the **While Loop** body will never get executed. However, since LabVIEW tests the condition *after* starting the loop, the **While Loop** *always* executes at least once. (This is similar to the DO . . . WHILE statement.)

*The **While Loop** in LabVIEW tests the condition after starting the loop; therefore, the **While Loop** always executes at least once.*

This is an important notion to recognize so you can avoid damaging data acquisition systems with **While Loop**. For example, suppose that your VI application acquires voltage signals from a system that generates high voltage if the system breaks down. In this instance, your application uses a **While Loop** to acquire a voltage signal as long as the signal remains under a certain threshold. The threshold exists in order to avoid any damage to the data acquisition system. Unfortunately, the problem here is that since the LabVIEW **While Loop** tests the condition *after* starting the loop, the first run will completely damage your system if it was already malfunctioning. After running the system after the first run of the **While Loop**, the loop stops! This can be avoided by testing the voltage once independently without passing it to the main data acquisition system *before* starting the **While Loop**. Therefore, you should keep this important difference in mind whenever you attempt to use the **While Loop** in LabVIEW. On the other hand, the **For Loop** tests the condition if the input to the iteration input terminal is positive *before* starting the loop.

*The **For Loop** in LabVIEW tests the condition to the iteration input terminal before starting the loop.*

Figure 4.4 illustrates a LabVIEW **While Loop** that corresponds to the preceding pseudo code, displaying **x** while incrementing it until **x** is greater than or equal to 10. The comparison LabVIEW function **Less?** can be found in the **Comparison** subpalette within the **Functions** palette and it returns a TRUE or FALSE Boolean value. (You can also identify the output type by the green colored wire.) When you create the VI in Figure 4.4, remember to use the automatic input creation feature of LabVIEW by right clicking on the input and selecting either **Create Constant**, **Create Control**, or **Create Indicator**. In this example, selecting **Create Indicator** after right clicking on the index terminal creates the indicator. Actually, the last value of **x**, which is 10, is the same in both the pseudo code and the diagram window. However, LabVIEW will display the last value whereas the pseudo code will display only up to 9 since the while loop in the pseudo code will exit when the test fails.

Figure 4.4 introduces a new terminal that is called *conditional terminal*. It expects a Boolean input since its color is green, and the **While Loop** will repeat until the input to the conditional terminal becomes false. The iteration terminal and **Wait (ms)** have been explained in the **For Loop** discussion section. After you complete the diagram shown in Figure 4.4, go to the front panel and run it to see the increment of **x**.

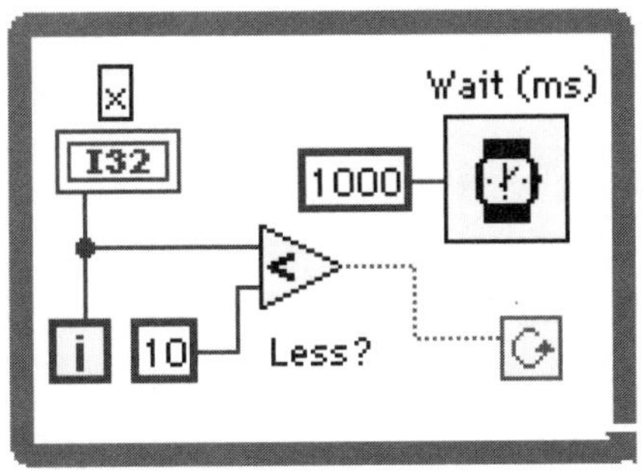


```
This corresponds to
the following pseudo code:

x = 0;
while (x < 10)
  display x;
  pause 1 second;
  x = x + 1;
end;
```

Figure 4.4
Conversion of conventional while loop into LabVIEW **While Loop**. The last value of **x** is the same in both, but LabVIEW will display the last value whereas the pseudo code will display only up to 9.

4.3 Case Structure

The **Case** structure in LabVIEW is analogous to the if and else statement in text-based programming languages. However, realization of nested if and else statements in LabVIEW requires more attention since LabVIEW provides graphical conditional selection in a two-dimensional diagram window. The nested if and else statements can be realized by nested **Case** structure or multiple frame **Case** structure. You can also specify a range for each case. The following example shows a pseudo code and its conversion to a proper LabVIEW code with **Case** structure. Consider the following if and else pseudo code:

```
if (x < y)
    z = −1;
else
    z = 1;
end;
```

In this simple example, you have two options, where only one of them will be executed based on the condition test; therefore, you need two frames for the **Case** structure, as shown in Figures 4.5 and 4.6.

When you place a **Case** structure in the diagram window, it comes with True and False frames that you can toggle by clicking the arrow next to the label. When you finish the diagram shown in Figure 4.5, you will realize that the **Run** button is broken (: Broken Run Arrow). This indicates that the VI is not complete and LabVIEW cannot compile it. It is because the indicator **z** is expecting a value based on two conditions (either **x** is greater than **y** or not), so an output for the False case also must be defined, as illustrated in Figure 4.6.

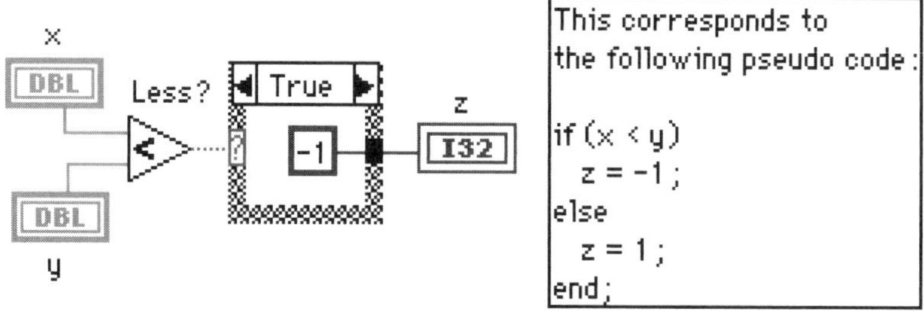

Figure 4.5
Conversion of conventional if and else statement into LabVIEW **Case** structure.

Figure 4.6 Completing **Case** structure by specifying an output for each case.

If you go to the other False frame and wire a **Numeric Constant** 1 to the white square *tunnel*, it will turn black as shown in Figure 4.6, and now the VI is complete. It is important to specify an output (if any) for both cases of **Case** structure as most errors in the use of **Case** structure result from unspecified output.

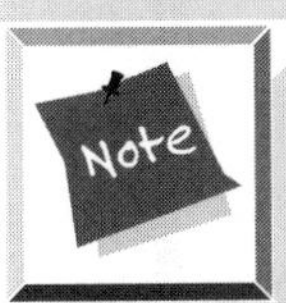

*With **Case** structure, you must specify outputs for all of the frames if there is any output at all. Any missing output at the tunnel can be detected by the presence of a white tunnel. However, this rule does not apply to input (data entering the **Case** structure).*

If you complete the False frame as in Figure 4.6, the Broken Run Arrow should become a **Run** button indicating that the VI is ready to execute. Go to the front panel, enter some value in the controls **x** and **y**, and observe the value in the indicator **z**.

4.4 Case Structure with Multiple Frames

In the previous section, the **Case** structure had True and False frames with Boolean input to select either one. However, the **Case** structure will also accept numeric values or string inputs to allow the user to select from more than two frames. Examples will be used to illustrate each case.

4.4.1 Case Structure with Numeric Input Control

Figure 4.7 uses both G-language and text-based language to illustrate an example of the **Case** structure with more than two frames. You can specify the range by using two period symbols (..) in the label of the frame. (You can change the label using the Text Editor.) Also, you can specify a default frame from the pop-up menu of any frame. Note that you must have all of the

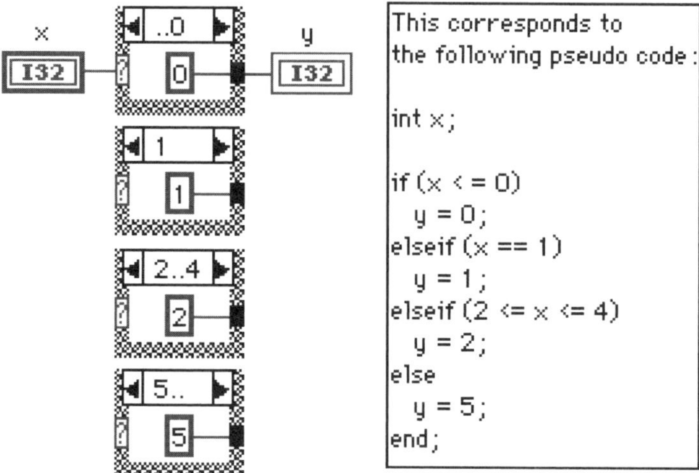

Figure 4.7
Example of the **Case** structure with more than two frames. Numeric value is used to select a frame. This figure shows a single **Case** structure with its frames shown together.

frames cover all of the possible ranges. Failing to do so will result in the Broken Run Arrow.

4.4.2 Case Structure with String Input Control

The **Case** structure can also take string variables as its input to select frames. An example is shown in Figure 4.8. This example performs the same task as that in Figure 4.7, but their frame labels are different. As the labels indicate, each frame is chosen by a string input instead of a numeric value. Note that you can still specify a range of numbers in string format using two period symbols (..). Also, note the difference in the label of the first frame such that you do not have the label **.."0"** for the first frame in Figure 4.8, whereas in Figure 4.7, you have two periods before zero for the first frame. If you use the label **.."0"** for the first frame with string input control, LabVIEW will return the Broken Run Arrow. This is because the input is string type and the string range **.."0"** does not cover the rest of the range in string format, thus violating the requirement in the coverage by labels. However, by labeling the first frame Default, the entire range is now covered and LabVIEW will be able to execute the VI.

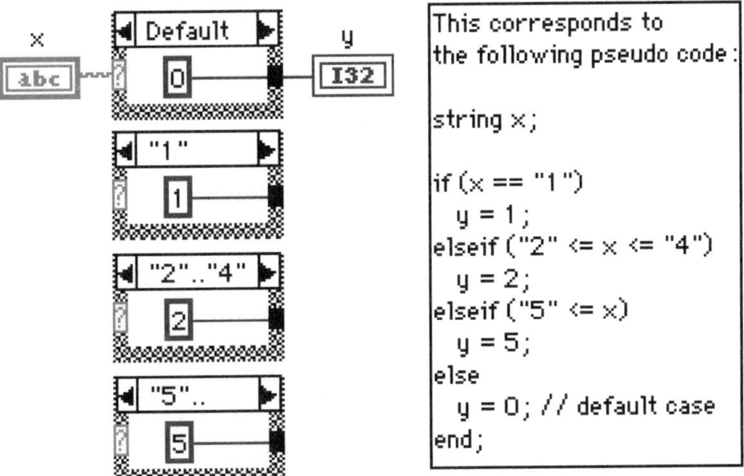

Figure 4.8
Example of the **Case** structure with more than two frames. String input is used to select a frame. This example performs the same task as Figure 4.7 does except that input values are now in string format instead of numeric format. This figure shows a single **Case** structure with its frames shown together.

4.5 Sequence Structure

In text-based programming languages, the top-to-bottom flow of execution is easily understood. However, LabVIEW must specify the order of execution in a two-dimensional diagram window. In LabVIEW, the execution order can be specified through wires or the **Sequence** structure. The first method is based on the simple mechanism such that if a control **x** is wired to an indicator **y**, **x** will be executed first while generating a value that will subsequently be passed to **y**. Therefore, the order of execution is **x** followed by **y**, which makes the direction of data flow, from **x** to **y**.

Figure 4.9 illustrates an example of using wires in LabVIEW to specify the execution order. Note the wire from the control **x** to the **While Loop**. Even though the **While Loop** is not using the value coming through the wire from **x**, this is a legitimate way of programming in LabVIEW to specify the order of execution. You can also apply this rule to other elements in LabVIEW, including the **For Loop**, **Case** structure, and **Sequence** structure. Therefore, the order of execution in Figure 4.9 is **x** first followed by the **While Loop** and **y**, since the order between **y** and the **While Loop** cannot be

Figure 4.9 Specifying execution order through wires. Control **x** will be executed prior to the execution of **y** and the **While Loop**.

defined. (Actually, the *content* in the **While Loop** starts after the value in **x** is written to **y** since the value in **x** reaches the *boundary* of the **While Loop**, not its content. To see such a behavior, try the light bulb debugging tool to see the signal flow. Therefore, the accurate order of execution is **x -> y -> While Loop**.) The second method (specifying the execution order using the **Sequence** structure) could be preferred by some users since the **Sequence** structure explicitly indicates the order of execution.

You can consider the **Sequence** structure as a collection of movie frames. When you play a movie, the player will project the images sequentially from frames 0, 1, 2, and so on. You can put a **Sequence** structure in the diagram window in the same way you create a **For Loop** or a **While Loop**. When you place it in the diagram window it will come with only one frame, but you can add more by right clicking on the border and selecting **Add Frame After** or **Add Frame Before**. If you add two more frames, for example, you will have frame numbers 0, 1, and 2, as shown in Figure 4.10.

In Figure 4.10, **Sequence Local** is introduced and is a channel through which parameters created in a certain frame can be made available to subsequent frames. Right clicking on the border of the frame and selecting **Add Sequence Local** will create it. As shown in the figure, the data you generate in a certain frame will not be available in the previous frames. (The corre-

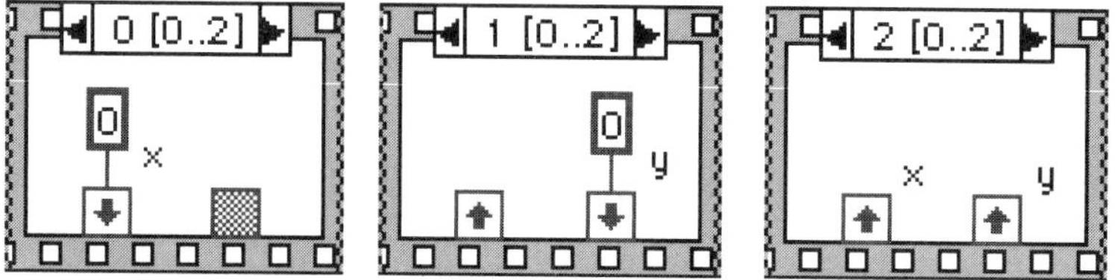

Figure 4.10
Sequence Local of **Sequence** structure. **x** is created in frame 0 so that it is available in all of the subsequent frames; however, **y** is created in frame 1 so that it is not available in frame 0, but is so in frame 2. Grey **Sequence Local** indicates its unavailability.

sponding **Sequence Local** is grayed out.) The order of execution is frame 0, frame 1, and so forth. Note that the **Sequence** structure should be used with discretion since manually specifying the execution order by the **Sequence** structure can degrade the speed performance of your VI.

4.6 Global Variable and Local Variable

GLOB LOCAL First, start with **Local Variable**s. In C/C++ codes, everything you have in each function is local unless it is declared as global. Therefore, the default in C/C++ is local. In LabVIEW, the same is true because once you close your VI, all of the variables in that VI become unavailable until you open it again. Therefore, you may ask why LabVIEW has **Local Variable**s if everything is already local. Note that **Local Variable**s in LabVIEW have a slightly different meaning than conventional local variables in text-based languages. Consider the following example.

Suppose you have a VI with many variables (controls and indicators) connected through complicated wires. Naturally, you will use a larger diagram window to see more objects on the screen. Also, suppose you have an object **x** at one corner of your screen and you need to access it at the other end, but between them are a multitude of wires and objects. Of course, you could wander through objects and wires to make your connection, but this may not be the best way to program. One solution would be to use a **Local Variable** of **x**, which is a copy of **x**, to access its value or even to update its value. In other words, you can read or write from or to **Local Variable**s. This implies that you are no longer limited to the rule that controls can only generate (write) values and indicators can only receive (display) values.

__Local Variable__s of an object (either a control or an indicator) are identical copies of the original objects. Using __Local Variable__s in LabVIEW, you can read from indicators or write to controls. To create one, right click on the object and choose __Create >> Local Variable__, or choose __Functions >> Structures >> Local Variable__ and select a corresponding object from its pop-up menu.

Consider a few examples where the use of **Local Variable**s is unavoidable. In Figure 4.11, you are trying to stop two **While Loop**s using the **QUIT** Boolean control. When you toggle the Boolean switch **QUIT** to generate

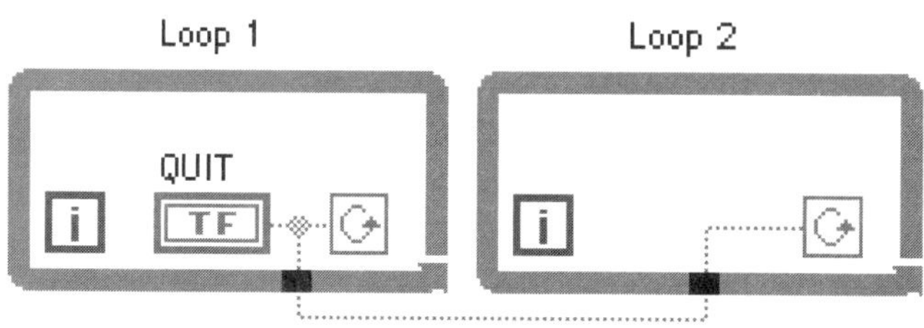

Figure 4.11
Controlling two **While Loops** without using **Local Variable**s.

FALSE, what would happen to both **While Loop**s? There will be no problem in controlling Loop 1 because it will stop when **QUIT** generates FALSE. However, Loop 2 will never stop because of the inherent characteristics of the **While Loop**. That is, once it starts, nothing goes in or comes out until the loop stops. Once Loop 2 starts, no other external value can get in the loop so that the conditional terminal in Loop 2 will keep receiving TRUE forever! Let us look at an alternative as shown in Figure 4.12. If you use a **Local Variable** of **QUIT** inside Loop 2, you can now stop the second loop when the first one stops because the **Local Variable** of **QUIT** will reflect the action of the original Boolean switch **QUIT**.

Another good example of **Local Variable** is incrementing a variable repeatedly. Suppose you want to perform the following:

k = k + 1;

Note that *k* on the left side is an indicator since it is receiving a value, but *k* on the right side is a control since it is generating a value. Also, note that you cannot use one object for both control and indicator without using a **Local Variable**. Therefore, beyond this point there is no such restriction that controls can only write and indicators can only display. In order to perform $k = k + 1$, you can use a **Local Variable** for either one of two *k*'s. Changing

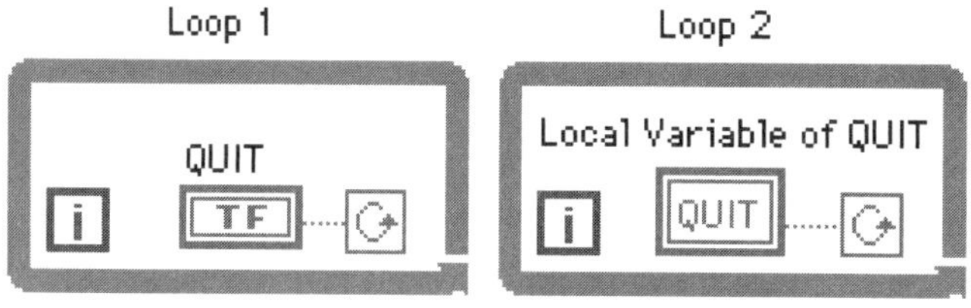

Figure 4.12
Controlling two **While Loop**s using **Local Variable**s.

the **Local Variable**s to indicators or controls can be simply done. Right click on the **Local Variable** and choose either, **Change to Read Local**, or **Change to Write Local**. Now, look at the borderline; if it is thick, it is a control, and if it is thin, it is an indicator. When you use **Local Variable**s, be aware of the update order of each copy, since each one will be updated sequentially. Therefore, certain copies can still contain old data.

As for **Global Variable**s, the concept is identical to conventional programming. For example, global variables are those that can be accessed from different functions in C/C++. In LabVIEW, **Global Variable**s are those that can be accessed from different VIs. Since they are defined outside VIs, they look just like a VI except that they do not have a diagram window. To create a **Global Variable**, choose **Global Variable** under the **Structures** subpalette in the **Functions** palette. Place it in the diagram window and double click on the icon; this will bring up the front panel of the **Global Variable** without a diagram. Put any objects you would like to use as **Global Variable**s and save the front panel. The file name of the VI you just saved will be the name of **Global Variable** that can be used for the objects it contains. Since the use of **Global Variable**s can cause some degree of confusion, readers should use them with discretion.

An example of the use of a **Global Variable** is as follows: Suppose you have multiple VIs running independently but they share the same data source. For example, suppose you have the two VIs **APP1.vi** and **APP2.vi** running independently, where **APP1.vi** acquires data and saves them to a **Global Variable** (update a **Global Variable**) and **APP2.vi** retrieves the data from the **Global Variable** for processing. As you can see, the timing must carefully be managed in order to process the correct data. If the update of a **Global Variable** occurs too fast for the second VI to keep up with, the second VI will miss the data acquired by the first VI.

4.7 Formula Node

Even though LabVIEW itself provides a full functionality for any possible arithmetic expression, there may be situations where you prefer to use equations by typing in the formula. The **Formula Node** is for such purposes. The most useful and common situations where the **Formula Node** is used would be the linearization process of thermocouple, resistance temperature detector (RTD), or strain gauge readings. In these processes, equations are nested versions of the same expression with different coefficients,

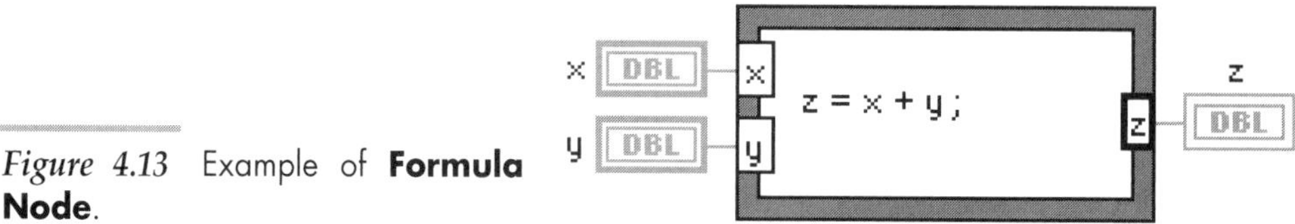

Figure 4.13 Example of **Formula Node**.

and recursive calls can simplify such computations. However, LabVIEW does not support recursive calls; therefore, nested equations will need to be entered manually using the **Formula Node**. Figure 4.13 shows a simple example of how to use the **Formula Node**.

First, put a **Formula Node** from **Functions >> Structures** in the diagram window. Second, right click on the boundary and select **Add Input** or **Add Output**. Conventionally, inputs are put on the left side and outputs on the right side. One thing to remember is that you have to label each input and output, as shown in Figure 4.13. Then use the Text Editor to enter the equation in the box. The variety of expressions is limited, but most trigonometric functions and logical expressions are supported. For the complete list of expressions that are supported by the **Formula Node**, refer to the LabVIEW Users Manual.

4.8 Autoindexing and Shift Register

One of the unique features of LabVIEW is *autoindexing* and *shift registers* with loops. It is natural to consider indexing elements and using shift registers in **For Loop** or **While Loop** because you repeat the same operations. First, consider autoindexing.

4.8.1 Autoindexing

Suppose you are acquiring a single temperature reading in a loop (either a **For Loop** or a **While Loop**) at each iteration, and you want to collect each reading in an array. Or, you already have an array, and you want to parse each element out at each iteration. LabVIEW allows you to do both by enabling or disabling the autoindexing feature at the boundary of the loop. Figure 4.14 shows the **For Loops**, Loop 1 and Loop 2. In Loop 1, a single random number is generated at each iteration, and it repeats 10 times. The two

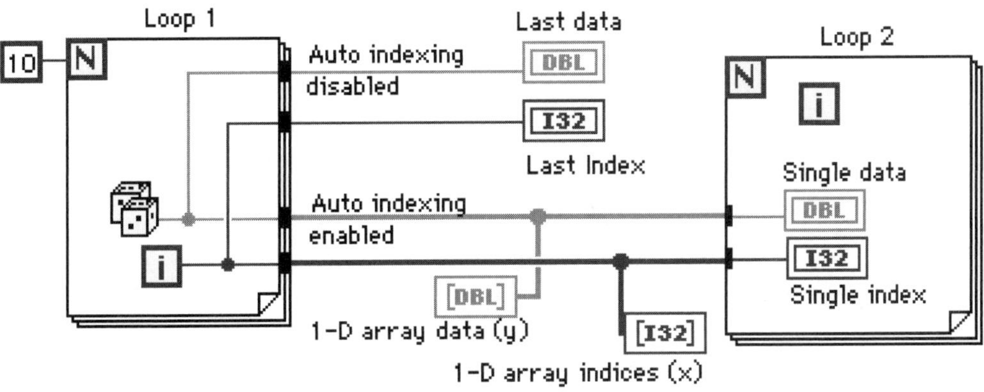

Figure 4.14
Autoindexing with **For Loop**s.

lower wires leaving Loop 1 are using autoindexing, and they are thicker, indicating the increase of dimension to a 1-D array of 10 elements from a single number. In Loop 2, a 1-D array enters and each element gets parsed out due to autoindexing. This explains autoindexing for the data entering the loop.

If you disable the indexing by right clicking on the tunnel and selecting **Disable Indexing**, the data leaving Loop 1 will be the *last* element generated, which will be the tenth element since the total number of iterations is 10 in Figure 4.14. The same thickness of the two upper wires leaving Loop 1 indicates that the dimension of the data has not been increased. Also, note that you do not have any input wired to the count terminal, [N], of Loop 2, but it will still iterate 10 times due to the size of the two 1-D arrays entering Loop 2 with the indexing enabled. This is another unique feature of the **For Loop** of LabVIEW. If you disable the indexing, the second **For Loop** will not be able to know how many elements there are in the input array. Then, you must provide an input to the count terminal to specify the total number of iterations. Autoindexing is a useful feature in array manipulation, and arrays will be discussed further in Chapter 6.

If a 1-D array is wired to the border of the ***For Loop*** *with indexing enabled, it will automatically iterate N times, where N is the size of the array, and it will ignore any number connected to the count terminal. However, if the indexing is disabled, you must wire an input to the count terminal,* [N].

4.8.2 Shift Register

While dealing with loops, you may have situations where the tracking of previous data elements is needed. LabVIEW provides shift registers to accomplish such tasks. You can create shift registers by right clicking on the border of either **For Loop** or **While Loop** and selecting **Add Shift Register**. This will create a pair of shift registers at both the left and the right borders of the loop, with a down arrow on the left and an up arrow on the right. The left one can be dragged down to include more than one element. You can also increase the number of shift registers on the left side by right clicking on the left shift register and selecting **Add Element**. An example is shown in Figure 4.15.

Figure 4.15 shows a **For Loop** with a shift register that has three elements, and $x[i]$ represents the data at the ith iteration. At the $(i+1)$th iteration, $x[i]$ will show up at the top element on the left side. At the following iteration, the top element will be pushed down to the second and the content in the second will be pushed down to the third, and this process will continue. Therefore, you will have a lag of one iteration between each element on the left boundary. The one leaving on the right side is the last one, $x[9]$. This feature can be used for moving average calculation for thermocouple signals or some noisy signals. If you average the three shift register elements on the left, the averaged value will become a moving average of three data elements. The moving average technique is studied again in Chapter 14.

Lastly, it is recommended that all of the elements of the shift registers on the left side be initialized with an initial value. In Figure 4.15, you have initialized three elements with zero in order to avoid any possible confusion, especially when the starting value is different from the default value of the data type wired to the shift registers. LabVIEW will not return the Broken Run Arrow even if no initial value is assigned to the shift registers, and you

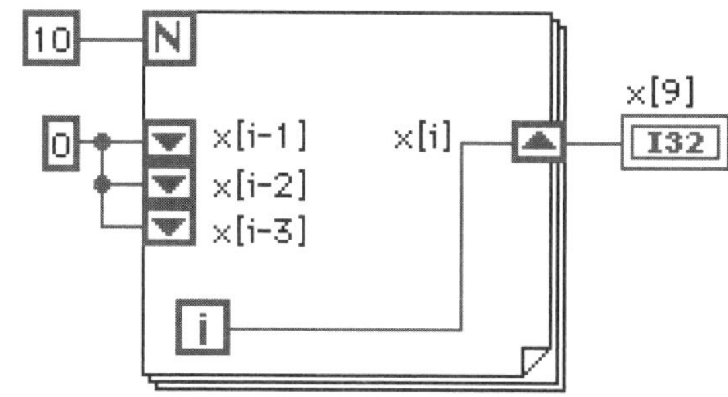

Figure 4.15 Shift register with three elements.

will not be able to realize the mistake until the VI returns wrong values. Therefore, it is always a good idea to initialize the shift registers manually.

PROBLEMS

4.1 Create a sub VI **P04_01.vi** as shown in Figure P4.1. The input of the sub VI is **x**, and the output is **y**.

(a) How many times will the **While Loop** iterate?

(b) What does this sub VI do? Hint: Run the VI twice with two different values for **x**.

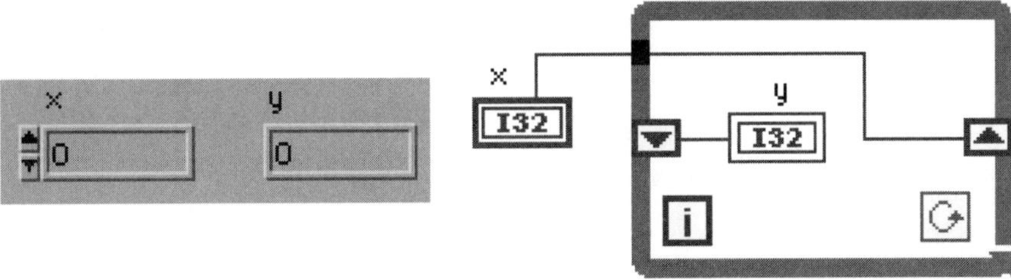

Figure P4.1

4.2 Create a VI **P04_02.vi** that calls the sub VI **P04_01.vi** three times simultaneously with three different input values. The diagram window and the front panel are shown in Figure P4.2.

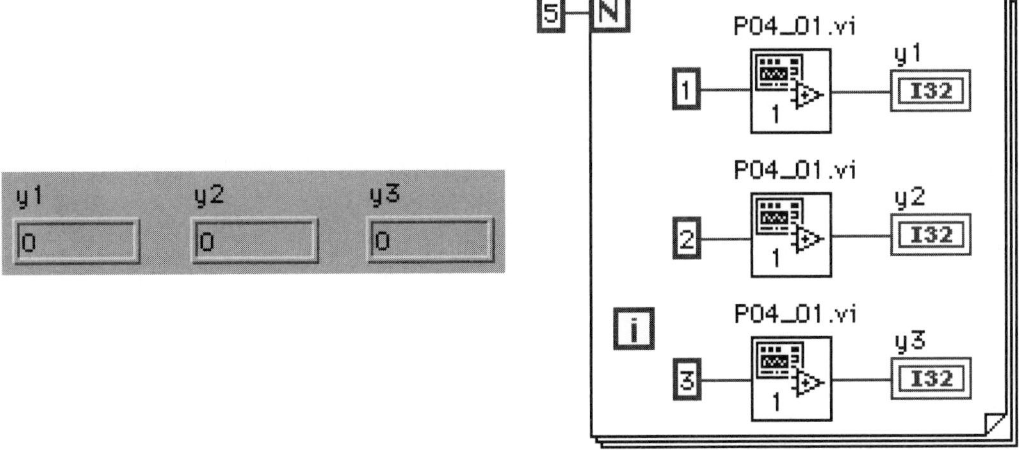

Figure P4.2

(a) Based on the answer in Problem 4.1(b), what are the expected values in **y1**, **y2**, and **y3** at each iteration? (Note that the total number of iterations is 5.)
(b) Run the VI **P04_02.vi** with the **Highlight Execution** light bulb on in the diagram window, and observe the three output values at each **P04_01.vi**. Why do the values displayed in **y1**, **y2**, and **y3** not agree to the answers in part (a)?

4.3 Select the **Reentrant Execution** option for **P04_01.vi** and run **P04_02.vi**, which you created in Problem 4.2, with the **Highlight Execution** light bulb on in the diagram window. Do the values of **y1**, **y2**, and **y3** match the answers in Problem 4.2(a) now? Explain why or why not.

4.4 Create a VI **P04_04.vi** and place the VIs shown in Figure P4.4. Using the Text Editor, or **String Constant**, list the path of each VI in the diagram window. Then create controls and indicators to each VI using the technique mentioned in subsection 4.1.1. After creating them, make sure to arrange the controls and the indicators nicely in the front panel.

Write To Spreadsheet File.vi

Simple Error Handler.vi

AI Config.vi

AI Start.vi

Figure P4.4

4.5 An array is a data structure that contains one or multiple elements of the same data type. Using the **For Loop** and autoindexing feature, create the following array **x** of 21 elements in **P04_05.vi**:

$$\mathbf{x}[k] = e^{-\frac{k^2}{10}}, k = -10, -9, \ldots, 9, 10$$

You will need to use **Exponential** from **Functions >> Numeric >> Logarithmic**.

4.6 Create a VI that turns on a LED when a random number generated by **Functions >> Numeric >> Random Number (0-1)** is greater than 0.5, and save it as **P04_06.vi**. Use **Functions >> Time & Dialog >> Wait (ms)** to iterate the **While Loop** at the rate of 500 ms. The complete diagram window is shown in Figure P4.6.

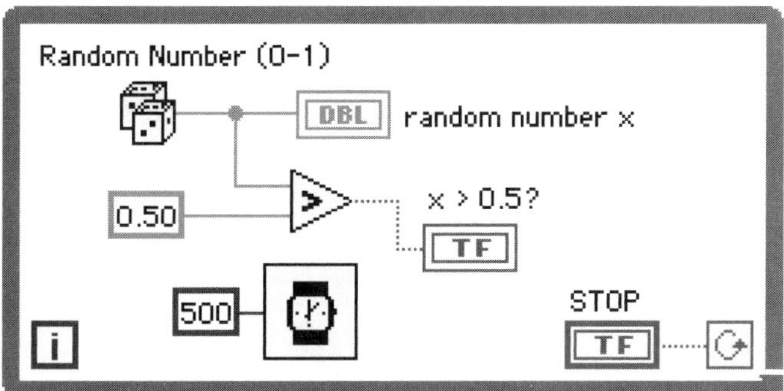

Figure P4.6

4.7 Figure P4.1 shows how a VI can have memory using the **While Loop** without any Boolean control wired to the conditional terminal. This is because the **While Loop** always executes at least once such as do while statement in C++. In other words, the **While Loop** in LabVIEW checks the condition *after* executing one iteration. Keeping that in mind, consider Figure P4.7, which is the incomplete diagram window of the pseudo code given in this problem. Find out why the VI needs to be modified, correct the error, and save it as **P04_07.vi**. (Repeat executing the VI to see the incompleteness of it.) When the VI stops after correcting the error, *k* should be 10, and *sum* should be 55 *every time* you run it.

```
k = 0; sum = 0;
while (k < 10){
     k = k + 1;
     sum = sum + k;
     display k and sum for 1 second;
}
```

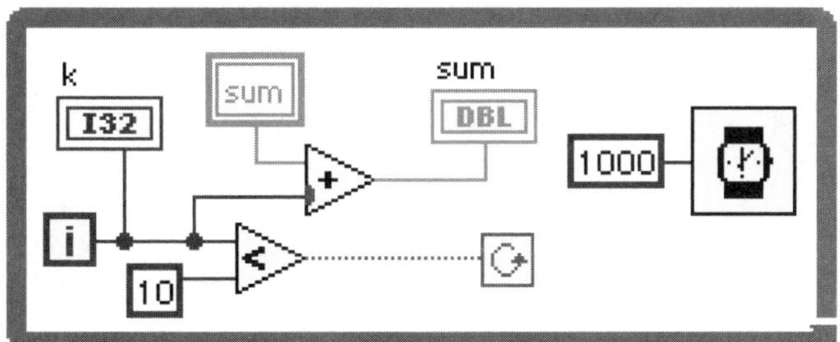

Figure P4.7

4.8 Rebuild the VI in Problem 4.7 using a shift register instead of a **Local Variable**, and save the VI as **P04_08.vi**.

4.9 Realize the following pseudo code using a **Case** structure with multiple frames instead of nested case frames. Save the VI as **P04_09.vi**.

```
if (k = 0)
     display a string "Case 0";
else if (k = 1)
     display a string "Case 1";
else if (k = 2)
     display a string "Case 2";
else
     display a string "Default case";
```

4.10 Suppose that data samples are streaming into your computer, and you want to average a certain number of samples continuously. This averaging scheme is often said to be *moving average* and is used to smooth noisy input data, such as temperature or strain gauge readings. Create a VI **P04_10.vi** that performs the moving average with four samples using a shift register: current sample and three previous samples. Use **Functions >> Numeric >> Random Number (0-1)** to generate a single data sample at each iteration in a **While Loop**. This is often said to be moving average of window size 4. See Figure P4.10 for details.

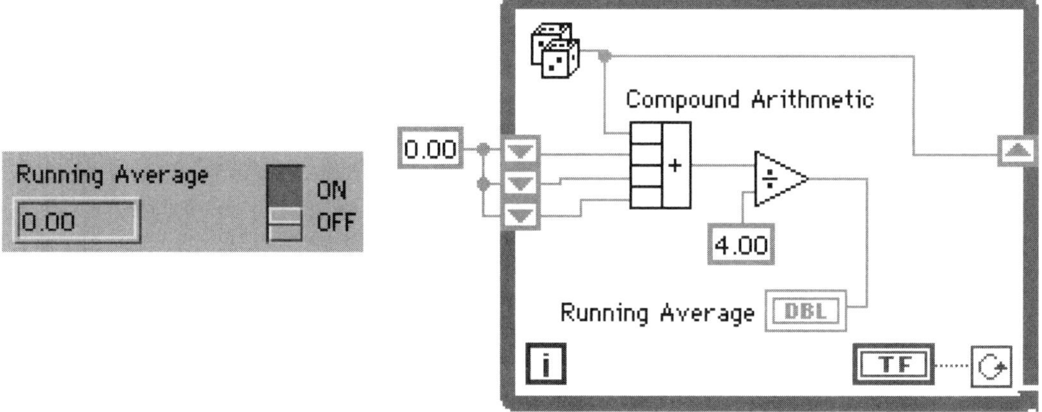

Figure P4.10

Data Display

5

Data display is an essential aspect of almost any application. In fact, visualization capability is so important that the whole purpose of programming may be meaningless without it. Since LabVIEW is a virtual instrumentation tool that concentrates heavily on data acquisition, displaying data naturally becomes an important part of G-programming.

In LabVIEW, there are two methods for displaying data: either with charts or with graphs. Within each method, there are different types. There are two types of charts: **Waveform Chart** and **Intensity Chart**. There are three types of graphs: **Waveform Graph**, **XY Graph**, and **Intensity Graph.** Since the majority of applications utilize **Waveform Chart**s and **Waveform Graph**s, they will be discussed first, in depth. Discussion about the remainder can be found in the latter part of this chapter. Let us begin with the **Waveform Chart**.

5.1 Waveform Chart

All of the charts and graphs can be obtained from the front panel. The **Waveform Chart** can be found under **Controls** >> **Graphs**. Before proceeding further, you need to consider two important questions about displaying data in LabVIEW: (1) Do you want to display the data online or offline? and (2) Do you want to display a single line or multiple lines of data (multiple channels) on a graph or a chart?

Generally, charts are used for online data display, and graphs are for offline display.

If you want to display real-time data acquisition, use a chart. If you want first to acquire data, save them, and then display them after acquisition is complete, use graphs. Consider the following example using the **Waveform Chart** in Figures 5.1 and 5.2. Build the example following the diagram shown in Figure 5.2. You have a **While Loop** that is controlled by a Boolean switch **Quit**, and the **Waveform Chart** receives a random number from the random number generator at every iteration. In the front panel in Figure 5.1, change the maximum value of the y-axis from the default value of 10 to 1.

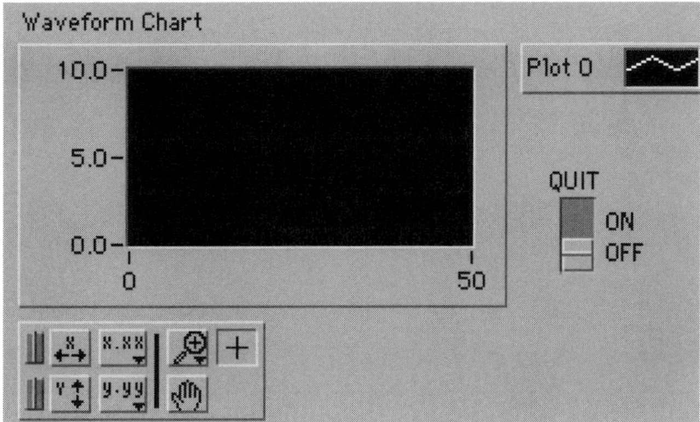

Figure 5.1 Front panel of an example VI with a **Waveform Chart**.

Since the default value of a Boolean switch is FALSE, you will have to set the Boolean switch to TRUE before you run the VI. Here, you will be introduced to the different settings of Boolean switches. Go to the front panel as shown in Figure 5.1, right click on the Boolean switch **Quit**, and select **Mechanical Action** from its pop-up menu. Choose **Latch When Released** and then set the switch **Quit** to the true position. Next, go to the **Operate** pull-down menu and select **Make Current Values Default**. This will cause the mechanical action **Latch When Released** to latch the switch back to the true position after generating FALSE when it is pressed. At the same time, **Make Current Values Default** will set TRUE (On) as the default state of the Boolean switch **Quit**. (Actually, **Make Current Values Default** will set the entire objects in the front panel to their current settings as their default values. To change the default value of an individual object, use **Data Operations** >> **Make Current Value Default** from its pop-up menu. Note that the word *value* is singular in the pop-up menu.) Then, the next time you open this VI, it will be ready to run since the switch is already at the true position. This simple combination of options may help you run a **While Loop** with the fewest preparation steps.

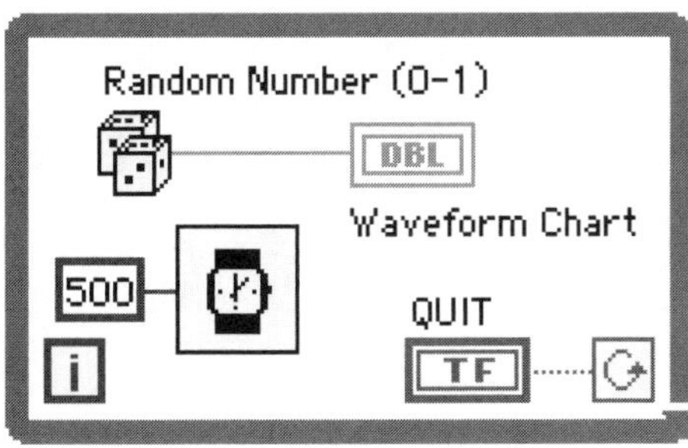

Figure 5.2 Diagram window of the VI in Figure 5.1.

*If you are using a **While Loop** with a Boolean switch to control it, first set the Boolean control to true. Select **Mechanical Action** >> **Latch When Released** from its pop-up menu, and choose **Operate** >> **Make Current Values Default**. Then save the VI to preserve the settings.*

More details about the rest of the mechanical action options are introduced in the next subsection. Assuming you have finished the VI in Figures 5.1 and 5.2, run it to view the single channel data. The **Wait (ms)** is in the **While Loop** in Figure 5.2 to slow down the display of single-channel data.

If you repeatedly run and stop the VI, you will realize that the chart does not refresh the screen. This means that the data from the previous run remain on the chart display. You can clear the display by selecting **Data Operations** >> **Clear Chart**. However, this can be done programmatically with Attribute Node. (See Section 5.1.2 for the detailed discussion of the Attribute Node.) Before starting a discussion on multiple channel data display, the different mechanical actions of Boolean switches will be discussed.

5.1.1 Mechanical Actions of Boolean Switches

Open an example from the following path: LabVIEW folder >> **examples** >> **general** >> **control** >> **booleans.llb** >> **Mechanical Action of Booleans.vi**. The front panel of the example is shown in Figure 5.3. This example has an

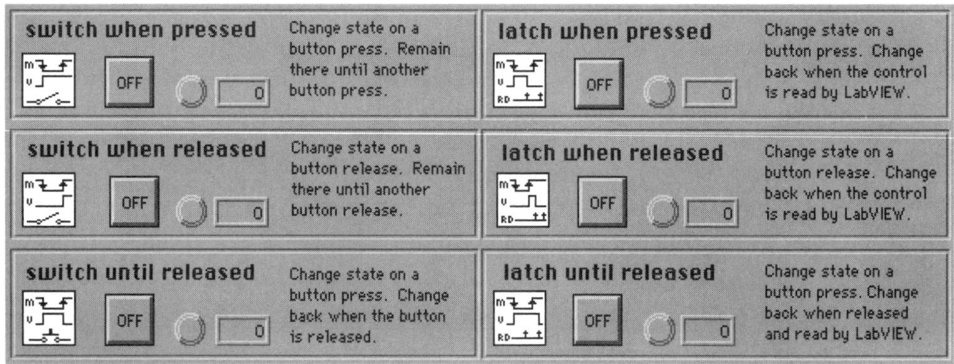

Figure 5.3
Front panel of **Mechanical Action of Booleans.vi**.

excellent description of the six different mechanical actions you can set on Boolean switches. One of the options, **Latch When Released**, is useful with the use of **While Loop**, as discussed in the previous section, in combination with the option **Make Current Values Default** under the pull-down menu **Operate**, or the option **Make Current Value Default** from the pop-up menu of Boolean switches. Execute the VI to see the differences between actions.

5.1.2 Attribute Node

You may encounter situations where you want to customize and control objects in the front panel. For example, you may want to have a Boolean switch blink when a data reading, such as temperature, exceeds a preset threshold. You may want to make certain option buttons disappear if you do not want the end users to be able to see them depending on different situations. Or, as mentioned in the previous section, you may want to clear the screen of the **Waveform Chart** every time you run the VI.

In LabVIEW, almost every object in the front panel has an Attribute Node with which you can change the object properties. As for the actions you can take with Attribute Nodes, there are two options available: (1) You can *read* the current setting of an object from its Attribute Node, or (2) you can *set* an object to desired settings. Suppose, for example, you want to programmatically position a Boolean switch in the front panel. To do so, you must first read the switch position to determine if repositioning is necessary. If it is, you would then change its coordinates to the desired position.

Some of the commonly used Attribute Nodes will be listed here, but you are encouraged to explore the others, too. Due to the wide variety of Attribute Nodes, it is not practical to discuss all of them. Therefore, only the most commonly used ones will be discussed.

Consider again the **Waveform Chart** example shown in Figure 5.2. Right click to pop up the menu on the orange **Waveform Chart**, with DBL indicating that it is displaying double precision type data. Select **Create** >> **Attribute Node** to create an Attribute Node, as shown in Figure 5.4. Using the Hand Tool, click and hold on the icon to review all of the options available. A partial list of the options is as shown in Figure 5.5. As seen in Figure 5.5, there are quite a number of options available just for the **Waveform Chart**. Since most of the options are quite useful, they will be explained in the order they appear in Figure 5.5.

The first option, **Visible**, allows users to programmatically make the chart visible or invisible. Note that displays in the front panel in LabVIEW impose

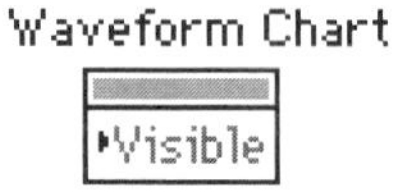

Figure 5.4 Attribute Node of the **Waveform Chart**.

overheads onto the performance of VIs. Excessive use of overheads can significantly slow down the VI execution speed. Therefore, it is recommended to reduce the number of on-screen displays and display updates of graphs, charts, and numeric indicators. If you must have multiple **Waveform Charts** in the front panel, you may optimize your VI by making charts visible only when you need to view them using Attribute Nodes. Therefore, using a Boolean switch in the front panel to control the visibility of charts can increase the execution speed of your VI.

An example of the next option, **Disabled**, is shown in Figure 5.6. A few aspects of this example need to be pointed out. First, note that you are using the **Sequence** structure to make sure that the **Disabled** setting becomes effective *before* the main task (displaying data) starts. Since LabVIEW is a G-programming language, the mere presence of objects in the diagram window will not specify the execution order. The order of execution is defined when objects are wired or a **Sequence** structure is used. (If two objects are connected through a wire, the one expecting data will *wait* until the data arrives.) To see the significance of the order of execution, consider a case where the Attribute Node refreshes the screen of the **Waveform Chart**. This option is **History Data**, which is the last option, and will be discussed again

Figure 5.5 A partial list of options of the Attribute Node of the **Waveform Chart**.

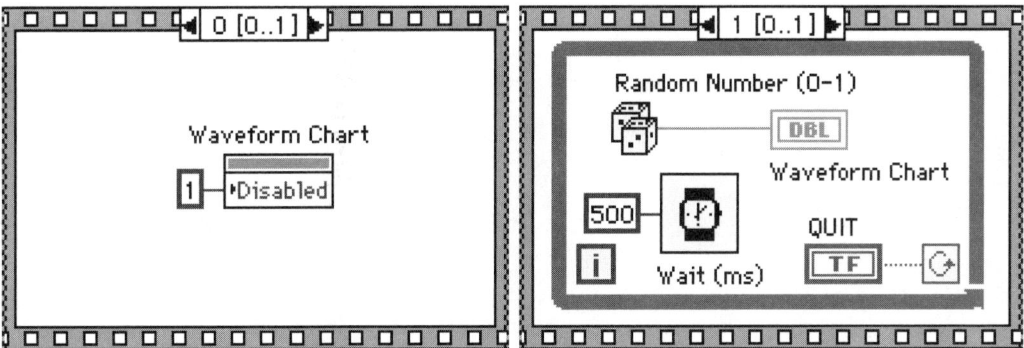

Figure 5.6
Attribute Node **Disabled** of the **Waveform Chart**.

shortly. Then consider the following question: What would happen if the **Waveform Chart** displayed the data followed by calling the Attribute Node that will refresh the chart display? All of the painfully acquired data would disappear because the execution order is not correct! Such a situation can be avoided by using a **Sequence** structure to clarify the execution order.

The next aspect that needs attention is the arrow on the left side of the Attribute Node, where a **Numeric Constant** 1 is wired in Figure 5.6. If the arrow is on the left side, it implies that the Attribute Node is expecting an input setting. In other words, you can *set* a property to the object of the Attribute Node. If the arrow is on the right side, it means that you can *read* the current setting of the object that the Attribute Node belongs to (in this case, the **Waveform Chart**). You can change the direction by right clicking on the Attribute Node and choosing **Change To Read** or **Change To Write**.

The third aspect you must consider pertains to the creation of a correct input or output to the Attribute Node. Due to the wide variety of options of Attribute Nodes, it is impossible to remember all of the correct input or output data types for the Attribute Nodes. However, creating the correct inputs and outputs can be achieved easily by right clicking on the terminal and choosing either **Create Constant**, **Create Control**, or **Create Indicator**. This method was already introduced in Chapter 4 during the discussion of coercion dots. (See Section 4.1.1.) Without using this method, it would be difficult to find out that the **Disabled** option in Figure 5.6 expects an unsigned 8-bit integer data type.

Finally, note that you can control more than one property with a single Attribute Node. To see this, go to frame 0 of the **Sequence** structure in Figure 5.6 and change the cursor to the Arrow Tool. Point the Arrow Tool to the right bottom corner of the Attribute Node box until it turns to a reversed double L shape, as shown in Figure 5.7. Now, click and hold the mouse button and drag it down to increase the number of elements, then release the button to complete the process. In Figure 5.7, there are four options selected, but there is no limit to the number of options you can have. The order of execution of these options is from top to bottom, and the same item can appear in the list more than once. Also, each item can have different input or output directions so that the combination of reading and setting can be done in a sequential manner.

The next option in Figure 5.5 is **Key Focus**. This option selects the object of the Attribute Node as default, but it does not apply to the **Waveform Chart** or the **Waveform Graph** since you do not enter inputs to them using keyboard keys. A good example of the use of **Key Focus** would be Boolean switches. If you select **Key Focus** for the option of the Attribute Node of a Boolean switch and wire a Boolean constant TRUE to its input, the switch will automatically be selected when the Attribute Node is executed. Therefore, you would not need to move your mouse to select the Boolean switch. Instead, you can just hit the Return key since the switch would already be selected. This is a useful option if you have multiple switches in the front panel and want to have one of them selected as a default.

The next option, **Blinking**, will control the blinking of an object. For example, it can be used with a **Case** structure for displaying a warning message. If a certain condition happens, set an object to blink; otherwise, do nothing. The option **Position** allows users to programmatically change the

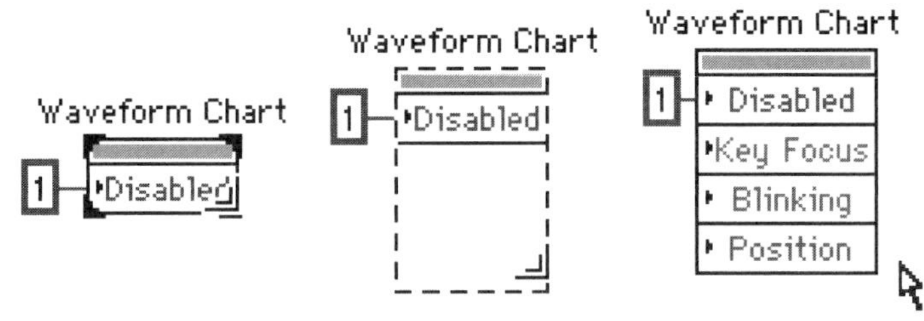

Figure 5.7
Multiple options with a single Attribute Node. A different direction can be assigned to each individual option.

position of an object in the front panel. **Plot Area Size** will control the plot size so that you can organize the screen more efficiently if you have a number of objects. **X Scale Info** and **Y Scale Info** are options that allow users to set the minimum, maximum, and increment of *x*- and *y*-axes of **Waveform Charts** or **Waveform Graphs**. The last item that is commonly used is **History Data**. Figure 5.8 is an example of this option.

To begin, modify frame 0 of the **Sequence** structure in the diagram window shown in Figure 5.6 to resemble Figure 5.8. In Figure 5.8, the input is a one-dimensional (1-D) array constant that is created by selecting **Create Constant** from its pop-up menu. In this example, LabVIEW created a 1-D array of orange floating-point numbers because you have one channel of data to be displayed. If, however, you had multiple channels of data wired to the **Waveform Chart**, LabVIEW would have automatically adjusted the dimension of the input array to 2-D. For a detailed discussion of arrays, see Chapter 6.

Figure 5.8 shows that the input to the Attribute Node **History Data** is an empty array. Setting the history of **Waveform Chart** to an empty array results in deleting all of the traces of the previous data in the chart. Run the VI to see the effect. The display should be refreshed before graphing new data every time the VI executes.

The alternative method to using the **Sequence** structure to specify the execution order is illustrated in Figure 5.9. Even though the **While Loop** does not expect any data, the wire from the Attribute Node completes its end on the boundary of the **While Loop** without resulting in the Broken Run Arrow. The simple act of attaching the wire to the **While Loop** from the Attribute Node specifies that the **While Loop** will not run until the Attribute Node executes. (This trick also works with the **For Loop**.) Therefore, Figure 5.9 has the same effect as Figure 5.6: executing the Attribute Node first. To

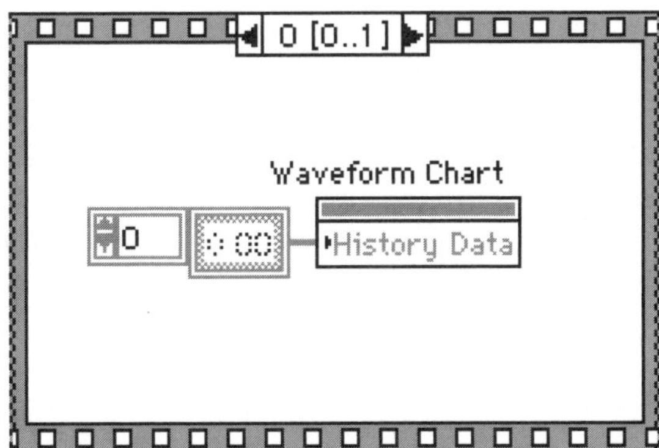

Figure 5.8 Attribute Node **History Data** of the **Waveform Chart**.

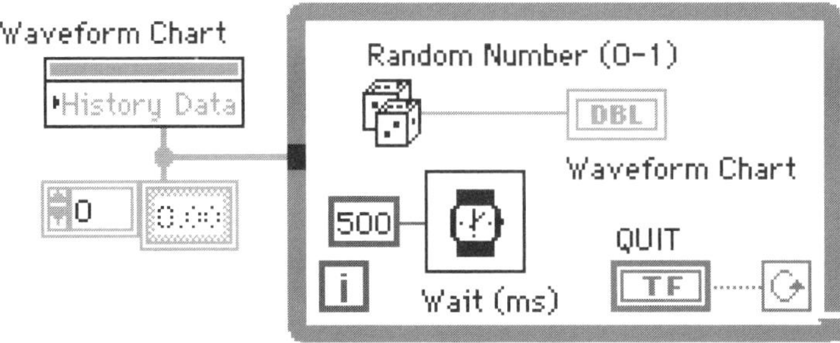

Figure 5.9
Alternative way to specify the execution order.

verify this, use the **Highlight Execution** to run the VI, and you will see that the array constant reaches both the Attribute Node and the *boundary* of the **While Loop** simultaneously, but the *content* in the **While Loop** does not start until the constant gets written to the Attribute Node.

As for the rest of the options in Figure 5.5, you are encouraged to explore them for yourself. When you do, you will likely find that some of them are self-explanatory and some are not. In order to find out what other options do, create a simple VI to test the effect of them before implementing them. Online help is also a good place to look for more information.

5.1.3 Update Mode of the Waveform Chart

There are three different update modes for the **Waveform Chart**: **Strip Chart**, **Scope Chart**, and **Sweep Chart**.The **Strip Chart** is the default setting, and the screen scrolls as the display reaches the right end of the chart. The **Scope Chart** refreshes the display once the chart screen is full, but the **Sweep Chart** does not refresh the screen. Therefore, **Sweep Chart** mode will display new data over the old data display. In summary, the **Scope Chart** and the **Sweep Chart** both start from the beginning of the chart screen when the screen is full, but the **Scope Chart** does not keep track of the previous data (refresh its screen). Since the mode **Scope Chart** does not trace the previous data (does not have memory), it is the fastest and has the most efficient memory usage. The **Sweep Chart** mode will be the next fastest, and the **Strip Chart** mode will be the slowest since it has to remember previous values to display. Changing the display mode can easily be done. Simply right click on the **Waveform Chart** and go to **Data Operations** >> **Update Mode** to choose one of the three modes. This can also be done programmatically with the Attribute Node of the **Waveform Chart**.

In LabVIEW, the number of displays or display updates is closely related to the amount of overheads; therefore, updating many of the **Waveform Charts** can consume quite a bit of memory and require significant execution time. To optimize your VI in terms of execution speed and overheads, you should minimize the number of image updates in the front panel. For example, suppose you have 1000 data samples to display. In this case, displaying them all at once will be much faster than displaying 100 points 10 times. A rule of thumb for optimizing overhead is that you should simplify the front panel and its operation as much as possible. This can be done by simplifying the number of updates in the front panel and hiding objects that need not be shown all of the time.

5.1.4 Multiple Channels of Data on the Waveform Chart

To understand how to plot more than one channel of data, a brief discussion of a new data type *bundle* will follow. (It will later be presented in detail in Chapter 6.) Consider Figures 5.10 and 5.11. The new data type bundle is

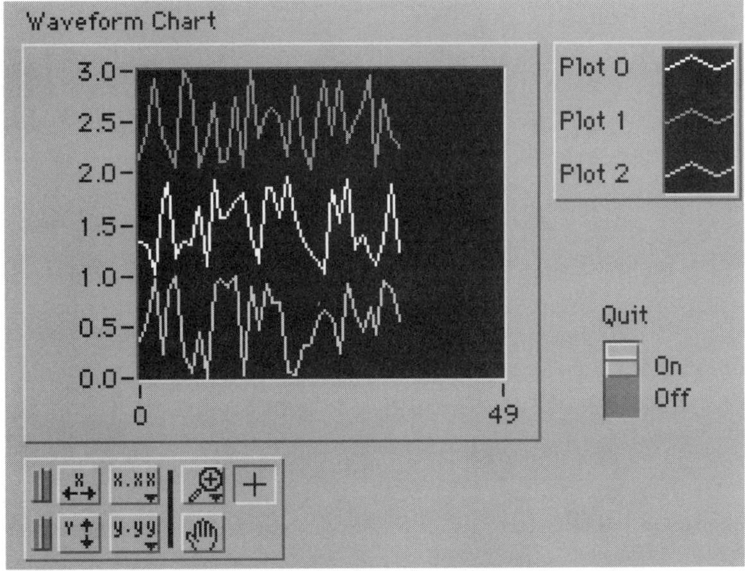

Figure 5.10
Displaying multiple channels of data on a **Waveform Chart** (front panel).

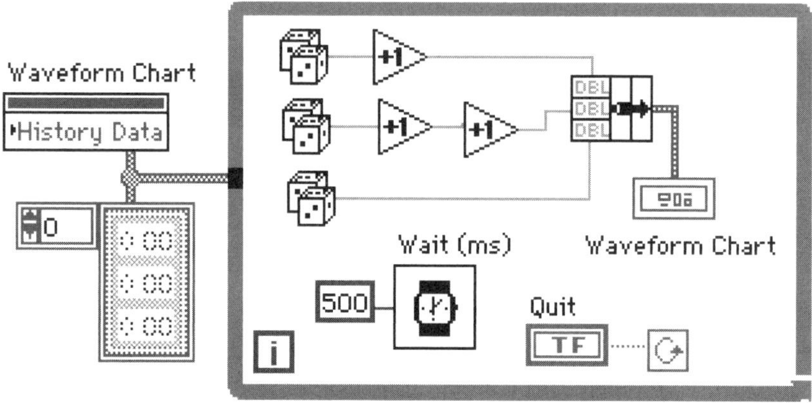

Figure 5.11
Displaying multiple channels of data on a **Waveform Chart** (diagram window).

introduced in Figure 5.11. In this example, a wire specifies the execution order between the **While Loop** and the Attribute Node of the **Waveform Chart**. As you can see from the constant wired to the input of the Attribute Node, its type is different from the empty array you saw in Figures 5.8 and 5.9. Three random number generator functions are generating three data points simulating data from three different channels. In this example, you can see how **Bundle** combines them and sends them to the **Waveform Chart**.

The front panel in Figure 5.10 shows the screen shot of the VI while it is running. Note that **Waveform Chart**s display a 2-D array of data (multiple channel data) columnwise. Consider the following 2-D array (matrix) **X** of size M by N, which is the collection of multiple channels of data:

$$\mathbf{X} = [\mathbf{x}_1 \ldots \mathbf{x}_N] \tag{5.1}$$

LabVIEW returns data from multiple input channels in a 2-D array format as in (5.1). (This is discussed again in the chapters about data acquisition.) **X** is a matrix where each column $\mathbf{x}_i$ is the data vector of size M from the ith channel. Then the ith line in the **Waveform Chart** will correspond to the ith data vector $\mathbf{x}_i$. Therefore, Plot 0 displays the vector $\mathbf{x}_1$, Plot 1 does $\mathbf{x}_2$, and Plot 2 corresponds to $\mathbf{x}_3$ in Figure 5.10 if **X** in (5.1) is wired to the **Waveform Chart**.

5.2 Waveform Graph

Waveform Graphs are different from **Waveform Charts** in many ways. First, **Waveform Graphs** are generally used for offline display. In other words, **Waveform Graphs** are mostly used to display data that are already acquired and saved, whereas **Waveform Charts** are mostly used to display data in real time. Second, you can specify the initial point and interval of the x-axis with **Waveform Graphs** using **Bundle**, whereas **Waveform Charts** do not allow it without using the Attribute Node. This is useful if you need to customize the x-axis scales when, for example, you are displaying the result of Fourier transform. (This is discussed in detail in Chapter 14.) Third, you use **Build Array** to combine multiple channels of data to display them together on a single **Waveform Graph**, whereas **Bundle** is used for multiple channel data display on a single **Waveform Chart**. Fourth, **Waveform Graphs** display data rowwise, not columnwise. In other words, if you have a matrix **X** that is the collection of multiple channels of data, as in equation (5.1), you must *transpose* it before you feed the matrix to the **Waveform Graph**. This is because each graph will correspond to each row vector instead of column vector in **X**.

Often, LabVIEW programmers forget to transpose the data matrix before displaying it on the **Waveform Graph**, so extra attention must be paid to the correct format of data. Transposing a matrix can be done easily by using **Transpose 2D Array**, found under the **Functions** >> **Array** subpalette. An alternative method to the use of **Transpose 2D Array** would be to select the option **Transpose Array** from the pop-up menu of the **Waveform Graph**. To do so, right click on the **Waveform Graph** and select **Transpose Array**. Specifying the initial value and interval of the x-axis is illustrated in Figure 5.12. The **Numeric Constant** zero is the initial value of x-axis, 0.5 is the interval of each data sample, and the array constant is 1-D data to be displayed.

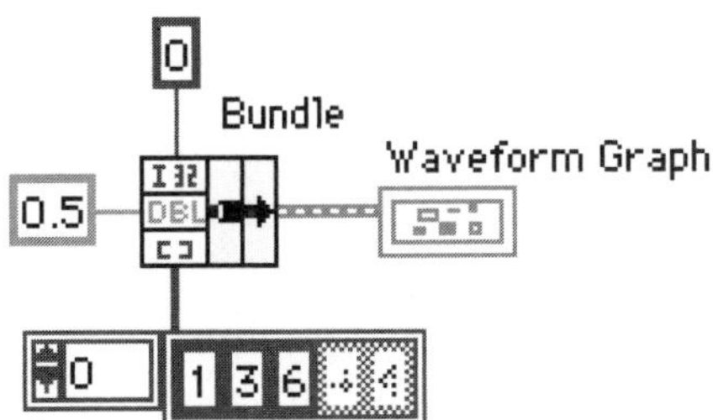

Figure 5.12 Specifying the initial value and the interval of x-axis with the **Waveform Graph**.

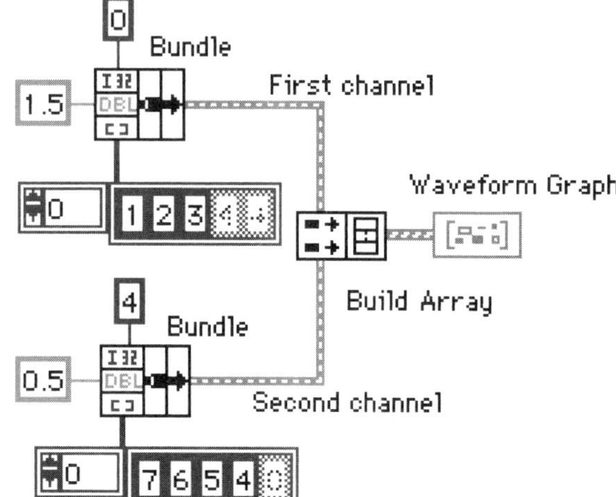

Figure 5.13 Multiple-channel display with individual initial and interval values for *x*-axis. The lengths of the two data arrays from each channel can be different.

5.2.1 Multiple Channels of Data on the Waveform Graph (I)

You can display multiple channels of data where the initial value and the interval can be individually specified. An example is shown in Figure 5.13.

The first channel has 0 and 1.5 for its initial value and interval of *x*-axis, and the second one has 4 and 0.5. The bottom input of each **Bundle** is a 1-D array data. Notice that the thickness of the output wire of **Build Array** increases as you combine more input arrays. Thicker wire indicates the increment of dimension. If you have multiple channels of data with the same

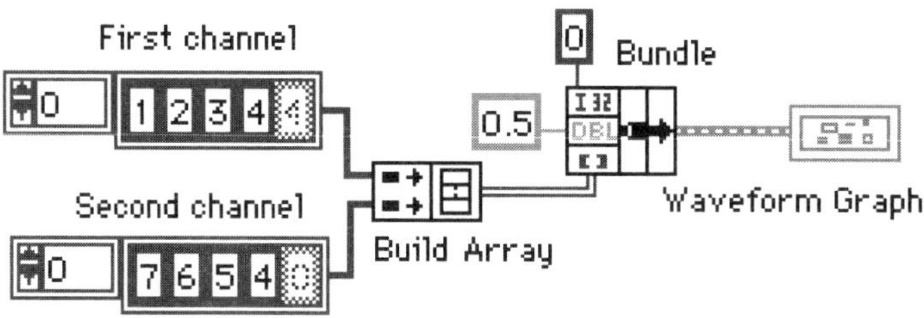

Figure 5.14
Multiple-channel display with the same initial and interval value for *x*-axis. The lengths of the two data arrays must be the same. Otherwise, the smaller one will use 0 (zero) to match its length to the bigger-size data array.

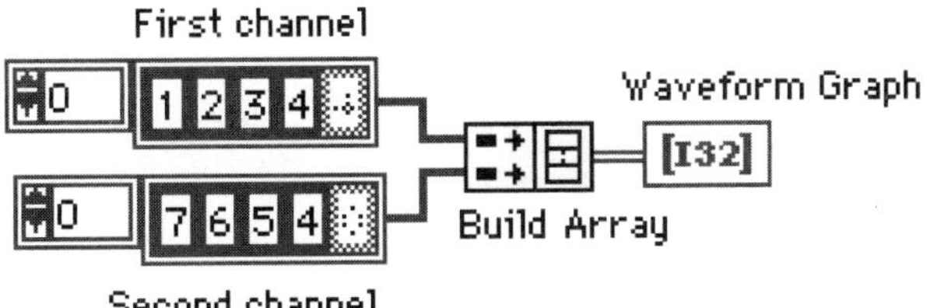

Figure 5.15 Multiple-channel display with default *x*-axis. The lengths of the two data arrays must be the same. Otherwise, the smaller one will use 0 (zero) to match its length to the bigger-size data array.

initial value and interval, you may display them using both **Build Array** and **Bundle**, as shown in Figure 5.14. **Build Array** and **Bundle** can be found in the **Array** and the **Cluster** subpalettes in the **Functions** palette. More details about arrays and clusters are presented in Chapter 6.

5.2.2 Multiple Channels of Data on the Waveform Graph (II)

The previous subsection discussed how to display multiple channels of data on a **Waveform Graph** while specifying the initial value and the interval of x-axis. This subsection will discuss how to perform the same task with default x-axis instead. The default x-axis starts at zero with interval 1, and Figure 5.15 shows the diagram window to display two channels of data without specifying the initial value and the interval of x-axis. Note that this will provide a plot similar to that in Figure 5.10, but **Waveform Charts**, as in Figure 5.10, are usually for real-time display.

5.3 XY Graph

If you need to plot data for given x-axis values, the **XY Graph** can be used. For example, you want to plot sin(x) versus cos(x), as shown in the diagram window of Figure 5.16. Note that **Bundle** is used for a single-channel display using the **XY Graph**. In case of multiple channels of data, **Build Array** needs to be used after **Bundle**, as shown in Figure 5.17. Figure 5.17 shows an example of plotting sin(x) versus cos(x) and $\sin^{-1}(x)$. The x-axis whose values are sin(x) is shared for both plots.

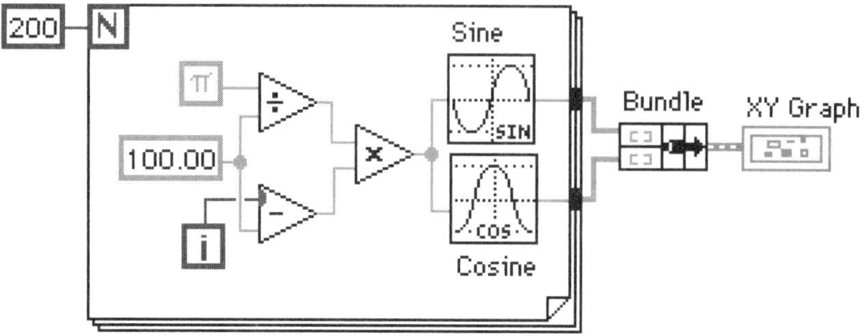

Figure 5.16
Single-channel display with the **XY Graph**.

5.4 Intensity Chart and Intensity Graph

The relationship between the **Intensity Chart** and the **Intensity Graph** is very similar to that between the **Waveform Chart** and the **Waveform Graph**. The major difference is the existence of memory: **Intensity Chart**s have memory whereas **Intensity Graph**s do not. The **Intensity Chart** and the **Intensity Graph** are especially useful if you have a 2-D array data (a matrix) where each element represents the density of a pixel of an image. The mapping between a data matrix **X** and the pixel index of the **Intensity Graph** is shown in Figure 5.18.

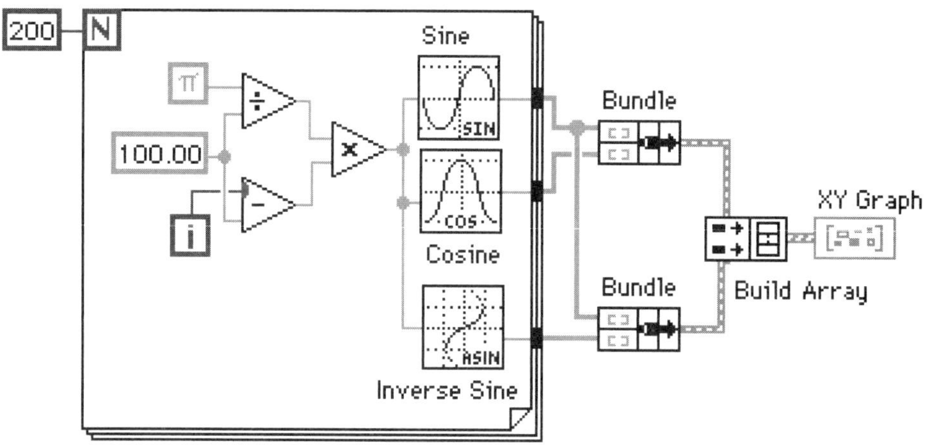

Figure 5.17
Multiple-channel display with the **XY Graph**.

$$\mathbf{X} = \begin{bmatrix} x_{11} & x_{12} & x_{13} & x_{14} \\ x_{21} & x_{22} & x_{23} & x_{24} \\ x_{31} & x_{32} & x_{33} & x_{34} \\ x_{41} & x_{42} & x_{43} & x_{44} \end{bmatrix}$$

x_{14}	x_{24}	x_{34}	x_{44}
x_{13}	x_{23}	x_{33}	x_{43}
x_{12}	x_{22}	x_{32}	x_{42}
x_{11}	x_{21}	x_{31}	x_{41}

Figure 5.18
Index mapping between the **Intensity Graph** and a 2-D array **X**.

The **Intensity Chart** and the **Intensity Graph** expect 2-D arrays as their input. You can also change the color for the value, and a couple of excellent examples of this are shown in the accompanying hands-on exercise manual. Interested readers may refer to it for examples of **Intensity Graph**.

PROBLEMS

5.1 Create a VI **P05_01.vi** that performs the following:
  (a) A chart **RN Display** displays a random number between 0 and 1 every 50 ms.
  (b) **P05_01.vi** stops when you press a Boolean button **QUIT**.
  (c) The default state of **QUIT** is TRUE.
  (d) When **QUIT** is pressed, it depresses by itself when you release the mouse button.
  (e) y-axis scale of the chart is between 0 and 1.

5.2 Modify **P05_01.vi** so that the display of **RN Display** refreshes every time you execute the VI using the Attribute Node **History Data**. Note that the option **History Data** is only available for **Waveform Charts**, not for **Waveform Graphs**. Save the VI as **P05_02.vi**.

5.3 Create a VI **P05_03.vi** that has two **Waveform Charts**, **RN Display 1** and **RN Display 2**, and each one performs the five tasks listed in Problem 5.1. Using the Attribute Node option **Visible** for both charts, create two Boolean buttons **Show 1** and **Show 2** to make each chart visible or invisible while displaying a random number continuously. Also, make sure that both charts refresh their screen whenever you restart the VI.

5.4 Using a **Waveform Chart**, display two channels of random numbers every 50 ms. One channel displays a random number between 0 and 1, and the other, between 1 and 2, in different color. Label the chart **RN 2Ch Display**, and adjust the y-axis scale to 0 and 2. Be sure to use the Attribute Node to refresh the screen whenever you execute the VI, and change the mechanical action and the default state of the Boolean switch **QUIT** as described in Problem 5.1. (Refer to Figure P5.4.) Also, label each channel **CH 0** and **CH 1** in the legend. Save the VI as **P05_04.vi**.

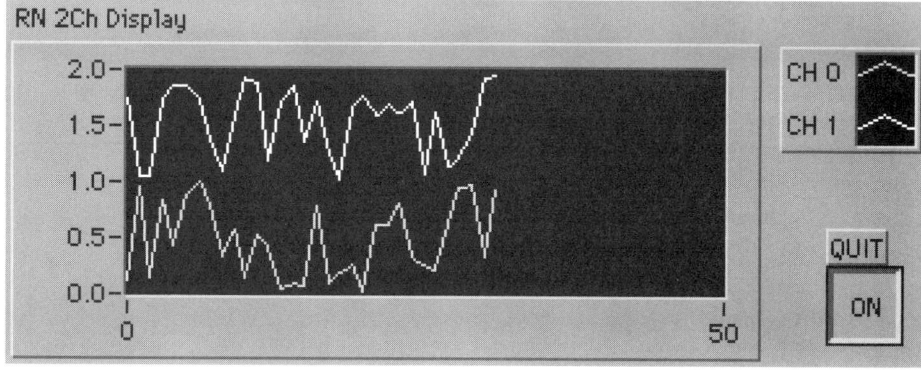

Figure P5.4

5.5 Using a **Waveform Graph**, display two channels of data, where each channel contains 10 random numbers. (Use autoindexing with a **For Loop**.) One channel has random numbers between 0 and 1, and the other, between 1 and 2. Use 0, 1, . . . , 9 for x-indices for both channel data. Label the graph **RN 2Ch Display**, and display the data from each channel using different color. Also, label each channel **CH 0** and **CH 1** in the legend. The complete front panel is shown in Figure P5.5. Save the VI as **P05_05.vi**.

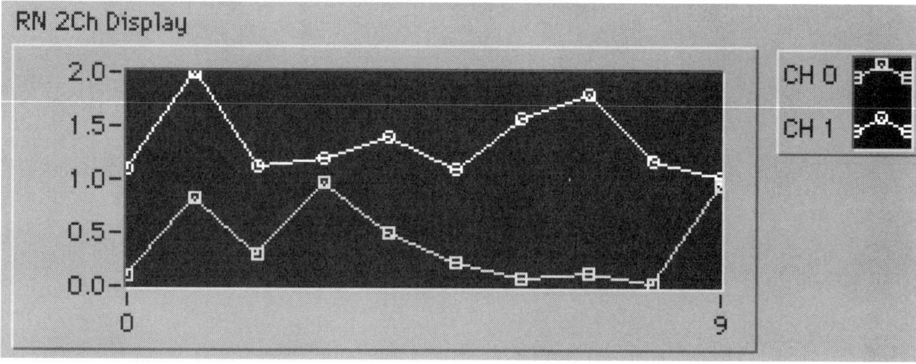

Figure P5.5

5.6 Modify **P05_05.vi** such that x-axis indices for both channel data are now 1, 3, 5, 7, . . . , 19 and all of the other information is identical. The complete front panel is shown in Figure P5.6. Save the VI as **P05_06.vi**.

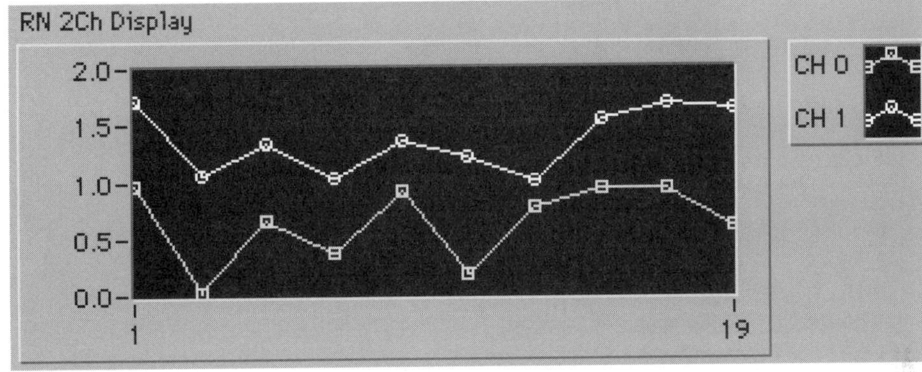

Figure P5.6

5.7 Modify **P05_05.vi** such that x-axis indices for each channel are different. The indices are 1, 3, 5, 7, . . . , 19 for **CH 0**, and 3, 6, 9, 12, . . . , 30 for **CH 1**. All of the other information is identical, and the complete front panel is shown in Figure P5.7. Save the VI as **P05_07.vi**.

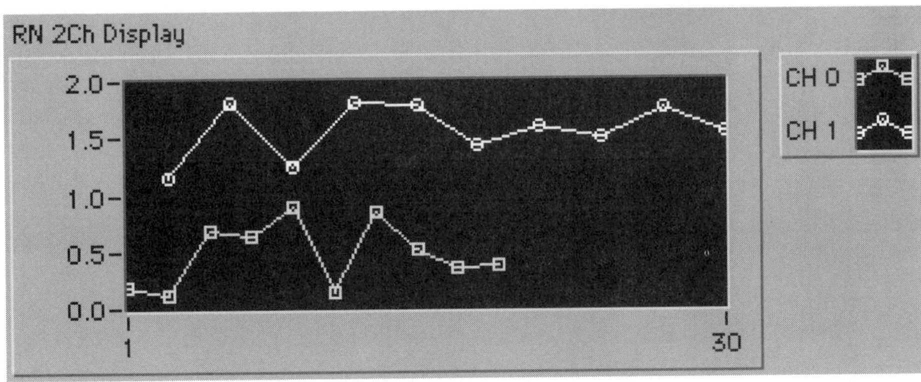

Figure P5.7

5.8 Modify **P05_07.vi** using a **XY Graph** in lieu of **Waveform Graph**; all of the other information is identical. Therefore, the complete front panel should be the same as Figure P5.7. Save the VI as **P05_08.vi**.

Arrays and Clusters

6

You briefly encountered the use of arrays and clusters during the discussion of **Waveform Chart** and **Waveform Graph**. This chapter will explore them in detail and provide examples.

6.1 Arrays

An array is a data type that has one or more elements of the *same* type, such as integer, single or double precision, or Boolean. As long as the type of each element is identical, the array structure can be used to contain multiple elements. Using mathematical terms, vectors and matrices are arrays where the former are one dimensional, and the latter are two dimensional. One advantage of the array data structure is that you do not have to carry all of the elements individually, as referring to the name of the array provides full access to each element through indexing.

Consider the following example to see an advantage of the array structure: When you create a sub VI, you need to create terminals for the objects in the front panel of the sub VI. (See Chapter 3 for the complete discussion of sub VIs.) A large number of objects can create a problem since a large number of terminals will make the wiring difficult. However, if the type of objects is identical, you may use a single terminal for an array, which consists of all of the objects of identical data type. For example, suppose that you have 10 numeric indicators, which will return 10 numeric values to another VI. Instead of having 10 terminals, you can create an array indicator of those 10 numeric indicators and assign a terminal to the array. If the data type of objects is different, you will need to consider a data type cluster, which is the topic of Section 6.2.

If you want to create an array from multiple objects of identical data type, you will need a function that combines them into an array, and **Build Array** does that. In order to access each individual element or a portion of the array, use **Index Array** or **Array Subset**. Since the array data structure is important enough to be emphasized repeatedly, each function in the **Array** subpalette will be discussed in Section 6.1.2. The next subsection discusses how to create an array.

6.1.1 Creating an Array

During the discussion of arrays, lowercase letters in boldface will be used to indicate a one-dimensional (1-D) array (or a vector). Upper case letters in boldface represent a 2-D array (or a matrix). Therefore, **X** is a matrix and **x** is a column vector.

$$\mathbf{X} = \begin{bmatrix} x_{11} & x_{12} & \cdots & x_{1,M} \\ x_{21} & x_{22} & \cdots & x_{2,M} \\ \vdots & \vdots & \ddots & \vdots \\ x_{N,1} & x_{N,2} & \cdots & x_{N,M} \end{bmatrix} \qquad \mathbf{x} = \begin{bmatrix} x_1 \\ x_2 \\ \vdots \\ x_N \end{bmatrix}$$

In C or C++, arrays are declared as follows:

```
int x[100], X[100][100];
```

The preceding expression declares a 1-D array (a vector) and a 2-D array (a matrix), and each element is accessed by its index. LabVIEW declares arrays in a different way, however, and there are two steps in creating (declaring) them:

Step 1: Place an *array shell* in either the front panel or the diagram window. If the array will be either a control or an indicator, place it in the front panel. If the array will be a constant, place it in the diagram window.

Step 2: Declare the type of the array by dropping a proper data type of a control, an indicator, or a constant in the array shell.

Arrays may be of any data type, such as Boolean or numeric. They can also be either controls or indicators depending on the type of element in the array shell. Step 3 in Figure 6.1 shows an array of numeric control so that you can *enter* data instead of displaying them. Step 1 in Figure 6.1 is an empty array shell obtained from the **Array** subpalette in the front panel. Step 2 shows how it will appear when you get a numeric **Digital Control** from the **Numeric** subpalette in the front panel. When you place it in the empty array shell, it will show a dotted line to indicate that the element is locked in

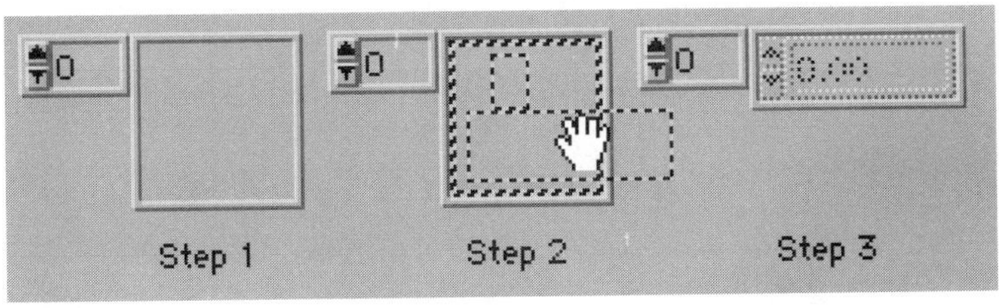

Figure 6.1
Steps to create an array. Step 3 shows an array of numeric control type.

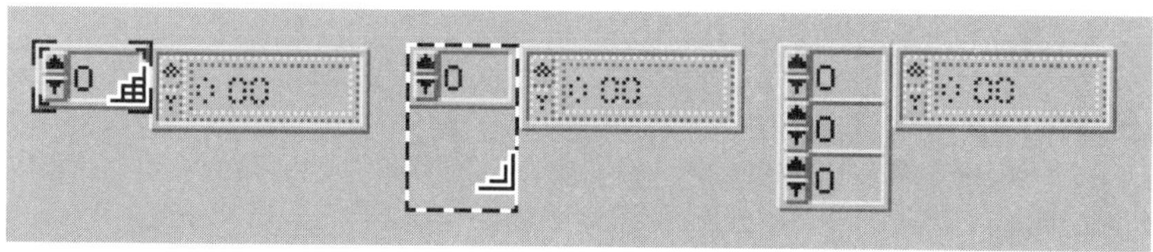

Figure 6.2
Increasing the dimension of an array from 1-D to 3-D.

the position. Releasing your mouse button will complete the step, and you will have an array of numeric control, as shown in Step 3.

The up and down arrows of the shell are to increase or decrease the start index of display. In other words, if you have 3 as the index of the shell, the first element has the index 3, making it the fourth element in the array since indices start at zero in LabVIEW. To increase the dimension of array, drag the index box down to show more indices. Figure 6.2 shows the steps in changing the dimension of an array.

Now that you know how to create an array, you must learn how to manipulate it with actual data. In LabVIEW, there are two ways to build an array: concatenation and indexing. In Chapter 4, you learned about indexing during the discussion of **While Loop** and **For Loop**.

For the sake of convenience, the indexing example is again shown in Figure 6.3. The thickness of the wire before and after the boundary of the two **For Loops** shows the dimension of the data. In Loop 1, the upper two wires

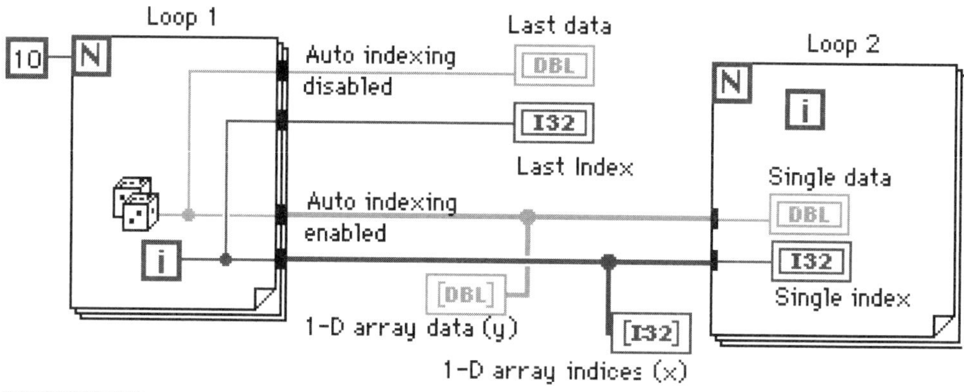

Figure 6.3
Indexing with the **For Loop**.

leaving the **For Loop** have the same thickness since the indexing is disabled. Indexing can be disabled by right clicking on the black dot and selecting **Disable Indexing**; however, the lower two wires leaving Loop 1 are thicker, indicating that the dimension of data has been increased. In this case, the data leaving Loop 1 at the bottom are two 1-D arrays constructed from 10 data elements. This indexing method is the fastest way to create an array and requires the least amount of overhead. In other words, in LabVIEW, creating arrays with the indexing feature is optimized.

Another way of creating an array is by concatenating data of the same type. This means that you manually combine multiple elements to create an array using **Build Array**. Consider the example shown in Figures 6.4 and 6.5. In Figure 6.4, you have two column vectors, **x** and **y**, of numeric *control* type. This means that you can *enter* or *set* values. Their size is identical and is

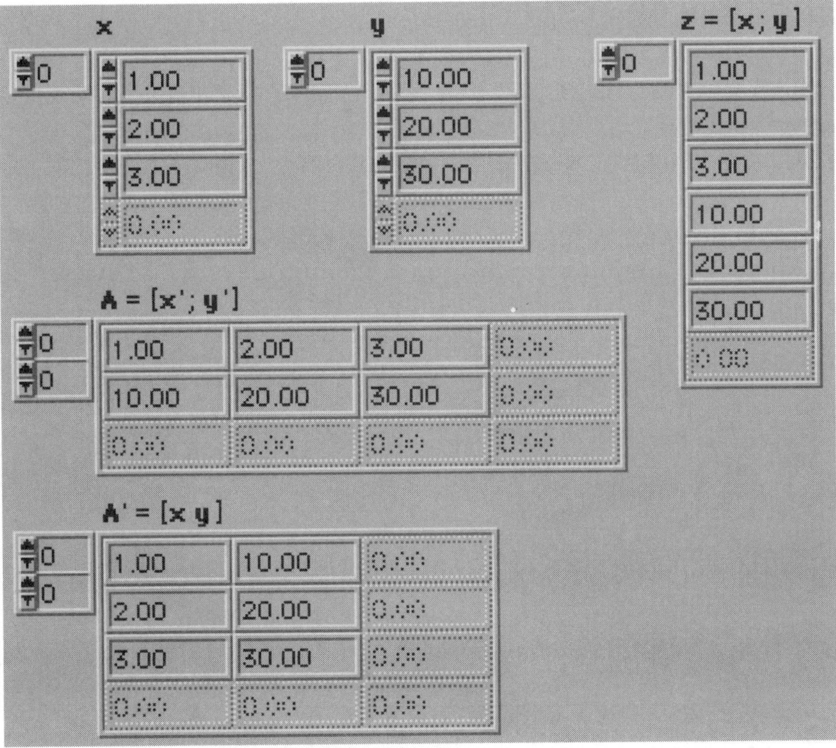

Figure 6.4
Building different arrays using **Build Array** (front panel).

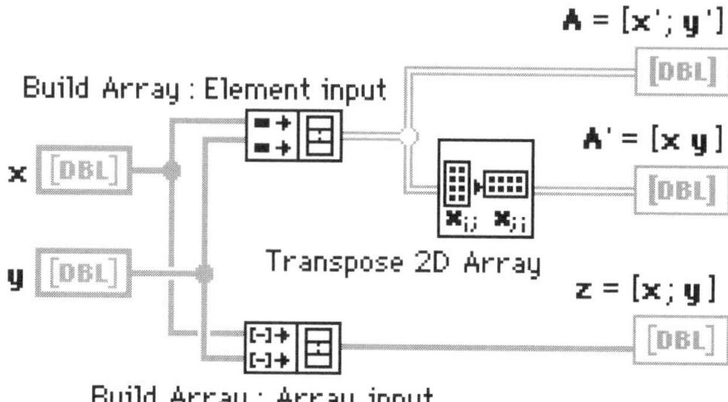

Figure 6.5 Building different arrays using **Build Array** (diagram window).

3 by 1. As shown in Figure 6.4, you can build another vector **z** from **x** and **y**, or create a matrix **A** with its columns or rows using **x** and **y**. (**A'** indicates the transpose of **A**.) In order to build another vector or matrix using **Build Array**, as shown in Figures 6.4 and 6.5, you need to understand the two options with **Build Array**: element input and array input.

In Figure 6.5, two different types of input to **Build Array** are shown. You may have realized that the shape of the input to two **Build Array**s is different: one with two black square dots, and the other with two dashes [-]. In order to change the type of input to **Build Array**, right click on the input to pop up its menu, and select either **Change to Array** or **Change to Element**. With the two different types of input, the output of **Build Array** will be either a new array with increased dimension or a new array of the same dimension. The thickness of the wire at the output of two **Build Arrays** shows the dimension of the output array. If you use the element input type, this will increase the dimension of the input array by one; therefore, using a 1-D array (vector) as its input will create a 2-D array (matrix), which is **A** in Figure 6.5. If you use the array input type, however, this will not increase the dimension but will create an output array of the same dimension, which is just another 1-D array (vector) **z** in Figure 6.5. The notation [**x y**] has been used to indicate two columns, where each column is the vectors **x** and **y**, and [**x;y**], one column with **x** stacked on **y**. One last important reminder about **Build Array** is that it treats 1-D array inputs as *row* vectors instead of column vectors. Therefore, the output of **Build Array** with the element input type using 1-D arrays **x** and **y** as its input will be a matrix **A**, and its first row will be **x** and the second row, **y**. To transpose **A** to have **x** in the first column of **A** and **y** in the second column, use **Transpose 2D Array**, which can be found in the **Functions** >> **Array** subpalette.

6.1.2 Array Functions

This subsection will examine each LabVIEW function in the **Array** subpalette in the **Functions** palette.

Array Size This function returns the dimension or the size of the input array. If the input is one dimensional, the output will be a scalar. If the input is two or higher dimensional, the output will be a 1-D array. If the input is a 2-D array, the first element in the output 1-D array corresponds to the length of row, and the second element to the length of column.

Index Array This function provides access to each element in the input array and has two input types: the array, and the index or indices. To change the number of index input terminals, point the arrow cursor to the right bottom corner until the shape of the arrow cursor turns to a reversed L shape, and drag it up or down. The number of index input terminals should be the same as the dimension of the input array. This also allows you to index certain entire row or column elements. For example, consider a 2-D array **X** of size 3 by 4 as follows:

$$\mathbf{X} = \begin{bmatrix} 1 & 4 & 7 & 10 \\ 2 & 5 & 8 & 11 \\ 3 & 6 & 9 & 12 \end{bmatrix}$$

In order to index any entire column of **X**, you need to disable the indexing of row input. If you want to index any entire row of **X**, the indexing of column input needs to be disabled. You can disable or enable the indexing of row or column input by right clicking on it and selecting **Disable Indexing** or **Enable Indexing** from the pop-up menu. If it is disabled, it will appear as a white square dot, and if it is enabled, its shape will be a black square dot (Figure 6.6). The default setting is enabled indexing (black square dot). Note that since **X** is a 2-D array, you must have two index input terminals.

If the 2-D array **X** is the input to **Index Array** in (b) and (c) in Figure 6.6, (b) will return a 1-D array whose elements are 4, 5, and 6, and (c) will return a 1-D array whose elements are 3, 6, 9, and 12.

Replace Array Element If you want to replace any individual element in an array, this function can be used. You specify the original array to be modified, the value for replacement, and its index as the input to this function. To increase the number of index input terminals, follow the method stated in the description of **Index Array**.

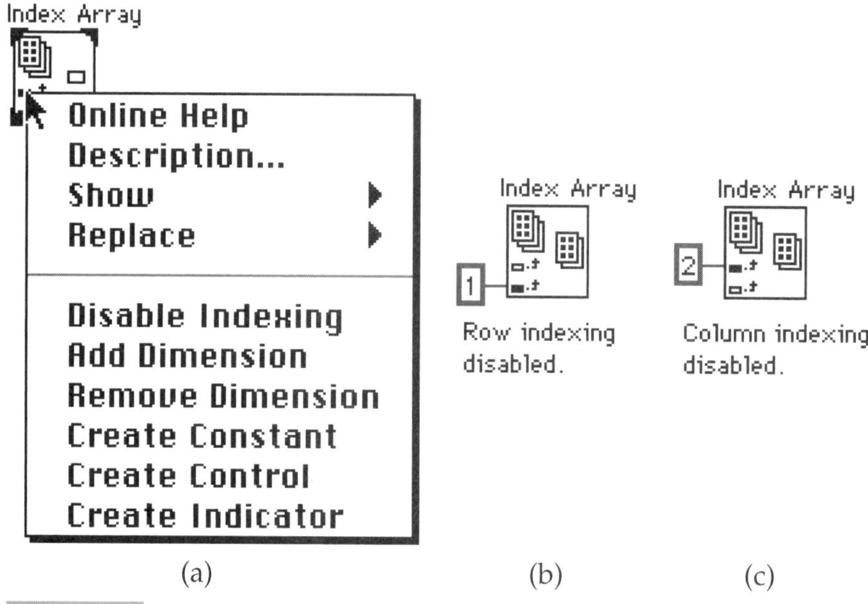

Figure 6.6
Disabling or enabling index input terminals of **Index Array** to index an entire row or column of a 2-D array input. **Index Array** in (b) will return the second column of the 2-D input array since the index starts at zero. The output of **Index Array** in (c) will be the third row of the 2-D input array.

Array Subset This function can extract a range of elements from the input array. The inputs are the original array and the starting index and the length of the range to be extracted.

Reshape Array This reshapes the input array. For example, if the input is a 1-D array of six elements, and if you reshape it with 2 for the input **Dimension Size**, it will truncate the original array to one of size 2. If you increase the number of **Dimension Size** input terminals to two and specify 2 and 3 for their input values, it will return a 2-D array of size 2 by 3 from the original 1-D array of six elements. When you increase the dimension of the output array, the number of elements in the input array and that in the output array must be the same. If the number of elements in the input array and that in the output array are not the same, the **Dimension Size** input value that is smaller than the length of the input array will truncate the input array, and the larger value will zero pad the input array. This can be useful in FFT data processing when zero padding is desired. (See Chapter 14 for more discussion about zero padding.)

Initialize Array This function is to initialize an array. Dynamic increase in the size of array can result in inefficient memory usage in LabVIEW. You can avoid this by allocating a block of memory to arrays and not changing their size dynamically. Therefore, if the number of array elements will vary, initialize an array that is big enough to hold all of the possible elements instead of changing the size of the array during the execution of VI applications.

Build Array This creates another array using input arrays. The dimension of the output array depends on the selection of input array mode: Element input and Array input. More details about those modes have been discussed in the previous subsection.

Rotate 1D Array This rotates the elements in the input 1-D array. In other words, this function shifts the elements of the input 1-D array by the amount of *n*, which is the number of places (bits) to be shifted. If *n* is positive, the elements rotate to the right, and to the left for negative values. The right-most element will move to the left-most location for a positive *n*, and the left-most element to the right-most location for a negative *n* during the shifting process.

Reverse 1D Array This reverses the order of elements in the 1-D array input.

Transpose 2D Array This transposes its 2-D input array. It may have to be used when displaying data using the **Waveform Graph**. For more information about the **Waveform Graph**, refer to Chapter 5, where displaying data is studied.

Search 1D Array This has three inputs: a 1-D array, a value that is under search, and a search start index. This function will search for the value starting at the index in the 1-D array input and return the index of the value found. If it is –1, the value under search is not in the array, and if the value has more than one appearance in the array, the output is the index where it appears for the first time.

Sort 1D Array This function sorts the input 1-D array elements in ascending order but does not provide the indices of the elements.

Array Max & Min This returns the maximum and the minimum values in the input array with their indices. If the input array is

1-D, the output indices will be a scalar. If the input is a 2-D array, the maximum and minimum indices are returned in a 1-D array: The first element is the row index, and the second is the column index. If there are duplicates of the value, the output indices correspond to the element at the first appearance.

Split 1D Array This returns two portions of the 1-D input array where the index of the first portion is 0 to n –1. The index of the second portion is *n* to the end, where *n* is the value to the input terminal **index**.

Interpolate 1D Array The name of this function is somewhat misleading. This function takes a 1-D array and a fractional index as its input and returns an estimated scalar value at that fractional index input. This value is computed by using only the two values in the immediate neighborhood of the fractional index. It is not computed by first fitting all of the elements in the array into an equation and then evaluating it at the fractional index. For example, if the input is [0 1 4 9 16], which is the output of a simple quadratic function $y = x^2$, the value at index 3.75 should be 14.06 since $y(x = 3.75) = 14.06$; but this function returns 14.25, which is the result of first-order polynomial data fitting using two closest adjacent values 9 and 16 whose indices are 3 and 4. $(9 + (3.75 - 3) \times (16 - 9) = 14.25)$. Refer to Chapter 14 for a discussion of interpolation VIs.

Threshold 1D Array This function takes a 1-D array, a threshold value, and a starting index as its input. The name of this function is also misleading. The name suggests that this function would leave all of the values below the threshold that is connected to the input **threshold y** and saturate the others above it. However, this function performs the opposite operation to the **Interpolate 1D Array**. If the input array is [0 1 4 9 16], the input **threshold y** is 14.25, which was the output of the **Interpolate 1D Array** in the preceding example, and if the input **start index** is 0, the output **fractional index or x** returns 3.75. In order to obtain the correct fractional index in general, the **start index** input value must be less than the correct fractional index. If the elements in the input array increase monotonically, you may set it to zero, but if the **threshold y** value appears more than once in the input array, the function may return a wrong index unless the **start index** is chosen carefully.

Interleave 1D Array This interleaves each element from multiple input arrays alternatively. If you have **x** = [1 2 3 4] and **y** = [4 5 6] as two inputs to this function, the output array becomes **z** = [1 4 2 5 3 6]. If the size of any of the input arrays is bigger than the others, the output array will be constructed based on the minimal size, and all of the elements beyond the minimal size will be ignored. In this example, **x** is bigger than **y** by one element so that the last element of **x** is ignored in constructing the output array **z**.

Decimate 1D Array This function performs the operation opposite to **Interleave 1D Array**. If the input array is **z** = [1 4 2 5 3 6] and you have two outputs **x** and **y**, this function will return **x** = [1 2 3] and **y** = [4 5 6]. The length of output arrays is identical. The valid input array length is $N \times M$, where N is the number of output arrays and M is the length of each output array. If the length of the input array P is larger than $N \times M$, the last $P - N \times M$ elements in the input array will be ignored.

Array Constant This creates an empty array shell in the diagram window so that an array constant can be created by placing any type of constant in it.

Array to Cluster This takes an array as input and converts it to a cluster. (Clusters are discussed in the following section.) You can also adjust the number of elements in the cluster beforehand. Just right click on the **Array to Cluster**, choose **Cluster Size . . .**, and enter a number to change it.

Cluster to Array This reverses the process of **Array to Cluster**. Note that arrays can only have the same type of data as their elements; therefore, if the input cluster, which can have different types of data, does not have the same data element type, this function will return the Broken Run Arrow. This function is useful if you have a cluster of Boolean elements. For example, suppose you have multiple Boolean buttons as elements in a cluster in the front panel to control different processes, such as DATALOGGING, READ DATA, DISPLAY DATA, and so on. Though there are many different ways of doing so, you may monitor which button is pressed by first converting the cluster to an array of Boolean switches, and using **Search 1D Array** to find out if any one of the buttons is pressed. If none is pressed, the output from **Search 1D Array** will be negative one (–1). Based on such output, you may have **Case** structures to perform corresponding tasks.

6.2 Clusters

Clusters are very similar to arrays in that they can have multiple elements. However, clusters can have different data type of elements, whereas arrays cannot. You were introduced to **Cluster to Array** in the previous section, and now you may ask why you would want to use the cluster type even though the type of each element is the same. The answer is that you can have more flexibility with clusters, even with the same type of data elements. For example, suppose you have an array of Boolean switches, and want to have a different effect on each one such as different labels. If you change the label of any one of the Boolean switches, the rest will all change identically since arrays *must* have exactly the same type (even the labels) of data element. However, if you have a cluster of Boolean switches, you can change the labels individually since each element is independent. This implies that you can have different settings, such as color, label, mechanical action, or Attribute Node by using a cluster of Boolean switches. This should justify the use of cluster even with the same type of data elements for some occasions. The drawback in using clusters instead of arrays is more overheads, so discretion must be exercised in choosing which one to use with identical data type elements.

6.2.1 Creating a Cluster

The method of creating a cluster is similar to that of creating an array. First, you place an empty cluster shell and insert data elements one by one. Since you are dealing with different types of data elements, the *order* of inserting data elements in the cluster shell becomes important. For example, if you want to wire a cluster indicator to a cluster control, each element in the cluster indicator must match that in the cluster control including the creation order and data type. Figure 6.7 shows an example of creating a cluster whose elements are one numeric control and one Boolean LED control. The numeric control was created first in this example. Note that all of the elements in a cluster *must* be either controls or indicators, but not both; therefore, it may be said that clusters are similar to arrays such that all of the elements must be of the same type (control or indicator).

To change the order of cluster elements, right click on the boundary of the cluster shell as shown in Figure 6.8. Select **Cluster Order...** to display the order as shown in Figure 6.9. Then the cursor becomes a hand with the index finger pointing to the left. In this mode, you can change the order. After changing the order, click OK to exit the mode.

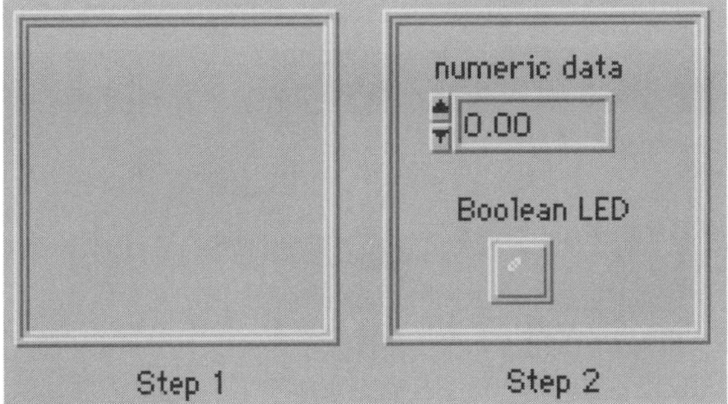

Figure 6.7 Steps to create a cluster.

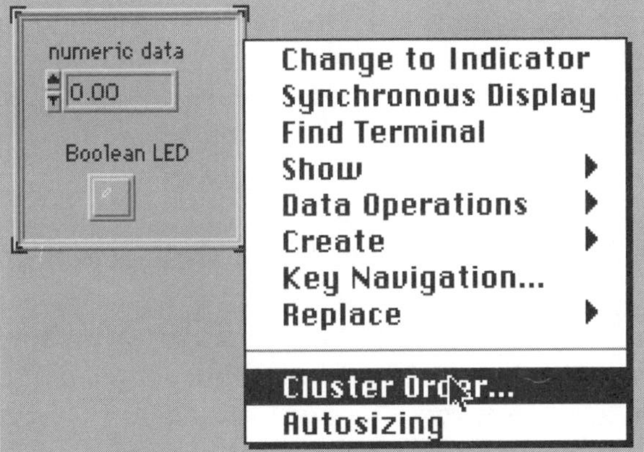

Figure 6.8 Displaying the cluster order.

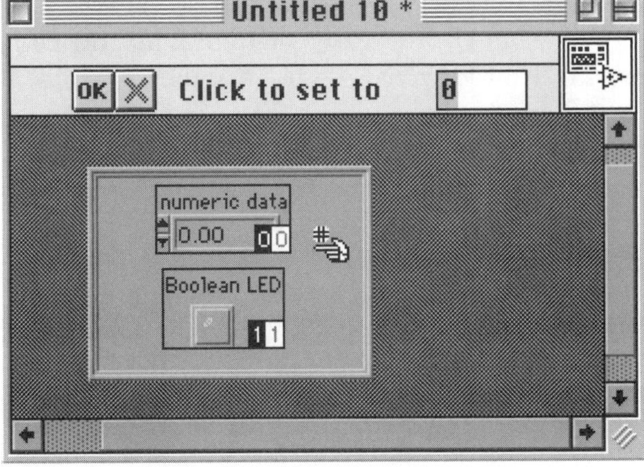

Figure 6.9 Changing the cluster order.

6.2.2 Cluster Functions

This subsection will discuss each LabVIEW function in the **Cluster** subpalette in the **Functions** palette.

Unbundle This returns each element in the input cluster. The number of output terminals *must* match that of elements in the input cluster; otherwise, it will return the Broken Run Arrow, indicating an error has occurred.

Bundle This function bundles up multiple input elements of different type and returns a cluster. The order of each input element is top first and bottom last. The middle input **Source Element** is optional and has two purposes: building a cluster and modifying a cluster. If you have a *template* cluster structure to build, you may wire it to the middle input. If you do so, the number and the order of input elements and their data type *must* match those of the elements in the template cluster.

The other purpose of the **Source Element** is to replace or modify any element in the cluster. The value to be replaced will be wired to the correct input terminal, and the cluster that needs to be updated will be wired to the **Source Element** terminal. Being similar to **Unbundle**, the number of input terminals *must* match the number of elements in the cluster that is wired to the **Source Element** terminal; otherwise, it will return the Broken Run Arrow.

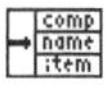

Unbundle By Name This function provides access to each element in the input cluster. If the elements have labels, each element can be accessed (retrieved) individually by selecting its label. This is the major difference from **Unbundle**. Note that **Unbundle** must have the same number of output terminals as the number of elements in the input cluster. If you have a cluster with 100 elements and want to access just the 99th one, you still have to have 100 output terminals just to access one element with **Unbundle**! However, if you use **Unbundle By Name**, you can have just one output terminal with the label of the 99th element selected. Therefore, it is a good practice to label each element in the cluster even though it is not displayed. It is also recommended that you use **Unbundle** to access the elements when there are relatively few elements in the cluster; otherwise, use **Unbundle By Name**.

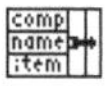

Bundle By Name This function is similar to **Bundle** since it also has two similar purposes. However, this *must* have middle

Source Element input wired. The first purpose is to simply create a cluster from a number of input elements of different data type. When you wire the template cluster to the **Source Element** input terminal, all of the labels of the input elements become available in the input terminals.Unlike the function **Bundle**, you do not need the same number of input terminals as the number of elements in the template cluster. In order to display all of the input labels, simply drag down the input terminals. The second purpose is to replace any cluster element. For example, suppose you have 100 elements in a cluster and you want to replace the 99th element. With **Bundle**, you must have the same number of input terminals (100) just to access the 99th one, but **Bundle By Name** allows you to have only one input terminal with the label of the 99th element. Of course, you should wire the cluster that needs to be modified to the **Source Element** input terminal to make all of the 100 labels available in the input of **Bundle By Name**.

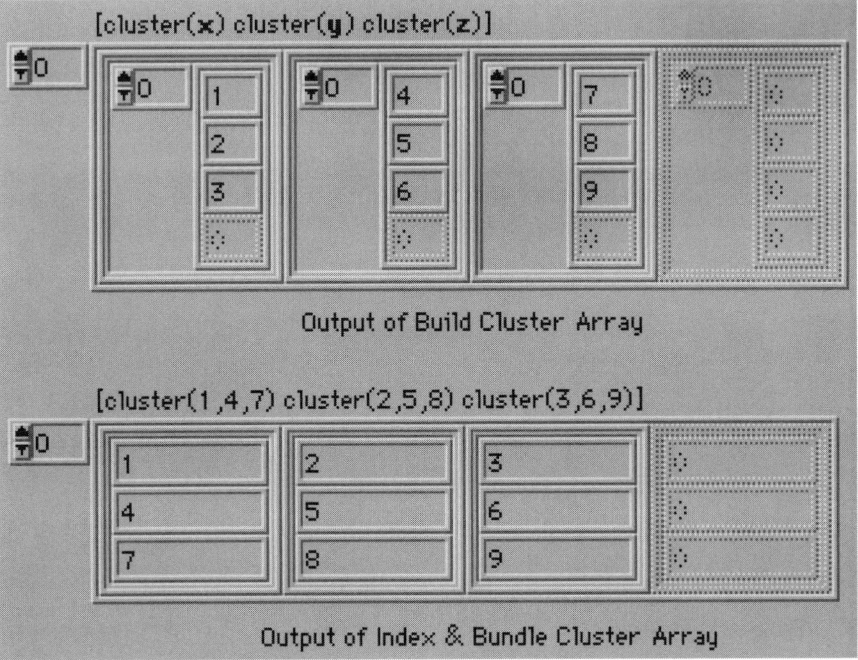

Figure 6.10
Outputs from **Build Cluster Array** and **Index & Bundle Cluster Array** with three inputs **x** = [1 2 3], **y** = [4 5 6], and **z** = [7 8 9].

Build Cluster Array This function builds an array whose elements are clusters that contain the input element wired to the input terminal of the function. For example, if you wire numeric constants 1, 2, and 3 to the function, the output will be [cluster(1) cluster(2) cluster(3)], where [ ] indicates an array, and cluster(x) indicates a cluster whose element is x. Therefore, if x is an array, this function will create an array of clusters where each cluster contains a single 1-D array **x**. For example, if three inputs to this function are **x** = [1 2 3], **y** = [4 5 6], and **z** = [7 8 9], the output will be [cluster(**x**) cluster(**y**) cluster(**z**)], as shown in Figure 6.10.

Index & Bundle Cluster Array This function is similar to **Bundle Cluster Array**, and it creates an array of clusters, but the element of the clusters is different and the input must be an array. Suppose you have **x** = [1 2 3], **y** = [4 5 6], and **z** = [7 8 9] for the input; then this function will return an array of clusters where each cluster has three elements as follows: cluster (1, 4, 7), cluster (2, 5, 8), and cluster (3, 6, 9). In other words, the output is [cluster (1,4,7) cluster (2,5,8) cluster (3,6,9)], as shown in Figure 6.10. Here, [ ] indicates an array, and cluster (a,b,c) indicates a cluster whose elements are a, b, and c.

Cluster Constant This constant is similar to **Array Constant** and provides an empty shell to create a cluster constant. You can place any type of constant in the shell to define its type.

6.3 Comparison of Functions

All of the functions for the array and cluster manipulation have been covered. Since their natures are somewhat similar to each other, they may cause some confusion. The following is a brief comparison and summary of the two types.

6.3.1 Build Array and Bundle (By Name)

These two functions are used to create an array or a cluster. You should be aware of the creation order of cluster elements. **Bundle** and **Bundle By Name** differ in terms of the number of input terminals required, and a cluster template can be wired to the input terminal **Source Element**.

6.3.2 Index Array and Unbundle (By Name)

These are to access each element in an array or a cluster. Be aware of the differences between **Unbundle** and **Unbundle By Name**.

6.3.3 Replace Array Element and Bundle (By Name)

These functions are to replace an element in an array or a cluster. As for the cluster, the **Source Element** input terminal *must* be wired, and such wiring is not optional. You must have the same number of input terminals as that of elements in the cluster to be modified with **Bundle**, and the cluster to be modified must be wired to the **Source Element** input terminal of **Bundle**. However, having the same number of input terminals does not apply to **Bundle By Name**.

6.3.4 Memory Issue

Dynamically resizing arrays will cause inefficient memory usage; this can be prevented by first allocating a sufficiently large block of memory to the arrays using **Initialize Array**. Clusters impose more overheads and can significantly slow down the process if they have complex structure elements. However, clusters provide excellent flexibility since elements are independent of each other. Note that all of the elements in a cluster must be either indicators or controls.

PROBLEMS

6.1 Generate two 1-D arrays **x** and **y** whose elements are as follows:

$$\mathbf{x} = [1\ 2\ 3\ 4\ 5],\ \mathbf{y} = [6\ 7\ 8]$$

Using **x** and **y** perform the same task as shown in Figure 6.4. Save the VI as **P06_01.vi**. How are the results different from those shown in Figure 6.4?

6.2 Create a 2-D array **X** of size 3 by 4 (three rows and four columns) with random numbers between 0 and 1. You may use two nested **For Loops** to generate such a 2-D array quickly. Complete the VI that can select a different column, and display the selected column **x** until you quit the VI using a **While Loop**. Note that you will need to create a 2-D array only once before the **While Loop** initiates. The complete front panel is shown in Figure P6.2. Save the VI as **P06_02.vi**.

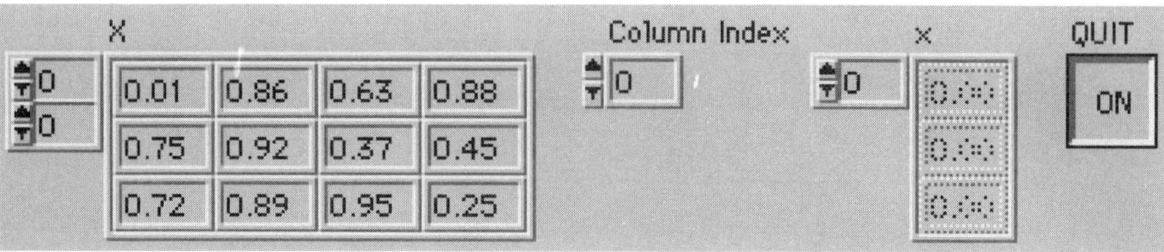

Figure P6.2

6.3 Expand the functionality of **P06_02.vi** such that the VI allows you to extract the entire column or row elements based on either column or row index and displays the result in **x**. Its front panel is shown in Figure P6.3. Save the VI as **P06_03.vi**.

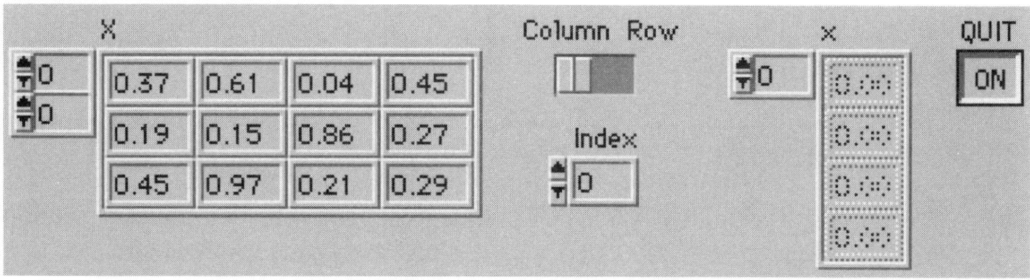

Figure P6.3

6.4 Build a VI that performs binary addition of two positive 1-byte-long integers; its complete front panel and diagram window are shown in Figure P6.4. The following are key functions used in this example:

```
Functions >> Array >> Reverse 1D Array
Functions >> Boolean >> Boolean Array To Number
```

Figure P6.4 (a)

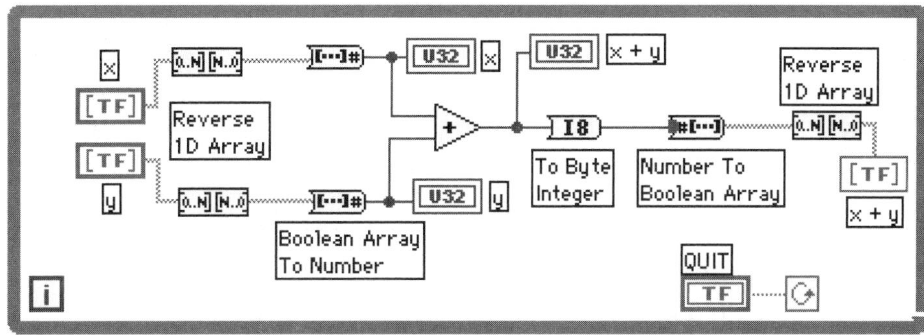

Figure P6.4 (b)

```
Functions >> Boolean >> Number To Boolean Array
Functions >> Numeric >> Conversion >> To Byte Integer
```

Save the VI as **P06_04.vi**, and explain the functionality of each VI. Why do you need to use **To Byte Integer**?

6.5 Create a cluster control **BIO Original** with three different types of data: **Name** (String), **Age** (U8 integer), and **Marital Status** (Boolean). Their order is top first, and bottom last in **BIO Original**. Complete the VI that updates the second element with the age in **New Age** and displays the result in the cluster indicator **BIO Updated**. Use **Bundle** to complete the example, and save it as **P06_05.vi**. The complete front panel is shown in Figure P6.5. How many input terminals of **Bundle** are needed to run the VI successfully?

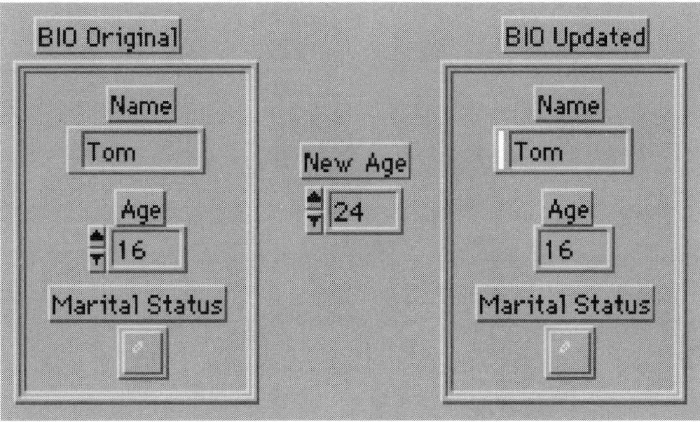

Figure P6.5

6.6 Repeat Problem 6.5 using **Bundle By Name** in lieu of **Bundle**, and save it as **P06_06.vi**. The front panel will be the same as Figure P6.5. How many input terminals of **Bundle By Name** are needed to run the VI successfully?

6.7 First, save **P06_06.vi** as **P06_07.vi**, and delete all of the labels of three controls in **BIO Original** in **P06_07.vi**. The **Run** button will turn into **List Errors** (Broken Run Arrow). Explain why, and suggest how the error should be fixed. Also, summarize your observation from Problems 6.5 through 6.7.

6.8 One application where clusters are commonly used is a database. Figure P6.8 shows a simple example with three entries, **Name**, **Age**, and **Marital Status**, in each record. Note that an array of cluster is used for **Database Array**. Build a VI as shown in Figure P6.8, save it as **P06_08.vi**, and explore its functionality with different entry values. Can you *add* more record to **Database Array** using the diagram in Figure P6.8?

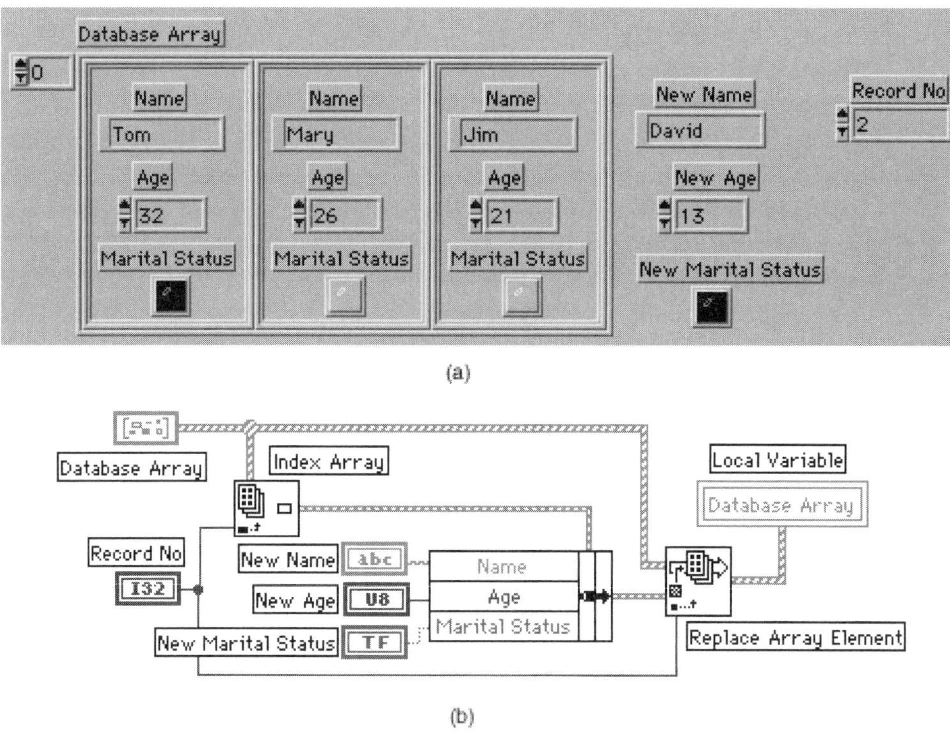

Figure P6.8

6.9 The database example in **P06_08.vi** is only to *update* the record in **Database Array**. In order to be able to *add* more record to **Database Array**, how would you modify **P06_08.vi**? Complete your suggestion, and save it as **P06_09.vi**.

Prerequisite for Data Acquisition

7

Even though LabVIEW can be used for regular programming purposes, it is ultimately designed for data acquisition, as the file extension implies: VI (Virtual Instrumentation). This chapter will discuss different data acquisition techniques, such as analog input and output, digital input and output, and counter operation.

Since the signals to be acquired in the real world are in different conditions—for example, very small magnitude voltage signals or extremely high voltage signals—they need to be preconditioned before being applied to an analog-to-digital (A/D) conversion board that plugs into the computer. Also, due to the sampling speed limitation on the A/D boards, you should be aware of the range of input signal frequency to avoid losing information about source signals. This chapter will first review the fundamental concept behind the sampling theorem and then discuss different input modes in data acquisition.

7.1 Sampling Theorem

The sampling theorem is formally stated as follows:

Sampling Theorem

If $x(t)$ is a bandlimited signal with its Fourier transform $X(\omega) = 0$ for $|\omega| > \omega_B$, $x(t)$ is uniquely determined by its sample values $x(kT)$, $k = 0, \pm 1, \pm 2, \ldots$, if

$$\omega_s > 2\omega_B \tag{7.1}$$

where $\omega_s = 2\pi/T$, t is time, ω, ω_B, and ω_s are frequencies in rad/s. Here, T is the sampling time and $2\omega_B$, which is the minimal sampling rate, is referred to as the Nyquist rate. ■

To understand the minimal sampling rate better, consider it in more depth using the impulse sampling scheme. Let $p(t)$ be an impulse train that is expressed as

$$p(t) = \sum_{k=-\infty}^{\infty} \delta(t - kT) \tag{7.2}$$

where $\delta(t)$ is the unit impulse function that is defined to be 1 if $t = 0$ and zero otherwise. Then, as shown in Fugure 7.1, the Fourier transform of $p(t)$, $P(\omega)$, is written as

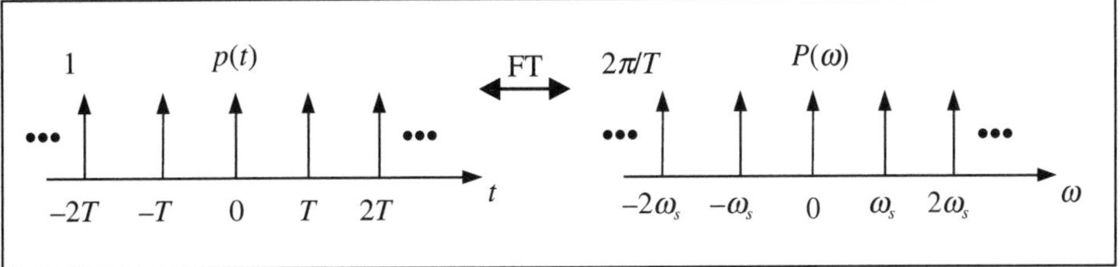

Figure 7.1
Fourier transform of an impulse train.

$$P(\omega) = \frac{2\pi}{T} \sum_{k=-\infty}^{\infty} \delta\left(\omega - k\frac{2\pi}{T}\right) \tag{7.3}$$

You can write a sampled signal $x_p(t)$ as

$$x_p(t) = x(t)p(t) \tag{7.4}$$

If you take the Fourier transform of $x_p(t)$,

$$X_p(\omega) = \frac{1}{2\pi} X(\omega) * P(\omega) \tag{7.5}$$

where $*$ indicates convolution. Substituting (7.3) in (7.5),

$$\begin{aligned} X_p(\omega) &= \frac{1}{2\pi} X(\omega) * \frac{2\pi}{T} \sum_{k=-\infty}^{\infty} \delta\left(\omega - k\frac{2\pi}{T}\right) \\ &= \frac{1}{T} \int_{-\infty}^{\infty} X(\tau) \sum_{k=-\infty}^{\infty} \delta\left(\omega - k\frac{2\pi}{T} - \tau\right) d\tau \\ &= \frac{1}{T} \sum_{k=-\infty}^{\infty} \int_{-\infty}^{\infty} X(\tau)\delta\left(\omega - k\frac{2\pi}{T} - \tau\right) d\tau \\ &= \frac{1}{T} \sum_{k=-\infty}^{\infty} X\left(\omega - k\frac{2\pi}{T}\right) \end{aligned} \tag{7.6}$$

Equation (7.6) shows that $X_p(\omega)$ is a replicate of $X(\omega)$ scaled by $1/T$ at every $\omega = k(2\pi/T)$, $k = 0, \pm1, \pm2, \ldots$. This is shown in Figure 7.2. Note that if $\omega_s < 2\omega_B$, both ends of each spectrum will start overlapping, and such a phenomenon is referred to as aliasing. Therefore, in order to avoid aliasing, sampling frequency must be greater than $2\omega_B$. This requirement is for signal

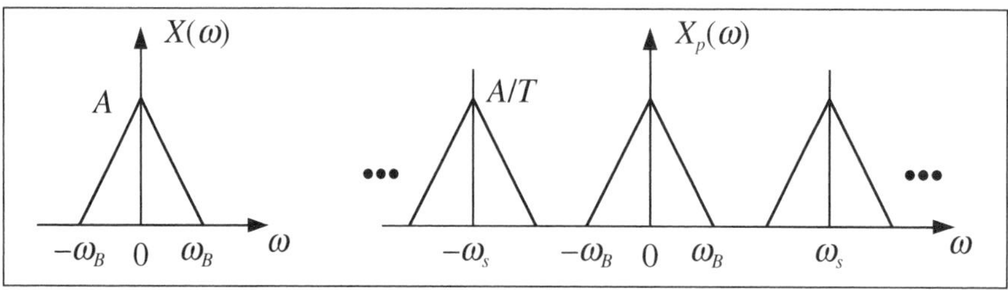

Figure 7.2
Original signal $X(\omega)$ and the spectrum of the sampled signal $X_p(\omega)$.

reconstruction purposes only, and if the display of sampled data is of interest, you should sample much faster than just twice the maximum frequency of the signal. In other words, if the waveform of signals in time is of interest, sampling frequency must be about $6\omega_B$ or faster.

To reconstruct the original signal $x(t)$, use a lowpass filter $H(\omega)$ to extract only the spectrum in the baseband; that is,

$$\hat{X}(\omega) = X_p(\omega)H(\omega) \tag{7.7}$$

where

$$H(\omega) = \begin{cases} T, & |\omega| < \dfrac{\omega_s}{2} \\ 0, & \text{otherwise} \end{cases} \tag{7.8}$$

This reconstruction procedure is shown in Figure 7.3, and $\hat{x}(t)$, which is the reconstructed signal of $x(t)$ and the inverse Fourier transform of $\hat{X}(\omega)$, is derived next. Since $h(t)$ is the inverse Fourier transform of $H(\omega)$,

$$h(t) = T\frac{\sin\left(\dfrac{\omega_s}{2}t\right)}{\pi t} = \frac{\sin\left(\dfrac{\omega_s}{2}t\right)}{\dfrac{\omega_s}{2}t} = \operatorname{sinc}\left(\frac{t}{T}\right) \tag{7.9}$$

and

$$x_p(t) = x(t)\sum_{k=-\infty}^{\infty}\delta(t-kT) = \sum_{k=-\infty}^{\infty}x(kT)\delta(t-kT) \tag{7.10}$$

where

$$\operatorname{sinc}(x) = \frac{\sin(\pi x)}{\pi x} \tag{7.11}$$

Then

$$\hat{x}(t) = x_p(t) * h(t) = \int_{-\infty}^{\infty} \sum_{k=-\infty}^{\infty} x(kT)\delta(\tau - kT)h(t-\tau)d\tau$$

$$= \int_{-\infty}^{\infty} \sum_{k=-\infty}^{\infty} x(kT)\delta(\tau - kT)\frac{T\sin\left\{\frac{\omega_s}{2}(t-\tau)\right\}}{\pi(t-\tau)}d\tau$$

$$= \sum_{k=-\infty}^{\infty} x(kT)\frac{T\sin\left\{\frac{\omega_s}{2}(t-kT)\right\}}{\pi(t-kT)} = \sum_{k=-\infty}^{\infty} x(kT)\frac{\sin\left\{\frac{\pi}{T}(t-kT)\right\}}{\frac{\pi}{T}(t-kT)} \tag{7.12}$$

$$= \sum_{k=-\infty}^{\infty} x(kT)\operatorname{sinc}\left(\frac{t-kT}{T}\right)$$

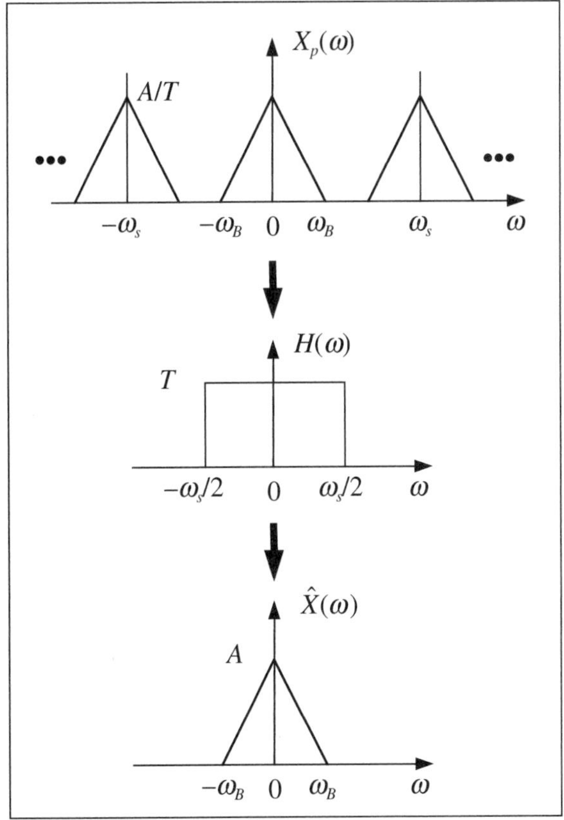

Figure 7.3 Reconstruction of the original signal from its samples.

To summarize, sampling frequency must be greater than the Nyquist rate, and the reconstruction of the original signal from sampled signal values is expressed by the sum of weighted data samples by sinc((t–kT)/T) for all k. However, sampling is indeed performed by pulse train of width W, as shown in Figure 7.4. Therefore, it can be shown that the Fourier transform $X_{pw}(\omega)$ of sampled signal $x_{pw}(t) = x(t)p_w(t)$ is

$$X_{pw}(\omega) = \frac{W}{T}\left[X(\omega) + \sum_{k=1}^{\infty} \text{sinc}\left(\frac{kW}{T}\right)\{X(\omega - k\omega_s) + X(\omega + k\omega_s)\}\right] \quad (7.13)$$

Then its reconstruction becomes

$$\hat{X}_w(\omega) = X_{pw}(\omega)H(\omega)\frac{1}{W} \quad (7.14)$$

And its inverse Fourier transform is

$$\hat{x}_w(t) = [x_{pw}(t) * h(t)]\frac{1}{W} = \frac{1}{W}\sum_{k=-\infty}^{\infty} x_{pw}(kT)\text{sinc}\left(\frac{t - KT}{T}\right) \quad (7.15)$$

Note that $x_{pw}(kT)$ is no longer an weighted impulse function but a signal of width W in time such that

$$x_{pw}(kT) = \begin{cases} x(t), & kT - \frac{W}{2} \le t \le kT + \frac{W}{2} \\ 0, & \text{otherwise} \end{cases} \quad (7.16)$$

Comparing (7.12) with (7.15), you can see that the only difference is the scaling factor, and the two equations essentially show the same result.

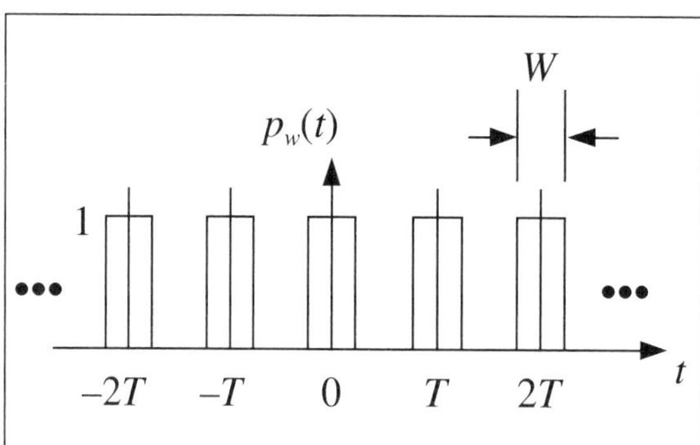

Figure 7.4 Pulse train for the pulse train sampling scheme.

7.2 Analog Input Signal

Acquiring signals that are in analog format requires extra attention before the acquisition is performed because of the measurement reference point. For example, when you say that you have a 1.5-V battery, you never really think about the meaning of 1.5 V. This section will define its meaning more precisely and consider different ways of representing analog input signals in data acquisition. It will also present correct measurement methods of various types of analog signals.

7.2.1 Referenced and Floating Source

To define a voltage rigorously, you would need to be introduced to electromagnetic theory. However, that is not the purpose of this subsection, and such a theory will not be discussed in this book. It will suffice to say that a voltage is the potential difference between two nodes (positive node and negative node) of a given source. For example, 1.5 V will be understood to mean that the potential difference between the positive node and the negative node is 1.5 V. Note that the negative node is not claimed to be grounded (0 V). Therefore, according to this definition, if a signal source has 35 V at the negative node and 36.5 V at the positive node, it will be understood that its capacity is 1.5 V, too. This simple example illustrates the importance of the reference point. In other words, the same quantity can be interpreted differently depending on the reference point.

A signal source whose negative node is wired to the ground is referred to as a *referenced* or *grounded* source. Then, for example, if a signal source is generated by a device that is plugged into a wall outlet, it is referred to as a referenced source, assuming that the generated signal shares the same ground of the power line (building ground) from the wall outlet. Otherwise, the signal source is understood as a floating source as in the case of battery, thermocouple, or strain gauge signals.

It is important to understand the nature of the signal source because its value or quantity can be interpreted differently depending on the reference point at the time of measurement. Having defined referenced sources and floating sources, you will now be introduced to three different measurement modes: differential (D), nonreferenced single ended (NRSE), and referenced single ended (RSE) modes.

7.2.2 Differential, (Non)Referenced Single Ended Modes

First, you need to understand how analog signals are acquired by data acquisition boards. For instance, you have a phenomenon to measure such as water temperature. You need to convert the temperature into a signal (temperature voltage) and bring it into a data acquisition board through a cable. On the data acquisition board, the temperature voltage is sampled and sent to an application such as LabVIEW, where the sampled data are analyzed. When the signal is sampled on the board, the key components you need to understand are the multiplexer and the instrumentation amplifier.

The instrumentation amplifier consists of three operational (OP) amplifiers with large input impedance and small output impedance, and amplifies the difference between the two inputs of the instrumentation amplifier. An ideal OP amplifier is a device with two input ports V+ and V–, and amplifies the difference between them. Since it amplifies only the difference in the inputs, it will have an effect of rejecting a signal that appears identically on the two inputs. Therefore, if you have an identical noise signal on both inputs, it will be rejected in the amplified output. Its input impedance is assumed to be infinite and output impedance is assumed to be zero. The output swing is limited to the two external power supplies such as ± 10 V.

The signal of interest (temperature voltage in the previous example) is multiplexed by a multiplexer, if the input signal is present at more than a single channel, followed by an instrumentation amplifier on the data acquisition board whose V+ and V– are connected to the input channels. The amplified and multiplexed signal is then sampled and quantized to be brought into computers where the data are analyzed by software. Usually, data acquisition boards have a single A/D converter so that multiple-channel sampling is performed in a sequential order. If simultaneous sampling is of interest, you should use a board that has an A/D converter for each channel.

The gain of an amplifier is often configurable either in hardware or in software. Combining different jumper switch settings on the board or setting (different) gains for different channels in LabVIEW usually does it. This will be discussed in detail in Chapter 8.

As for the three different input measurement modes, input signals are always measured with respect to the ground of the board, which is the same ground the computer shares (building ground). This ground is sometimes called earth ground since the computer is referenced to the ground of the wall outlet that is the earth. This is the common convention unless your data acquisition board specifies it differently.

It is a common practice that input signal is measured with respect to the earth ground by data acquisition boards unless otherwise specified.

Based on this notion, three different input measurement modes will be introduced. Two of them, D and NRSE, are essentially identical except for the number of input channels available. If a board provides 16 input channels, only half will be available in D mode since each differential channel measurement requires two physical input channels on the board.

In Figure 7.5, the person with a magnifying glass (person A) is measuring the height of the other person standing on a small hump (person B). If person A determines the height of person B by measuring the difference between where the head and the feet of person B are, it will be the correct height (D and NRSE). However, if person A measures where the head of

Figure 7.5
The height of a person can vary depending upon the point of reference. The person with a magnifying glass (person A) represents the data acquisition board, and the other on the hump (person B) does the input signal source. In this case, the input signal source is referenced source type, not floating, since person B is standing on a hump whose bottom is on the earth. For floating signal sources, remove the hump so that person B floats in the air.

person B is using the reference point of person A (RSE), the measurement will be incorrect as it sums the height of person B and the height of the hump he is standing on. The height of the hump is sometimes called ground loop and is usually the cause of 60-Hz AC or DC noise in acquired data. In Figure 7.5, the height of person B is analogous to the voltage of input signal source, and person A is analogous to the data acquisition board that measures the input signal source. Therefore, person B (signal source) is a referenced source (not a floating source) since he is standing on a hump whose bottom is the earth ground.

Now, what happens if the input source is floating? Then all three modes are possible in measuring input signal sources of floating type. The only problem is that since person A in Figure 7.5 will look at where the head and the feet of person B are in D and NRSE mode, person B should be within the sight of person A. In case of RSE, both person A and person B are referenced to the same point. Therefore, if person B is not within the sight of person A, it just means that person B is too tall for person A to measure the height. Therefore, in the case of D or NRSE modes, you may need to level person B up or down to bring him within the sight of person A. This can be done by implementing bias resistors between both inputs V+ and V– and the reference point of person A (the reference point of the instrumentation amplifier on the data acquisition board). The values of the two resistors must be identical and range between 10 KΩ and 900 KΩ to retain the balance between two inputs. However, you may need to fine tune the values of the resistors to provide the inputs V+ and V– with a DC path to bring the input signals up or down to the sight of person A. The range of the bias resistors is chosen such that it keeps the input source floating while not loading it to the resistors and keeps the input source within the range of the instrumentation amplifier on the board. If their values are too small, the resistors will no longer keep the input signals floated, and consequently the input signals will be loaded onto the resistors. If the values are too large, the bias resistors will keep the input signals floated, but they will not provide a return path for the bias currents injected by the instrumentation amplifier, which will keep the input signals out of range. However, if the input signal is DC coupled, only one resistor between V– and the ground would be sufficient to bring the signal down to a proper range, whereas an AC-coupled input source must be fed with two identical bias resistors.

Since it was mentioned that D and NRSE were basically identical except the number of input channels available, the following question may arise: When do you use D mode and when NRSE mode? The answer is as follows: Differential mode (D) is always the best since the amplifier only considers

If a signal source is floating and the measurement mode is either D or NRSE, the input signal can be out of the range of the data acquisition board instrumentation amplifier due to the bias current injected by the amplifier. To avoid this phenomenon, bias resistors must be used between the two inputs V+ and V– and the earth ground which is the reference of measurement on the board. For DC-coupled input sources, only one is needed between V– and the earth ground, whereas two identical resistors between 10 KΩ and 900 KΩ are required to maintain the input balance for AC-coupled input sources. Fine tuning of those resistor values may be necessary.

the difference between V+ and V–. If noise is present at both V+ and V–, it will be rejected in the amplified output. Also, NRSE and RSE are more susceptible to noise than D mode; therefore, use either NRSE or RSE if the signal-to-noise ratio (SNR) is high, because the signal will be relatively less sensitive to the noise due to the high SNR. Applying either NRSE or RSE depends on the type of input source: floating or referenced. Also, if you know that all of the input signals can share the same reference point, single-ended measurement modes can be applied. For instance, if you want to measure the potential of multiple batteries, you will connect the negative side of each one to the same reference point on the data acquisition board and apply either RSE or NRSE, noting that bias current must be taken into consideration in NRSE mode. The following algorithm summarizes how to select the correct signal measurement mode:

START: Is the input signal floating or referenced? If floating, go to **FLOAT**. If referenced, go to **REFERENCED**.

FLOAT: If SNR is high, use either RSE or NRSE. V– should be wired to the earth ground (usually indicated as AIGND on most data acquisition boards) in RSE mode and to a common reference (usually indicated as AISENSE on most data acquisition boards) in NRSE mode. NRSE mode must be used with bias resistors between the common reference and the earth ground. If SNR is not high, use D mode. D mode in this case must be used with bias resistors between V– and the earth ground, or both V+ and V– and the earth ground. See BIAS section below.

BIAS: If the source is AC coupled, use two identical resistors between 10 KΩ and 900 KΩ. If the source is DC coupled, one resistor between V– and the earth ground would suffice. Fine tuning of the resistor value may be necessary.

REFERENCED: Use either D or NRSE mode. RSE mode may cause ground-loop problems.

7.3 Data Acquisition Hardware and Driver Software

If you want to perform any type of data acquisition using a personal computer, you will need a plug-in card that performs signal conversion and a program that brings the converted signal into the computer. LabVIEW is the software that allows you to perform data acquisition at the highest user level, and actual signal acquisition occurs at the lowest level on the data acquisition board. The intermediate-level software that controls the communication between LabVIEW (highest level) and hardware (lowest level) is the driver software. On PC platforms, the driver software utilizes dynamic link libraries (DLLs), whereas the Macintosh platforms use Extensions and Control Panels. One important piece of common knowledge in regard to the driver software is as follows:

*It is recommended that you install the driver software on the computer **before** the hardware. Therefore, you should install LabVIEW first, the driver software second, and the hardware (data acquisition board) last. This correct installation order will prevent most of the common computer-related problems, such as the boards not getting recognized by the computers or inconsistent behavior of the boards.*

The expression *data acquisition* has been used loosely to refer to all types of tasks, such as analog input, analog output, digital input and output, and counter operations. Analog input refers to applications where you measure real-world phenomena, such as temperature, voltage, strain, and pressure, in terms of voltage or current. Analog output is the opposite of analog input, and an example is DC or AC signal generation, which is sometimes referred to as point generation or waveform generation, respectively. If you are generating or acquiring TTL-level signals, it is called digital input and output. This is usually for applications where controlling digital switching devices such as relays is the main goal. Another example of digital data acquisition is generating a pattern of TTL-level signals, which is often implemented for the communication between digital devices such as scanners, printers, and computers. The communication is often achieved through handshaking where two digital devices request and acknowledge certain tasks. Counter operation often confuses LabVIEW programmers with digital input and output since counter operation also deals with TTL-level signals; however, it is different from digital data acquisition such that more general applications are covered by counter applications whereas digital applications generally deal with switching devices and digital communication. Examples of counter operation are frequency measurement, pulse width measurement, period measurement, pulse train generation, and TTL pulse counting, which is often referred to as event counting.

The next few chapters will discuss each topic individually. When each topic is studied, the hardware must also be discussed. Therefore, proper discussion of the hardware (data acquisition board) will also be presented whenever the need arises. The ultimate goal of the next few chapters is not to present all of the details about each topic, but to present the fundamental concepts that are essential to understanding different techniques of data acquisition. Once the core concept is understood, its expansion can be done easily. Concentrating on small details of different techniques will prevent you from seeing the complete scope, which will eventually confuse you with too many details. Therefore, the following strategy is employed in the next few chapters: First, the fundamental concepts and the outlines of different methods are introduced. Then each method will be studied using *examples* that can be found in the example directory in the LabVIEW directory. Those examples are all ready to run and cover all different types of (basic and advanced) implementation; therefore, it would be waste of time to learn how to build those examples from scratch. Instead, learning *how* to use and expand them to fit your own application would be a more meaningful and more efficient way of learning data acquisition techniques with LabVIEW. Also, the discussion of the hardware will be generalized as much as possible so that it is not limited to the boards of a specific vendor. Once the basics are mastered, understanding different brands of hardware should be accomplished quickly since the fundamental concepts and terminologies are similar or often identical. Specification sheets of data acquisition boards and how to understand them will be discussed first.

7.4 Specification of Hardware

If you want to set up a data acquisition system, you will first need to decide which board(s) will be used for your applications. Selecting the most appropriate board(s) will optimize your system financially as well as performance-wise. Therefore, understanding your application as well as your hardware is important as the first step in setting up your data acquisition system.

As you may expect, the specification sheet of data acquisition boards would normally show a lot of information, and this section will cover only a few of them, which are the most common and important aspects, such as:

Input and output analog signal range
Gain for analog input
Resolution in analog-to-digital (A/D) conversion
Resolution in digital-to-analog (D/A) conversion
Resolution of counter chips
Maximum frequency for sampling and waveform generation
Triggering capability
Number of digital lines and ports

7.4.1 Input and Output Analog Signal Range

This information indicates the range of input signal that a data acquisition board can handle. If it is for analog input, you must have *a priori* information about the input signal, such as the range, reference type (referenced or floating), and approximate signal-to-noise ratio (SNR). Input range is closely related to the gain that a board can afford, which is the topic of the next subsection.

Signal range of analog output indicates the range of signals that a board can generate, so if waveform generation is of interest, you must consider this specification first. One aspect that is essential to waveform generation on PC platforms is the availability of direct memory access (DMA) channels as well as interrupt request (IRQ) channel. DMA especially is critical in waveform generation so that waveforms cannot be generated without any DMA resource channel assigned to the board; therefore, you must free up at least one DMA channel for the board to be able to generate waveforms. As for Macintosh platforms, the DMA issue does not apply to the latest peripheral component interconnect (PCI)-based boards. If you want to generate waveforms with 68K models, however, a separate board that externally provides DMA channels is required.

7.4.2 Gain for Analog Input and Resolution in A/D Conversion

Gain and input signal range are so closely related that one cannot be considered without the other. It seems to be a common misconception with the gain setting of a board such that the gain is considered as a volume setting of an audio system. In other words, if you read that a board can apply a gain of 10, you tend to think that if your input signal is 1 V, for example, the output with the gain would be 10 V. This type of misunderstanding should be corrected in order to understand the gain setting on the board properly. Different gain settings will only affect the minimum change that can be detected by a board in the input signal. In order to elaborate this issue further, this subsection introduces the resolution in A/D conversion first.

If a board is said to have 1-bit resolution in A/D conversion, it implies that the board can detect two different levels of variation in the signal level. For example, if the input signal range is between 0 V and 1 V with 1-bit resolution, the board will represent the input signal with two different levels, such as 0 and 1 only. Such a case would be that if the input signal is between 0 V and 0.5 V, it will be represented as level 0, and range between 0.5 V and 1 V will fall to level 1. However, if a board has 2-bit resolution that can represent four different states, it will give us better representation of the input signal than that with 1-bit resolution so that now the input signal between 0 V and 0.25 V is mapped to level 0, 0.25 V and 0.5 V to level 1, and so forth. Also, note that the resolution is an *invariant* aspect of a board, whereas the gain and the input range vary. In other words, the resolution of a board is given and cannot change, whereas the gain and the input range can.

Having discussed the resolution in A/D conversion, consider the correlation between the gain and the input range. Suppose your input signal varies between 0 V and 1 V, and the board has 2-bit resolution that is invariant. As previously mentioned, the gain does not affect the range of the input signal acquired but affects the *accuracy* in the input signal representation. It is also important to note that the input range and the gain are reciprocally related. In LabVIEW, you can apply different gains to different input channels by choosing different input range for each channel, and broader range setting in LabVIEW applies smaller gain since the capability of a board is fixed. For example, if the specification sheet states that gain 1 with –10 V and 10 V input range, and gain 5 with –1 V and 1 V, it implies that LabVIEW will apply the gain of 1 if you select –10 V and 10 V for the input range, and gain of 5 with –1 V and 1 V input range. Therefore, the gain and the input range are reciprocally related.

You must select a correct input range in order to obtain the maximum accuracy in the representation of input signals.

Suppose you have an input signal that varies, at most, between –1 V and 1 V, and a board in the computer that has two different gain settings: gain of 1 with –10 V and 10 V input range, and 5 with –1 V and 1 V. Then which gain or input range setting do you need to apply in LabVIEW? The answer is obviously the gain of 5, which is for the input of range –1 V and 1 V. What happens if you apply the other gain of 1, which is for the input range –10 V and 10 V? In order to answer this question, you must recall the previous discussion of resolution. If you assume that the board has 2-bit resolution for the sake of convenience, you know that the board will represent the input signal using four different levels. If you select the gain of 1, which is for an input varying between –10 V and 10 V, the board will divide 20 V swing (–10 V and 10 V) into four different levels. However, with gain of 5, which is for the –1 V and 1 V range, it will divide 2 V swing (–1 V and 1 V) into four different levels. It is clear to see that the latter case (correct input range setting) will give you higher accuracy in the representation of the input signal. Therefore, a correct input range setting must be chosen for the maximum accuracy in the signal reading in LabVIEW. Again, the input range setting is reciprocally related to the gain setting, which has direct influence on the smallest change that a board can detect.

7.4.3 Resolution in D/A Conversion and Counter Chips

Discussion of the resolution in terms of D/A conversion and counter chips is similar to that of the analog input case. If you have a counter chip with 2-bit resolution, the maximum number you can count would be 3 (0, 1, 2, and 3); therefore, higher resolution will allow you to be able to count larger number of events. As for the analog output case (D/A conversion), having higher resolution (more bits) will result in finer representation in the waveform that a board generates.

7.4.4 Sampling Frequency and Maximum Update Rates

The maximum sampling frequency of a data acquisition board must be at least two times higher than the bandwidth or the maximum frequency component of input signals to meet the sampling theorem requirement. It should be noted that the nominal maximum sampling frequency of a data acquisition board applies to a single input channel. This implies that the maximum sampling frequency for more than a single input channel is the nominal maximum sampling frequency that the board can support divided by the number of input channels in use. This is due to the number of A/D converters on the board, and if it has a single A/D converter (which is the usual case), you must divide the maximum sampling frequency by the number of input channels in order to compute the maximum sampling frequency for each channel. For example, if the maximum sampling frequency of a board is 100 KHz, each channel will be able to sample at 20 KHz at most if you are using five input channels. With a single A/D converter, this division is necessary since the converter samples each channel sequentially. As for waveform generation, the *update rate* should be taken into consideration to make sure the board of interest is capable of generating the waveform at a specified frequency.

7.4.5 Triggering Capability

In applications where the initiation of a certain task depends on another event, triggering techniques can be employed. Consider an example of triggering usage with musical instrument digital interface (MIDI). These days, musical instruments are computer controlled so that their automation becomes trivial, and a single musician can play multiple instruments simultaneously. For example, while a pianist plays the piano, he or she can have multiple instruments, such as drum rhythm box or keyboards, played at the same time. Usually, the other instruments, such as drum rhythm box or keyboards, will be waiting for the right moment to start, and the pianist will send a digital pulse to initiate them using a pedal while playing the piano. In other words, when the pianist presses the pedal, it will send out a digital pulse to each instrument to initiate it upon the reception of the pulse.

Common triggering techniques can be divided into two categories: digital triggering and analog triggering. Digital triggering is more basic than the analog one, so only the advanced boards usually provide analog triggering capability. Digital triggering is supported by almost all data acquisition devices. Digital triggering can be described as follows: A certain task will be initiated based on the reception of a square pulse. The initiation can be set when the pulse changes either from 0 V to 5 V, which is said to be rising edge, or from 5 V to 0 V, which is said to be falling edge. On the other hand, analog triggering initiates a certain task based on the property of an analog signal, such as slope or threshold. If the slope of an analog triggering signal exceeds a certain level, or if the analog triggering signal level exceeds a preset threshold, a certain task will be initiated on the data acquisition board.

7.4.6 Digital Lines and Ports

If your application involves a digital operation, such as turning on or off relay switches or reading or writing a single digital value or a binary pattern, you should make sure that your board has digital lines and ports. Ports are analogous to channels in analog data acquisition, and each port has multiple lines (usually eight lines). If you need to generate more than eight bits, you can group multiple ports to construct such binary patterns. If the digital operation includes handshaking with another digital devices, the board must support handshaking capability in order to communicate with other digital devices.

7.4.7 Data Acquisition VIs

Having the information that has been discussed so far, you are ready to study data acquisition VIs in LabVIEW. If you go to the **Functions** panel in the diagram window and select **Data Acquisition** subpalette, you will see six subpalettes, as shown in Figure 7.6.

The first one, **Analog Input**, contains all of the VIs for analog input, and the **Analog Output**, **Digital I/O**, and **Counter** subpalettes in the first row contain the VIs for waveform generation, digital, and counter operation. The two subpalettes at the bottom are **Calibration and Configuration** and **Signal Conditioning**. The **Calibration and Configuration** subpalette contains VIs for calibrating different types of data acquisition boards, and the name of each VI shows which board it can be used for. It used to be recommended that calibration be done twice a year or so, but now it is recommended that it be done every time you run your application. Using the VIs in **Calibration and Configuration**, you can reset internal settings of a board, such as voltage reference value, and readers who are interested in more details about calibration may refer to online help for each VI, which discusses correct parameter settings and external connections if necessary.

The **Signal Conditioning** subpalette contains VIs for preconditioning input signals, such as unit conversion of thermocouple, thermistor, RTD, or

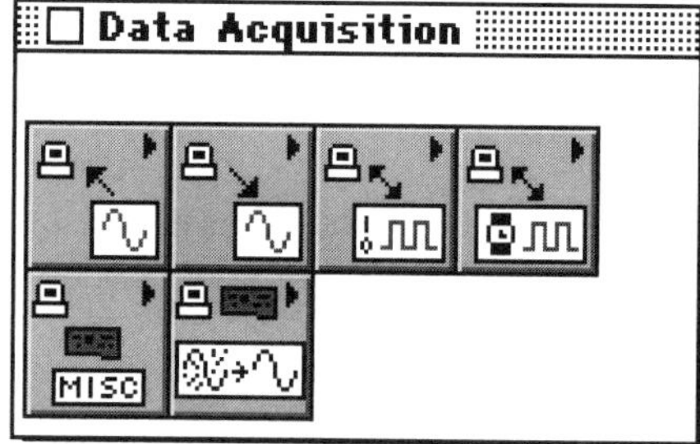

Figure 7.6 Data acquisition subpalette.

strain gauge voltage readings. Since those types of measurements return small voltage readings, conversion to corresponding units, such as Fahrenheit, Celsius, or unitless strain, needs be done. This also includes the linearization of the voltage reading since, for example, the response of the thermocouple probe to different temperatures is not linear. Furthermore, different types of signal conditioning are required, such as cold junction compensation for thermocouple readings, a correct bridge configuration and an external excitation voltage for strain gauge readings, and an excitation voltage for thermistor readings. Readers who are interested in the details of those applications are encouraged to refer to the manuals of corresponding signal conditioning devices for more specific configuration schemes.

The following few chapters will focus on the first four subpalettes in the first row in Figure 7.6, which are about analog input and output, digital input and output, and counter operations. The goal of those chapters is to understand the basic and the advanced techniques about different data acquisition programming. Once the fundamentals are mastered using the examples of LabVIEW, expanding them to accommodate your specific needs can be easily done with a minimal amount of effort and difficulty.

PROBLEMS

7.1 Equation (7.12) shows that the reconstructed signal from the samples $x(kT)$ is the sum of weighted $x(kT)$ by $\text{sinc}((t-kT)/T)$ for k from $-\infty$ to ∞. Show that the equation works for a 1 Hz sine wave for k from –50 to 50, and 101 Hz of sampling frequency.

$$x(t) = \sin(2\pi t)$$

A possible solution is shown in Figure P7.1.

7.2 What input mode do you suggest to be used for the following input signals? Provide your answer using either D, NRSE, or RSE, and justify it. Also, indicate how the signal should be wired using V+, V–, AISENSE, and AIGND. Will you need to use the bias resistors?

(a) Thermocouples
(b) Battery
(c) Signal from a function generator plugged into the wall outlet
(d) DC from a power supply plugged into the wall outlet

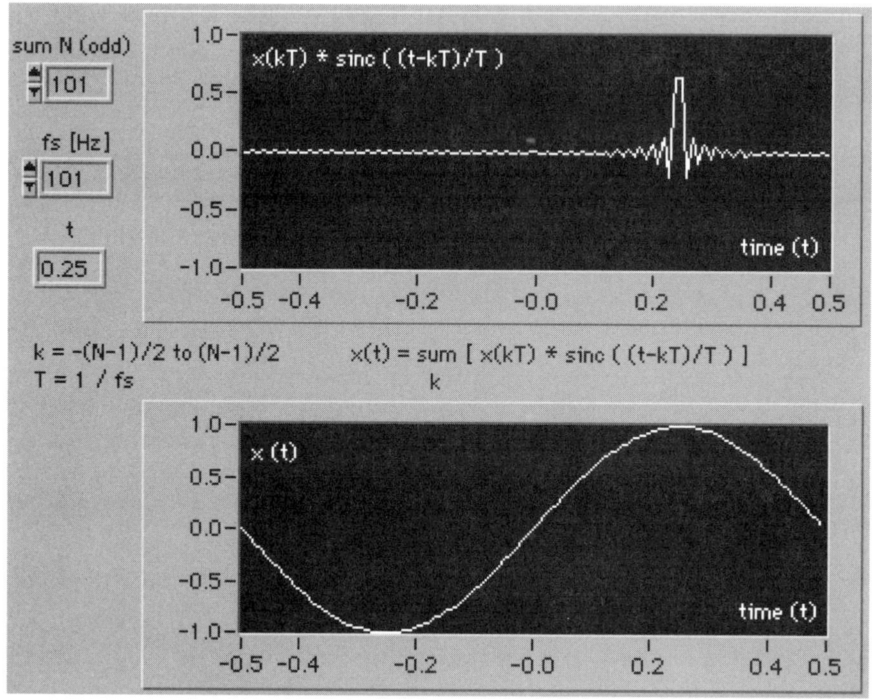

(a)

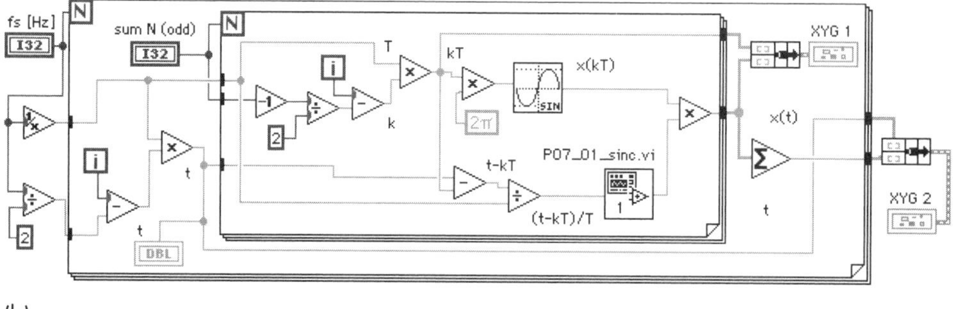

(b)

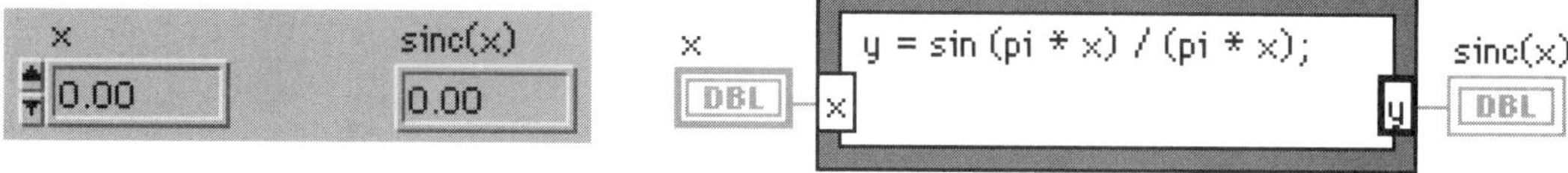

(c) sub VI **P07_01_sinc.vi**.

Figure P7.1

7.3 The smallest detectable change by a data acquisition board with gain G, input signal range ΔV, and resolution R is often called *code width* and is calculated by

$$\text{code width} = \Delta V \times (G \times 2^R)^{-1}$$

Compute the code width of the following five cases:

	Input Range	**Gain**	**Resolution [bits]**
Case 1	0 V to 10 V	1	12
Case 2	–1 V to 1 V	1	12
Case 3	0 V to 10 V	1	16
Case 4	–0.1 V to 0.1 V	100	12
Case 5	0 V to 5 V	10	16

7.4 Suppose that your data acquisition board has a single A/D converter with a maximum scanning rate of 100 KHz, and you are scanning four analog input channels. What is the maximum scanning rate for each channel? Also, what is the maximum frequency bandwidth that the input signal for each channel can have in order to avoid aliasing?

Data Acquisition: Analog Input

8

8.1 Sampling Signals

Figure 8.1 shows the VIs in the first subpalette in the **Data Acquisition** subpalette. First, consider the following general statement about the **Data Acquisition** subpalette:

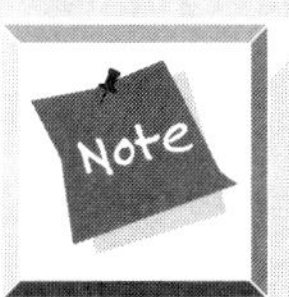

All of the four subpalettes, ***Analog Input****,* ***Analog Output****,* ***Digital I/O****, and* ***Counter****, in the* ***Data Acquisition*** *subpalette have similar VI organization. The first row contains high-level VIs, implying that they are ready to be used with limited controls. The second row contains intermediate-level VIs with more controls, and they are being used as a sub VI in the high-level VIs. Unless your application is very complicated, most applications can be written using only the VIs in the first two rows of each subpalette. The subpalette* ***Utility*** *contains VIs similar to the high-level ones, and these are useful for repeating processes using* ***For Loop****s or* ***While Loop****s. The subpalette* ***Advanced*** *contains advanced VIs, which give you the ultimate control, and accurate knowledge about your application will be critical to use them correctly.*

The following section will explain the high-level VIs with an example and discuss *how* you can expand them to fit your own application. Due to the variety of prewritten examples in LabVIEW and the range of their application coverage, the correct way of learning data acquisition programming with LabVIEW is to learn how to use those examples and customize them for your application in lieu of writing a new VI.

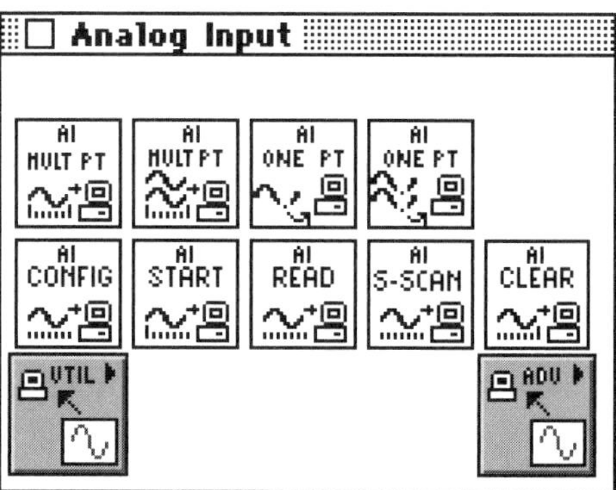

Figure 8.1 **Analog Input** subpalette.

8.2 AI Acquire Waveform.vi (High-Level VI)

Using high-level VIs is quite straightforward since those VIs are ready to be used, and specifying inputs is all you need to do. Therefore, **AI Acquire Waveform.vi** will be used to explain the input parameters for the high-level VIs since the parameters are very similar. **AI Acquire Waveform.vi** returns a one-dimensional (1-D) array of data from a single analog input channel. The following are its input and output parameters, with their data types indicated in the parentheses:

1. **device number** (I16): This is an unique ID number assigned to each data acquisition board in the computer; it is assigned in the driver software *NI-DAQ Configuration Utility*. Each different board, therefore, will have a different ID number. Referring to the ID number of a board of interest will do the addressing to the board. I16 indicates 16-bit integers.
2. **channel string** (string): You specify the input channel using this parameter. You may enter a channel number in string format or a name of the channel if *Channel Wizard* is used to assign a string to input channel(s).
3. **number of samples** (I32): This specifies how many data samples will be acquired from a single channel specified in the input **channel string**. I32 indicates 32-bit integers.
4. **sample rate** (SGL): Sampling frequency. Default is 1 KHz (1000 samples per second). The Nyquist rate criterion must be satisfied in order to preserve the characteristic of the original input signal. (See Chapter 7.) SGL indicates single precision floating numbers.
5. **high limit** and **low limit** (SGL): These control the input range, which determines the gain for each input channel. The gain or the input range determines the accuracy of signal representation (see Chapter 7); therefore, the closest range to the input signal must be chosen in order to maximize the accuracy in the representation of the input signal.
6. **actual sample period** (SGL): Actual sampling time, which is the inverse of actual sampling frequency returned by the data acquisition board. This may slightly be different from the input parameter **sample rate**.
7. **waveform** (SGL): One-dimensional (1-D) data array from a single channel specified in the **channel string** input.

Once proper input parameter values are set, the VI is ready to run and returns a 1-D array of data from a single channel. Note that high-level VIs use intermediate-level VIs, which provide you with more flexibility.

As mentioned earlier, the goal in this chapter is to learn the fundamentals and how to use the LabVIEW examples and expand them to each individual need instead of learning how to write a VI from scratch. Therefore, this section will discuss how the high-level data acquisition VIs can be modified. First, note that **AI Acquire Waveform.vi**, as well as the other high-level VIs, runs *once*, not continuously. This implies that if you want to perform continuous analog input data acquisition using high-level VIs, you will need to modify them properly. **AI Acquire Waveform.vi** will be examined in more detail before proceeding to the modification technique.

Open a new VI, and place **AI Acquire Waveform.vi** from **Functions >> Data Acquisition >> Analog Input** in the diagram window. Double click on the VI to show its front panel. Using the shortcut **Ctrl-E**, display the diagram window shown in Figure 8.2. It shows that **AI Acquire Waveform.vi** uses **AI Waveform Scan.vi**, which is one of the VIs in the subpalette **Analog Input Utilities**, which is located in the third row in Figure 8.1. The **Analog Input Utilities** subpalette contains VIs with slightly more control than high-level VIs, such as continuous acquisition capability. Double click on **AI Waveform Scan.vi** and go to its diagram window. Examine the diagram window to understand the VI better. When you open the diagram window, you will see a lot of information that can easily overwhelm you with complicated wires. When you see a VI with a complicated structure as

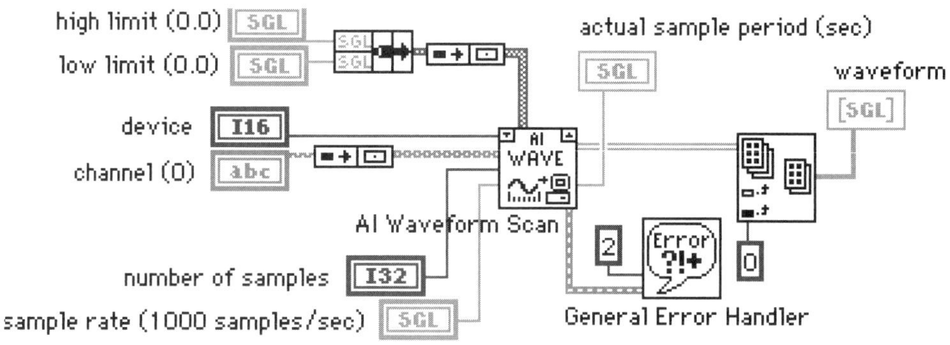

Figure 8.2
Diagram window of **AI Acquire Waveform.vi**.

shown in the diagram window of **AI Waveform Scan.vi**, remember the following statement:

When you see a complicated wiring structure in the diagram window of a data acquisition VI, you can always divide it into three stages—Configuration, Read or Write, and Reset. In the stage Configuration, LabVIEW initiates the communication with the hardware (data acquisition board) while setting input limits, sampling rate, channel strings, and so forth. In the stage Read/Write, LabVIEW acquires or generates the data, performing analog input or output, digital input or output, or counter operation. When the data acquisition stage is completed, LabVIEW terminates the communication with the hardware by releasing all of the parameter settings (the stage Reset). The rest of the wires and VIs are built upon these three stages.

Figure 8.3 shows a small portion of the entire diagram window of **AI Waveform Scan.vi**, which displays the three stages mentioned previously. The first stage contains, **AI Config.vi**, which establishes the communication with the board by configuring all of the parameters. The second stage contains **AI Start.vi** and **AI Read.vi**, where the data acquisition starts and the data are read. The last stage includes **AI Clear.vi**, where all of the parameter settings are reset and the communication becomes terminated.

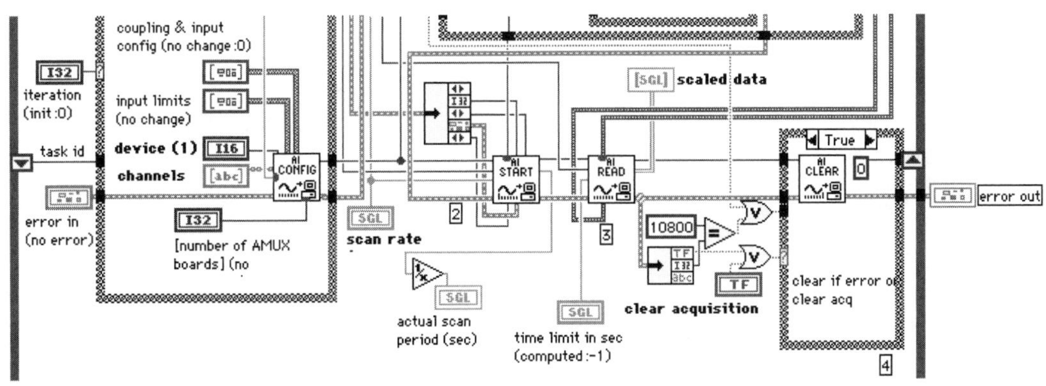

Figure 8.3
Three stages of data acquisition: Configuration, Read/Write, and Reset. The diagram window of **AI Waveform Scan.vi** is shown.

8.3 Modifying High-Level VIs

You have seen the three stages of data acquisition using **AI Waveform Scan.vi** in Figure 8.3. Among those three stages, it should be noted that the first stage, Configuration, is a time-consuming process compared to the others and does not have to be repeated if the same data acquisition board and parameters are to be used. Suppose that you want to perform continuous analog input waveform acquisition using one of the high-level VIs **AI Acquire Waveform.vi**. Since you need to repeat the acquisition knowing that **AI Acquire Waveform.vi** can be used without any complicated programming, you decide to include the VI in a **While Loop** calling the VI repeatedly. This argument is perfectly legitimate except for one problem. Since the high-level VIs perform all three stages in one VI, they configure the hardware repeatedly, which can slow down the continuous acquisition process. Then how do you modify **AI Acquire Waveform.vi** (the high-level VI) in order to configure the device only once at the beginning of the acquisition process and skip it afterward? The answer will be given in this section, and once you understand the technique, it can be applied to any other high-level data acquisition VIs.

Consider the diagram window of **AI Waveform Scan.vi** in Figure 8.3 again. You see a control **iteration** wired to a **Case** structure that contains **AI Config.vi**. Therefore, if the **iteration** is 0, it calls **AI Config.vi**, which configures the hardware, and if not, the first-stage Configuration will be skipped since **AI Config.vi** will not be called. If you move one VI level higher to **AI Acquire Waveform.vi**, as shown in Figure 8.2, you realize that a terminal is available to access the **iteration** input to **AI Waveform Scan.vi** (Figure 8.4). You can also see that in Figure 8.3 **AI Clear.vi** is called by a **Case** structure controlled by a Boolean control **clear acquisition**. Since these two controls (**iteration** and **clear acquisition**) are missing in the diagram window of **AI Acquire Waveform.vi**, you need a total of two additional input terminals to **AI Waveform Scan.vi** in Figure 8.2. However, it is important to save the VI

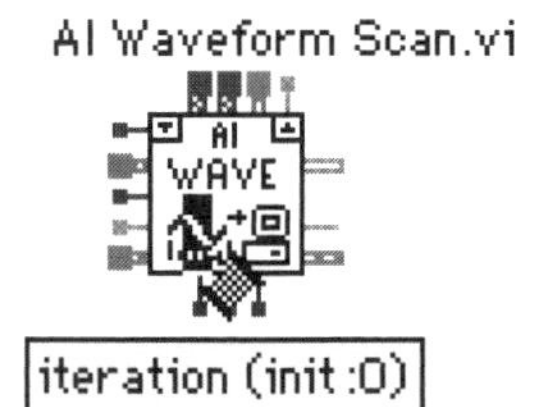

Figure 8.4 The input **iteration**.

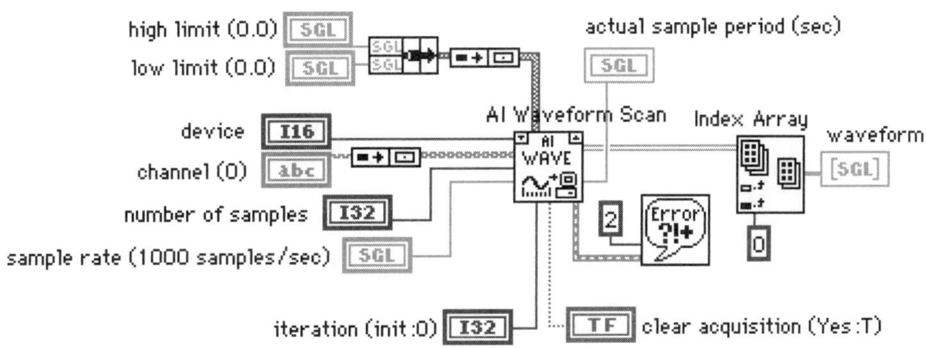

Figure 8.5
Diagram window of **My AI Acquire Waveform.vi**.

under a different name first since you do not want to change the original LabVIEW VI. So, first save the high-level VI **AI Acquire Waveform.vi** under a different name, such as **My AI Acquire Waveform.vi**, and add two controls, **iteration** and **clear acquisition**, to proper terminals. The complete modified version of **AI Acquire Waveform.vi** is shown in Figure 8.5, and Figure 8.6 shows continuous analog input acquisition using the modified high-level VI **My AI Acquire Waveform.vi** and a **While Loop**. Note that in Figure 8.6, the Boolean control **QUIT** is wired to **clear acquisition** through a **Not** gate since TRUE to **Clear Acquisition** will call **AI Clear.vi**, which will terminate the communication with the data acquisition board. Also, iteration

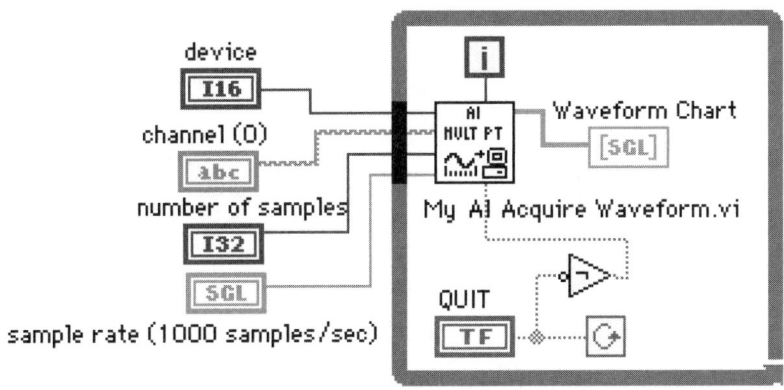

Figure 8.6
Continuous analog input using **My AI Acquire Waveform.vi**.

terminal [i] is wired to the input **iteration** of **My AI Acquire Waveform.vi**, which will call **AI Config.vi** only the first time the **While Loop** iterates.

The technique just studied can be applied to any other high-level data acquisition VI, and it gives you a great deal of flexibility after all. Similarly, you can easily customize any preexisting VI for your own applications. Then you may wonder how you know which input or inputs need to be additionally added or even removed. The answer is that you will need to debug the VI of interest to identify correctly what needs to be changed where. For example, in Figure 8.6, you could start with putting **AI Acquire Waveform.vi** in a **While Loop** only with the input **iteration** wired. If you run the VI, it will return an error message claiming that the device is invalid at the second iteration of the loop. Then you need to move one VI level lower to the diagram window of **AI Waveform Scan.vi**, and you will find that the error is coming from the **Case** structure, which contains **AI Clear.vi**. In order to control the **Case** structure, you need a Boolean control to **clear acquisition**. This will give us the complete version as shown in Figure 8.6. It needs to be reemphasized that complicated VIs should not frustrate you. Remember that all of the complex wires and VIs can be divided into the three stages described previously. Once you identify those three stages, miscellaneous sub VIs and wires can be analyzed one by one.

8.4 High-Level Analog Input VIs

The other three high-level analog input VIs are **AI Acquire Waveforms.vi**, **AI Sample Channel.vi**, and **AI Sample Channels.vi**. The only difference among them is indicated in their names: Waveform or Channel either singular or plural. If it is Waveform (singular), the VI works with a single input channel while returning multiple data samples in a 1-D array. If it is Waveforms (plural), it works with multiple input channels while returning multiple data samples in a 2-D array, where each *column* is the 1-D data array from each channel specified in the channel string input. If it is Channel (singular) instead of Waveform(s), the VI works with a single input channel while returning a single data sample. Similarly, if it is Channels (plural), the VI acquires a single sample from multiple input channels and returns a 1-D data array. Addressing multiple channels in the input channel string is done by listing channel numbers separated by a comma, such as 0, 1, 4 to address three input channels 0, 1, and 4. If a range of channels needs to be specified, the starting and ending channels can be separated by a colon; for example, 0:3 implies channels from zero to three, and 3:1 from channel three to one in a descending order.

8.5 Intermediate-Level Analog Input VIs

There are five VIs shown in the second row in Figure 8.1, and they will be explained using a LabVIEW example. Those intermediate-level VIs provide much more flexibility than high-level ones do, such as triggering, buffered or nonbuffered, and hardware-timed acquisition. The rest of this chapter will discuss each category using LabVIEW examples that are available in the following path:

`LabVIEW directory` **>> examples >> daq >> anlogin >> anlogin.llb**

Therefore, in the following sections the current directory will be assumed to be the one just listed unless it is specified otherwise; for example, if you refer to **Test.vi**, its full path would be LabVIEW directory **>> examples >> daq >> anlogin >> anlogin.llb >> Test.vi**.

8.6 Software Timing and Hardware Timing

Software-timed data acquisition means that you control the iteration of data acquisition using the timer of LabVIEW, which uses the CPU clock of the computer. This type of timing scheme does not provide a high order of accuracy due to the dependence on the CPU. An example of software timing would be a loop with **Wait (ms)** or **Wait Until Next ms Multiple**, which are shown in Figure 8.7.

On the other hand, hardware timing is when the iteration rate is controlled by the clock on the data acquisition board. This can be done by using the following two VIs: **AI Start.vi** and **AI Single Scan.vi**. Note that **AI Single Scan.vi** uses the on-board first-in-first-out (FIFO) buffer instead of the buffer in CPU memory, and it can return the most recent or the oldest data

Figure 8.7 **Wait (ms)** and **Wait Until Next ms Multiple** for software timing.

from the FIFO buffer depending on the input to the terminal **opcode** of **AI Single Scan.vi**. Whereas software timing controls the rate of sampling using CPU clock, hardware timing with **AI Single Scan.vi** uses the on-board scan clock and repeats sampling at the rate specified in the input **scan rate** to **AI Start.vi**. Points to remember with hardware timing are that (1) you can acquire (sample) signals at more accurate rate, but (2) since you are using the on-board FIFO buffer, which is relatively small, your scan rate must be relatively low in order to keep up with the size of the FIFO buffer. In other words, if you sample too fast, which will fill up the FIFO buffer quicker than LabVIEW can retrieve the data from it, you will start losing data samples. Therefore, use hardware timing if you do not have to sample fast (for example, less than 100 Hz) but need accurate sampling. If the high accuracy in sampling is not required, software timing can be used using the functions shown in Figure 8.7. Figure 8.8 shows an example of hardware-timed acquisition using **AI Start.vi** and **AI Single Scan.vi**, where **AI Config.vi** and **AI Clear.vi** are already assumed. You can also use buffers using CPU memory in lieu of FIFO on-board buffer with **AI Config.vi**, which is discussed in the next section.

__AI Single Scan.vi__ uses the on-board FIFO buffer so that hardware timing can be achieved, and the sampling rate can be specified by __AI Start.vi__. Without __AI Start.vi__, where you set the hardware timing rate (in other words, only with __AI Config.vi__, __AI Single Scan.vi__, and __AI Clear.vi__), it becomes hardware-timed acquisition at an unspecified rate, and __AI Single Scan.vi__ initiates a single scan using the fastest safe channel clock rate available.

A modified version of **AI Read One Scan.vi** is shown in Figure 8.9. (**AI Read One Scan.vi** can be found in the subpalette **AI Input Utility**.) It shows an example of using hardware timing at a specified rate set by **AI Start.vi**.

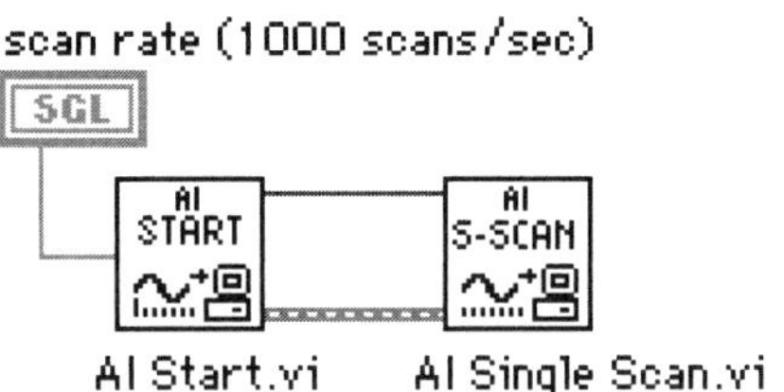

Figure 8.8 Hardware-timed acquisition with **AI Start.vi** and **AI Single Scan.vi**. The acquisition is performed at the rate specified in **scan rate** using the on-board scan clock, which is far more accurate than software timing, and depends on the on-board scan clock. Use this for slow sampling rate such as less than 100 Hz.

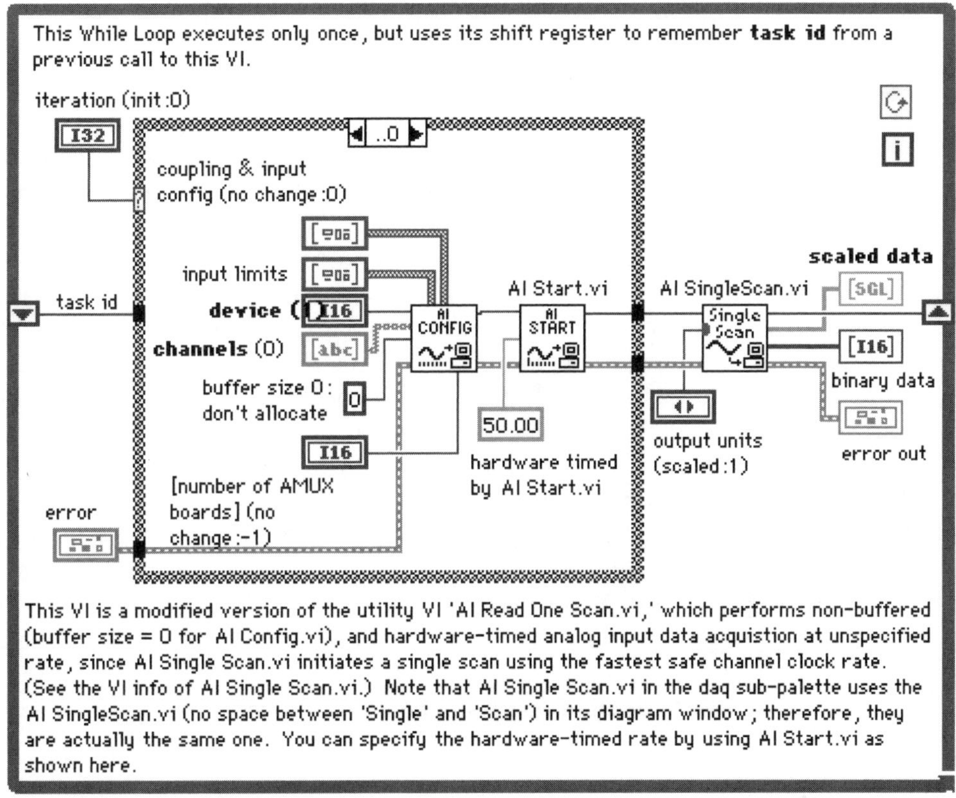

Figure 8.9
Modified version of **AI Read One Scan.vi** that can be found in the **AI Input Utility** subpalette. Note that **AI Read One Scan.vi** is hardware-timed at an unspecified rate, whereas this modified one is hardware-timed at a specified rate set by **AI Start.vi**. The sampling rate with hardware timing, however, must be relatively slow, such as less than 100 Hz.

8.7 Buffered and Nonbuffered Acquisition

LabVIEW can use buffers in two different locations: computer memory and data acquisition board. Single buffer and circular buffer are the types in the memory, and the FIFO buffer is the one on the data acquisition board. The single buffer is for single acquisition and the circular buffer is for continuous data acquisition. The FIFO buffer on the hardware can be accessed by the intermediate-level **AI Single Scan.vi**, and it can return the oldest or the most recent data from the FIFO buffer depending on the input **opcode** of **AI Single Scan.vi**.

Now, consider the following three examples: **Cont Acq&Chart (immediate) .vi**, **Cont Acq&Chart (buffered).vi**, and **Cont Acq&Chart(hw timed).vi**. Figure 8.10 shows the first two to compare nonbuffered and buffered continuous acquisition. In Figure 8.10(b), the three data acquisition stages are shown: Configuration, Read or Write, and Reset. Since you want to repeat the acquisition, only **AI Read.vi** is included in the **While Loop**. Similarly, **AI Single Scan.vi** is included in the **While Loop** in Figure 8.10(a) for nonbuffered continuous acquisition. Note that Figure 8.10(a) is not using hardware timing, which has been discussed in the previous section. (You do not have **AI Start.vi** to speficy the **scan rate**.) Instead, it is using software timing using **Wait Until Next ms Multiple**. Also, note that since **AI Single Scan.vi** uses the on-board FIFO buffer, the sampling rate (scan rate) must be slow, and the default value set in **Cont Acq&Chart (immediate).vi** is 3 Hz! The input **buffer size** of **AI Config.vi** is zero in Figure 8.10(a), indicating that it is nonbuffered acquisition, whereas it is 4000 in Figure 8.10(b), indicating that it is buffered acquisition. (Open the examples to see the default value in the front panel. Figure 8.10 shows only the diagram windows.)

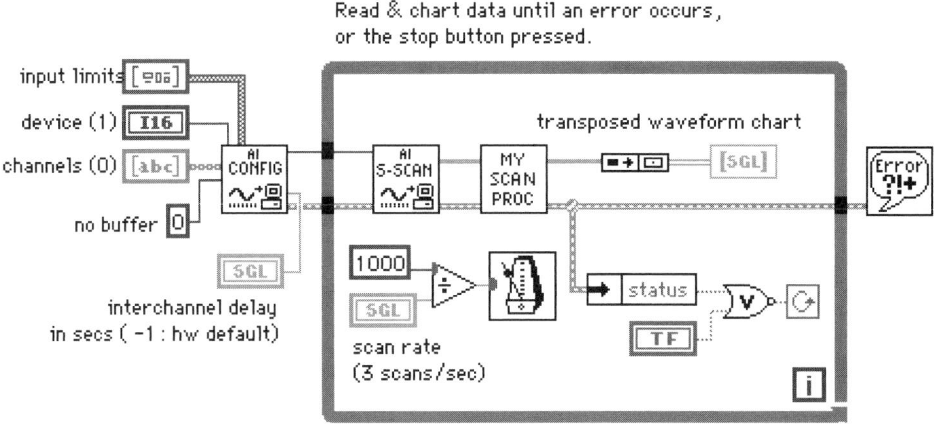

Figure 8.10 (a)
Diagram window of **Cont Acq&Chart (Immediate).vi**.

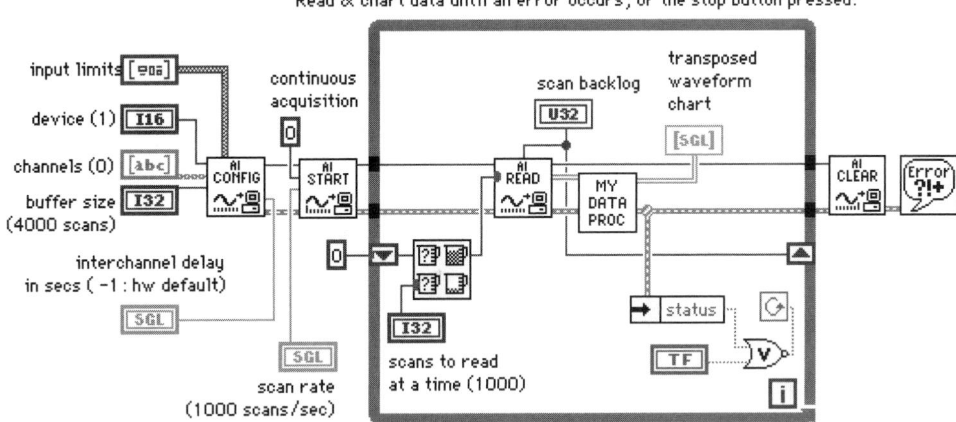

Figure 8.10 (b)
Diagram window of **Cont Acq&Chart (buffered).vi**.

*Examples of nonbuffered and buffered continuous AI acquisition are shown in Figure 8.10(a) and (b), respectively. Note that **AI Single Scan.vi** is used for the nonbuffered case, whereas **AI Read.vi** is used for the buffered case. Also, the input **number of scans to acquire** of **AI Start.vi** is zero to perform continuous acquisition in (b). hw stands for hardware in both diagrams.*

The last example, **Cont Acq&Chart(hw timed).vi**, shows nonbuffered continuous acquisition with hardware timing in Figure 8.11. Note the presence of **AI Start.vi** with **scan rate** 100 Hz. Also, the input **buffer size** of **AI Config.vi** is zero, indicating that it is nonbuffered acquisition. However, this example uses the on-board FIFO buffer since it is using **AI Single Scan.vi**, and you can use the buffer in the memory by setting a nonzero value to **buffer size**.

***AI Config.vi** controls the buffered acquisition. If the input **buffer size** is set to zero, it becomes the nonbuffered data acquisition, and if it is set to a positive integer number, it becomes the buffered data acquisition.*

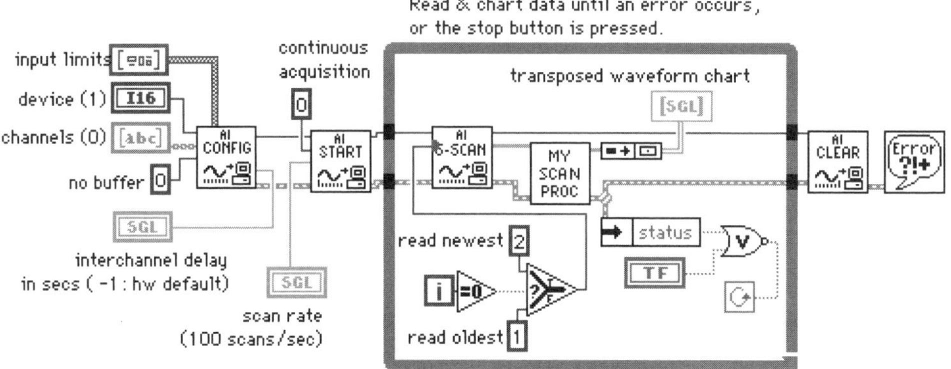

Figure 8.11
Nonbuffered, hardware-timed continuous acquisition using **AI Single Scan.vi**. Note the presence of **AI Start.vi** to set the sampling rate, which is somewhat limited due to the use of the on-board FIFO buffer by **AI Single Scan.vi**. Changing the input **buffer size** of **AI Config.vi** to a nonzero value will allow you to use the buffer in the memory. If you eliminate **AI Start.vi** and the **While Loop**, this example becomes **AI Read One Scan.vi**, which can be found in the subpalette **AI Input Utility**.

All of the previous three examples in Figures 8.10 and 8.11 use the circular buffer since they are all acquiring data continuously. If you eliminate the **While Loop**s in those examples, the type of buffer used becomes the single buffer. The circular buffer starts filling itself with data from the beginning, and when it reaches the end, it circles to the beginning of the buffer and continues to fill itself while replacing the old data. If the input **number of scans to read** of **AI Read.vi** is too small, LabVIEW will start losing data. In fact, there is no explicit equation among the inputs **number of scans to read** of **AI Read.vi**, **scan rate** of **AI Start.vi**, and **buffer size** of **AI Config.vi**. Usually, you should monitor the output **scan backlog** of **AI Read.vi** to adjust the values of those three parameters. A rule of thumb, however, is to set both **scan rate** and **number of scans to read** to the same value, and **buffer size** to about three to four times larger than **scan rate**.

8.8 Gain and Channel String for Analog Input

Chapter 7 briefly mentioned that the input range determined the gain for each input channel. This section will consider the mapping between the gain settings and the channels in the channel string input of **AI Config.vi**. As mentioned in Chapter 7, the gains for input channels are set by the input limits, and furthermore, different gains can be applied to different channels.

In Figure 8.12, the two inputs of **AI Config.vi** are shown: an array of string for channel inputs and an array of cluster for input limits. The reason why the input channels are of string type is due to the Channel Wizard, which is a new feature of LabVIEW since version 4.0. The Channel Wizard assists you in customizing the configuration and allows string-type channel names. If you label input channels with customized names, such as temp1, pressure, new_channel, etc., you may enter the names in the array of channel string input. As you can see in Figure 8.12, each element in the two arrays has a one-to-one correspondence. For example, Figure 8.12 shows that channels 0 to 2 have input limits –1 V and 1 V, and channel 4 has input limits –5 V and 5 V. These input limits determine which gain LabVIEW will apply before sampling input signals. Therefore, Figure 8.12 shows how you apply different gains to different channels. However, what would happen to the last channels 6 and 7? The third element in the input limit array is not defined. In such a case where there are not exactly the same number of

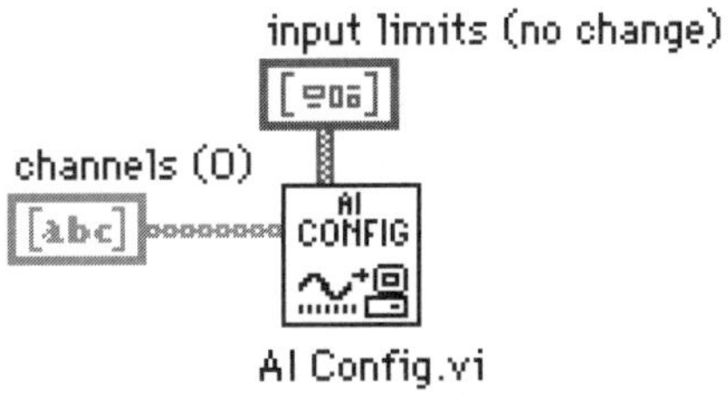

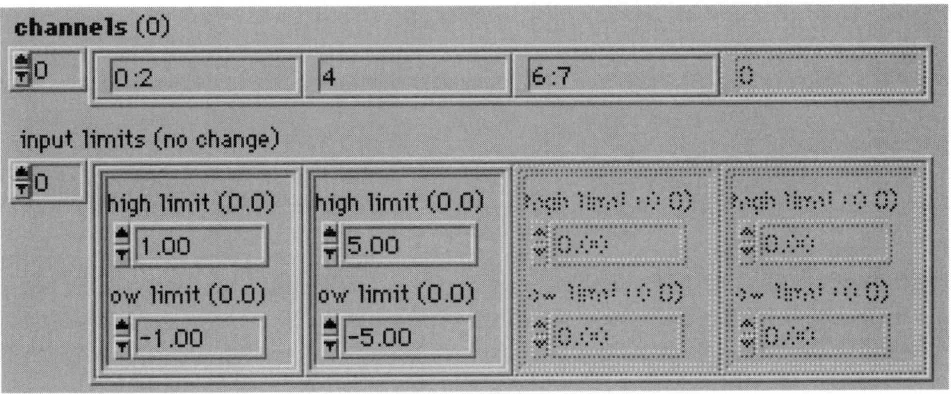

Figure 8.12
Channel string and input limit inputs of **AI Config.vi**.

elements in the two arrays, the last gain setting (the last input range) will be applied to the rest of the channels. Therefore, channels 6 and 7 will have the same gain as channel 4 in Figure 8.12.

8.9 Triggered Acquisition

There are two types of triggering in data acquisition: analog and digital triggering. Digital triggering is much simpler than analog triggering, and when a TTL-level triggering signal switches either from low to high or high to low, a certain operation begins. Generally, most data acquisition boards support digital triggering capability. On the other hand, the analog triggering initiates a certain task when the analog triggering signal level or slope exceeds a user-defined threshold. Since the analog triggering requires more sophisticated on-board curcuitry, not all of the boards support this capability, and only the relatively advanced ones provide this feature.

The following list summarizes key parameters that need to be specified along with key VIs and examples for each category.

1. *Digital hardware triggering*
   (a) Examples: **Acquire N Scans Digital Trig.vi** (for one-time acquisition), **Cont Acq&Graph (buffered) D-Trig.vi** (for continuous acquisition)
   (b) Parameters of **AI Start.vi**: **trigger type**, **pretrigger scans** (only for one-time acquisition, and not used for continuous case; not applicable (N/A) for single-point acquisition), **edge or slope**, **number of scans to acquire** (only for continuous acquisition with value 0)
   (c) Parameters of **AI Read.vi**: **read/search position** (only for one-time acquisition)
   (d) Trigger signal connection: STARTTRIG*, EXTTRIG*, DTRIG, PFI (*: falling edge triggerred)
2. *Analog hardware triggering*
   (a) Examples: **Acquire N Scans Analog Hardware Trig.vi** (for one-time acquisition), **Cont Acq&Graph (buffered) Hard A-Trig.vi** (for continuous acquisition)
   (b) Parameters of **AI Start.vi**: **trigger type**, **pretrigger scans** (only for one-time acquisition), **edge or slope**, **analog chan (-) & level**, **additional trig params**, **number of scans to acquire** (only for continuous acquisition with value 0)

(c) Parameters of **AI Read.vi**: **read/search position** (only for one-time acquisition)
(d) Trigger signal connection: trigger signal to the first channel listed in the channel string. This is not true if programmable function input (PFI) lines are used.

3. *Software triggering (only analog type)*
   (a) Examples: **Acquire N Scans Analog Software Trig.vi** (for one-time acquisition), **Cont Acq&Graph ExtScanClk Soft A-Trig.vi** (for continuous acquisition)
   (b) Parameters of **AI Start.vi**: **number of scans to acquire** (only for continuous acquisition with value 0)
   (c) Parameters of **AI Read.vi**: **read/search position, conditional retrieval** (from **Set up Conditional Retrieval.vi**)
   (d) Parameters of **Set up Conditional Retrieval.vi**: **hysteresis, trig slope, analog chan (-) & level, pretrigger scans**
   (e) Trigger signal connection: N/A

Figure 8.13 shows parameters and VIs for different triggering methods. The details about each parameter will not be covered in this book since doing so can easily divert your focus. Instead, this book will focus on the general understanding about where to go and which to use since it is a more efficient way of learning many different types of data acquisition techniques. Once you know which one to use, the details can be searched quickly and efficiently by referring to the online help of the examples, without getting confused by the variety of parameters.

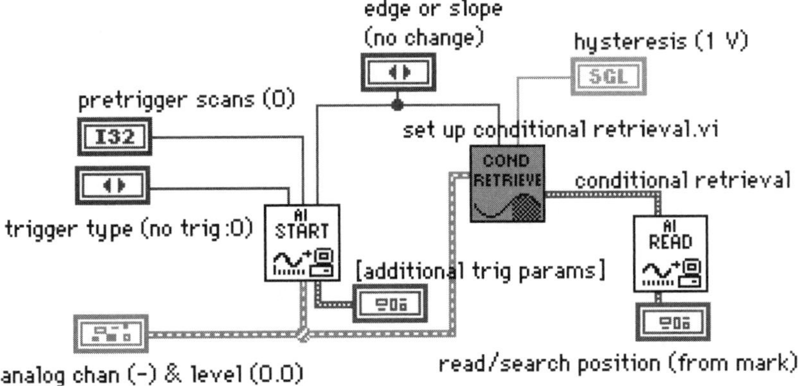

Figure 8.13
Parameters and VIs needed for different triggering scheme.

One interesting aspect about triggered data acquisition is that LabVIEW can also return the data *before* the initiation condition is met, and it is achieved via pretrigger scan values. Therefore, the following relationship can be established:

number of scans to acquire
= pretrigger scan + data acquired after triggering condition is met

If you set **pretrigger scan** to zero, the data acquisition board will start sampling input signals when the triggering condition is met. However, if **pretrigger scan** is set to a positive integer number, the board will continue to acquire signals even before the triggering condition is met while saving the data in the memory buffer. Once the condition is met, it will return the data of the number specified in the input **number of scans to acquire** of **AI Config.vi**. This applies to both analog and digital triggering methods.

8.10 Acquisition with External Clock

There are two types of clocks on data acquisition boards: a channel clock and a scan clock. The channel clock controls the time interval between adjacent samples within one sampling period, whereas the scan clock controls the sampling period for each channel. This section will explain the two clocks, using an example with Figures 8.14 and 8.15. Suppose you are sampling from three input channels 0, 1, and 2. Assuming that there is only one A/D converter on the board, the sampling circuitry will scan the output of the multiplexer, and the resulting data stream will be an interleaved version with the samples from each channel. The time interval between adjacent samples is the interchannel delay, which is controlled by the channel clock,

channels scanned
0 1 2 • • • • 0 1 2 • • • • 0 1 2 • • • • ... → *time*

Figure 8.14
Interchannel delay and scan delay (sampling time) with three input channels 0, 1, and 2. The time interval between the adjacent 0 and 1, or 1 and 2 is the interchannel delay, controlled by the channel clock. The time interval between 0 and 0, 1 and 1, or 2 and 2 is the scan rate, controlled by the scan clock. The • indicates a time instance with no sample acquired.

channels scanned
0 1 2 0 1 2 0 1 2 0 1 2 ... → *time*

Figure 8.15
Round-robin scanning with no scan clock. The time interval between adjacent samples is identical.

and the time interval between the data samples from the same channel is the scan rate, which is controlled by the scan clock. In other words, the channel clock controls the time interval between 0, 1, and 2, and the interval is said to be the interchannel delay. The scan clock controls the time interval between 0 and 0, 1 and 1, and 2 and 2, and this rate is set by the **scan rate** input of **AI Start.vi**. If a board does not have the scan clock, then the time interval between adjacent samples will be the same as that between the data sample from the last channel in the channel string list and that from the first channel in the channel string list. This is called round-robin scanning, and low-end data acquisition boards usually choose this type of scanning scheme. Note that if you externally provide a channel clock, the scan rate will become the channel clock frequency divided by the number of channels listed in the channel string input since it will become the round-robin scanning. If you provide an external scan clock, LabVIEW will apply the fastest interchannel delay in sampling data from each channel. You can also provide both an external channel clock and a scan clock, but this requires accurate timing be taken into consideration beforehand.

The following list summarizes which parameters need to be specified along with key VIs and a good example for each category.

1. *External channel clock*
   (a) Examples: **Acquire N Scans-ExtChanClk.vi**
   (b) Parameters of **AI Clock Config.vi**: **which clock**, **clock frequency** (set to zero to disable internal clock), **clock source**
   (c) Parameters of **AI Start.vi**: **scan rate** (set to zero to disable internal clock)
   (d) TTL level clock signal connection: EXTCONV*, PFI lines (*: falling edge scanning). Refer to the manuals of the boards.
2. *External scan clock*
   (a) Examples: **Acquire N Scans-ExtScanClk.vi**
   (b) Parameters of **AI Clock Config.vi**: **which clock**, **clock frequency** (set to zero to disable internal clock), **clock source**

(c) Parameters of **AI Control.vi**: **control code**
(d) **AI Start.vi** is not used.
(e) TTL-level clock signal connection: OUT2, PFI lines, OUT B1. Refer to the manuals of the boards.

8.11 Analog Input Utility VIs

There are three VIs in the subpalette **Analog Input Utilities**: **AI Read One Scan.vi**, **AI Waveform Scan.vi**, and **AI Continuous Scan.vi**. Those VIs use the intermediate-level VIs and are especially useful if you need to use the **While Loop** or the **For Loop** for repeated analog input acquisition. Unlike the high-level VIs, utility VIs have error input and error output, which provide you with lower-level error control capability. The following summarizes all three of them:

AI Read One Scan.vi: Hardware-timed acquisition at unspecified rate. The rate will be the fastest channel clock rate that a data acquisition board can support. If you use **AI Start.vi** between **AI Config.vi** and **AI SingleScan.vi**, the hardware-timed rate can manually be set; however, this rate should be relatively slow, such as less than 100 Hz, because **AI SingleScan.vi** from the subpalette **Advanced Analog Input** uses the on-board FIFO buffer. (See Figure 8.9 for an example.) There is an input terminal **iteration** so that it can be used in a loop.

AI Waveform Scan.vi and **AI Continuous Scan.vi**: These two VIs are almost identical except that **AI Waveform Scan.vi** provides input terminals for external triggering whereas **AI Continuous Scan.vi** does not. They both use the same intermediate-level VIs and have an input terminal **iteration** as **AI Read One Scan.vi** does. Therefore, they, too, can be used in a loop.

8.12 DAQ Occurrence

When you perform analog input data acquisition, LabVIEW first configures the hardware (**AI Config.vi**), initiates the acquisition (**AI Start.vi**), reads the data that are possibly stored in a circular buffer (**AI Read.vi**), and resets the hardware (**AI Clear.vi**). As shown in Figure 8.16, as soon as **AI**

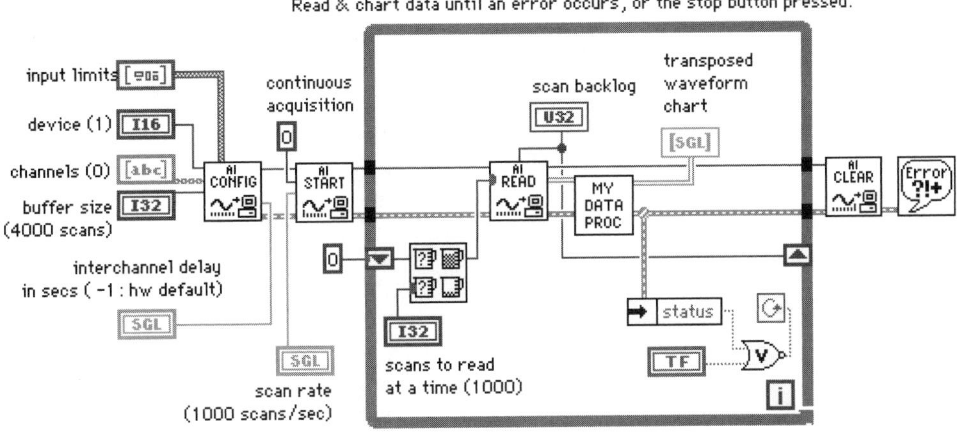

Figure 8.16
Continuous AI acquisition without using DAQ occurrence. The diagram window of **Cont Acq&Chart (buffered).vi** is shown.

Start.vi is called, **AI Read.vi** is called. If the process of acquiring data samples and storing them in the circular buffer is not completed when **AI Read.vi** is called, the VI will idle and other operation will be halted, too, because the CPU will be occupied during the acquisition process. However, if you can hold calling **AI Read.vi** until the complete number of data samples are acquired, and call it when the data are ready, this will free up the CPU time for other tasks to be executed.

The usage of data acquisition (DAQ) occurrence is shown in Figure 8.17, which is the diagram window of an example **Cont Acq&Chart (Async Occurrence).vi**. Note that **DAQ Occurrence Config.vi** between **AI Config.vi** and **AI Start.vi** generates a triggering signal when the condition, which is displayed right below the **While Loop**, is met. (The condition here is when a certain number of data samples are acquired.) Then the generated triggering signal triggers **Wait on Occurrence** inside the **While Loop**, which then generates FALSE. This will choose the False frame of the **Case** structure, which contains **AI Read.vi**. Otherwise, **Wait on Occurrence** will continue to generate TRUE, and the True frame of the **Case** structure is empty. Therefore,

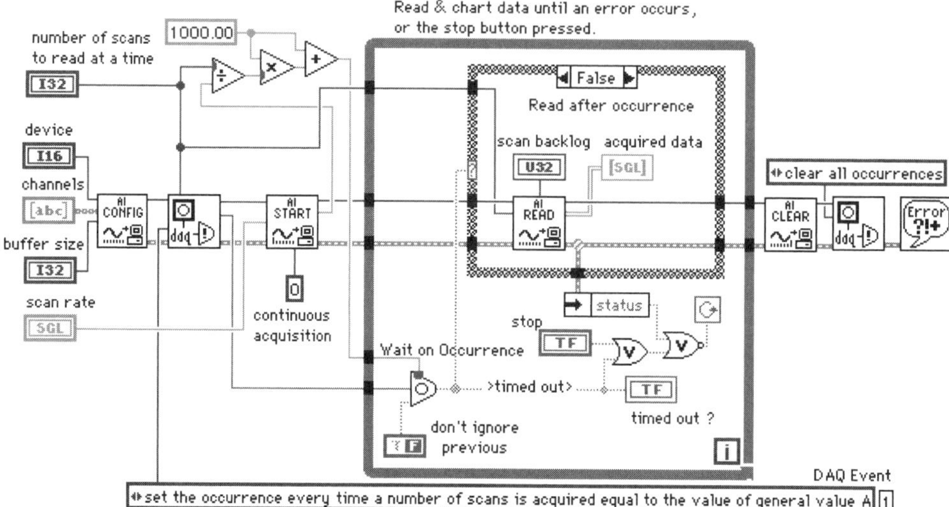

Figure 8.17
Continuous AI acquisition using DAQ occurrence. The diagram window of **Cont Acq&Chart (Async Occurrence).vi** is shown.

until a certain number of data samples are acquired and ready to be read, **AI Read.vi** will not be called, which will provide extra CPU time for other tasks to be executed. However, DAQ occurrence is not supported on Macintosh platforms, and the general-purpose occurrences will be discussed in Chapter 15.

PROBLEMS

8.1 Suppose that you want to acquire a waveform from a single channel as shown in Figure P8.1(a). As shown in Figure P8.1(b), you are trying to use the high-level VI **AI Acquire Waveform.vi** in a **While Loop** in order to perform continuous acquisition. However, the diagram window Figure P8.1(b) is not complete in order to achieve such a task. Modify the diagram so that the VI in the figure can continuously acquire waveform data. First save the high level VI **AI Acquire Waveform.vi** as **P08_01_AIAcqWave.vi**. Hint: You should not configure the hardware or clear the configuration at every iteration. Save your answer as **P08_01.vi**.

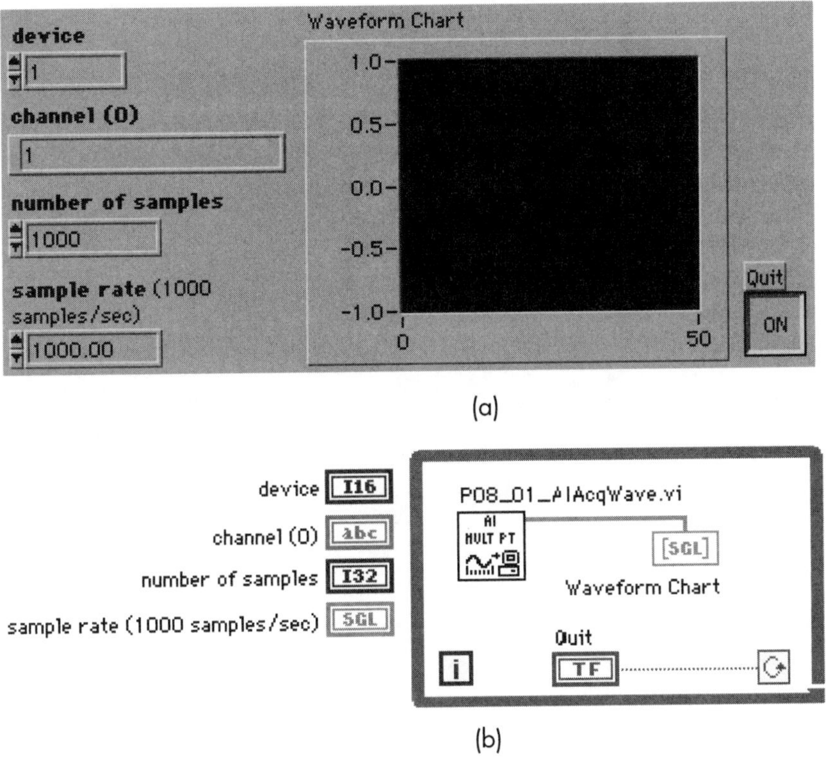

Figure P8.1

8.2 Repeat Problem 8.1 with the high-level VI **AI Acquire Waveforms.vi**. Save the VI as **P08_02_AIAcqWaves.vi** first. Since the diagram window will be similar to that in Problem 8.1, only the front panel is shown in Figure P8.2. Save your answer as **P08_02.vi**.

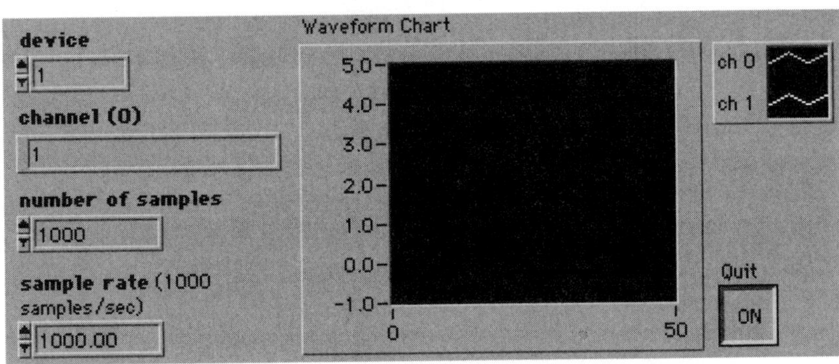

Figure P8.2

8.3 Open the example **Cont Acq&Chart (buffered).vi** from the following directory:

LabVIEW directory >> **examples** >> **daq** >> **anlogin** >> **anlogin.llb**

(a) In the front panel, there is a Boolean button **STOP**. Also, note that there is no **Abort Execution** in the menu bar. Explain why the button **Abort Execution** is hidden and the **STOP** button is in the front panel instead.
(b) Explain what this VI performs. Elaborate on the parameters of each sub VI in the diagram.
(c) Explain the condition when the **While Loop** stops.

8.4 Consider the example **Cont Acq&Chart (buffered).vi** again. (It was considered in Problem 8.3.) A portion of the diagram is shown. There is a sub VI with the label **MY DATA PROC** on it; save it as **P08_04.vi** first. Then modify it to process the 2-D input data as follows:
(a) First channel data will be increased by 1.
(b) Second channel data will be decreased by 1.
(c) All of the other channels will be ignored. This means that the output of **P08_04.vi** will have only two columns.

Be careful not to save **Cont Acq&Chart (buffered).vi** with the modified **MY DATA PROC**. You should modify only **MY DATA PROC**, and save it as **P08_04.vi**.

Data Acquisition: Analog Output

9

9.1 Signal Generation

There are applications where you need to control devices or generate a response with sinusoidal, square, or customized waveforms, such as controlling a power supply with a different level of DC signal, or generating a sine wave based on the output of an encoder. This chapter will discuss the fundamentals of waveform generation technique, which is said to be analog output data acquisition.

Signal generation is generally divided into two categories in LabVIEW: single-point (DC) generation and waveform (AC) generation. Before proceeding to the details about analog output data acquisition, it should be pointed out that there must be direct memory access (DMA) channels available on PC platforms in order to generate AC waveforms, but DC generation does not require the DMA channels. On Macintosh platforms, however, such an issue is not a concern for the PowerMac series, but the 68K series will need an extra board that provides DMA resources. You may contact the vendors of your board or refer to the manuals to find out the availability and the requirement of DMA channels for analog output data acquisition.

There are a lot of similarities between analog input and analog output data acquisition. First, the three stages of data acquisition are identical: Configuration, Read or Write, and Reset. As for analog output data acquisition, the middle stage will be Write or Update.

The organization of VIs is also identical. Comparing Figure 9.1 with the subpalette **Analog Input** in Chapter 8, the top row has high-level VIs, the second row has intermediate-level VIs, and the subpalettes **Analog Output Utilities** and **Analog Output Advanced** are in the bottom row. The next section will start with the utility VIs.

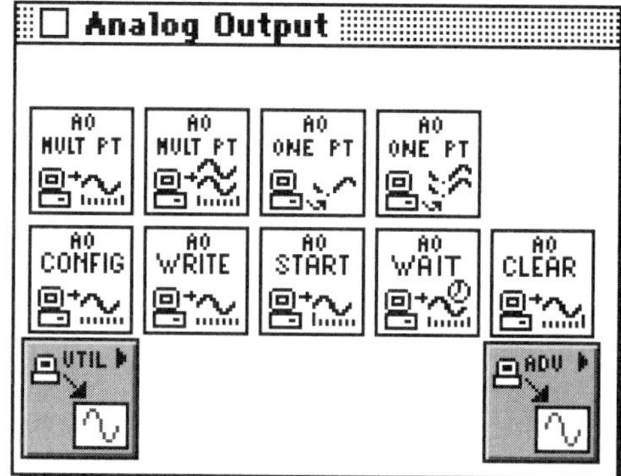

Figure 9.1 **Analog Output** subpalette.

9.2 Analog Output Utility VIs

Similar to the **Analog Input Utilities** subpalette discussed in Chapter 8, there are three utility VIs in the subpalette **Analog Output Utilities**: **AO Write One Update.vi**, **AO Waveform Gen.vi**, and **AO Continuous Gen.vi**. Again, these three analog output utility VIs have the input terminal **iteration** so that all of them can be used in a loop for continuous updates.

> **AO Write One Update.vi**: This VI writes a single data to an analog output channel or a group of analog output channels immediately without using the buffer in the memory. The input is a 1-D array since you can have a group of analog output channels. For example, if you have analog output channels 0 and 1 with 1.5 and 3 in the 1-D input array, this VI will update channels 0 and 1 with 1.5 V and 3 V. Only channels with data specified will be updated; that is, if the second data 3 V was not specified, only the channel 0 would be updated with 1.5 V. In order to use the buffer in the memory, set a positive integer number to the input **buffer size** of **AO Config.vi**. This VI is similar to the utility VI **AI Read One Scan.vi** in the subpalette **Analog Input Utilities** in such a way that both **AI Read One Scan.vi** and **AO Write One Update.vi** do not use the buffer in the memory, and read or write data immediately. **AO Write One Update.vi** uses **AO Single Update.vi** in the subpalette **Advanced Analog Output** whereas **AI Read One Scan.vi** uses **AI SingleScan.vi** in the subpalette **Advanced Analog Input**.
>
> **AO Waveform Gen.vi** and **AO Continuous Gen.vi**: Use these VIs to generate a waveform, instead of a single value, either single time or continuously in a loop. If you want to change the waveform while generating it, or if the waveform data are too long for the size of the buffer, use **AO Continuous Gen.vi**.

Figures 9.2 and 9.3 show the diagram window of the last two utility VIs. **While Loops** in the figures are used not to iterate but to use the shift register. (Since FALSE is wired to the condition terminal of the **While Loop**, it will run only once.) Therefore, those two diagrams should not be understood as a repeating process. If you place those two utility VIs (**AO Waveform Gen.vi** and **AO Continuous Gen.vi**) in a loop using the input **iteration**, the entire content in the **While Loop** will repeat; however, the VIs in frame zero of the **Case** structure will be executed when **iteration** is zero. This is to avoid redundant hardware configuration.

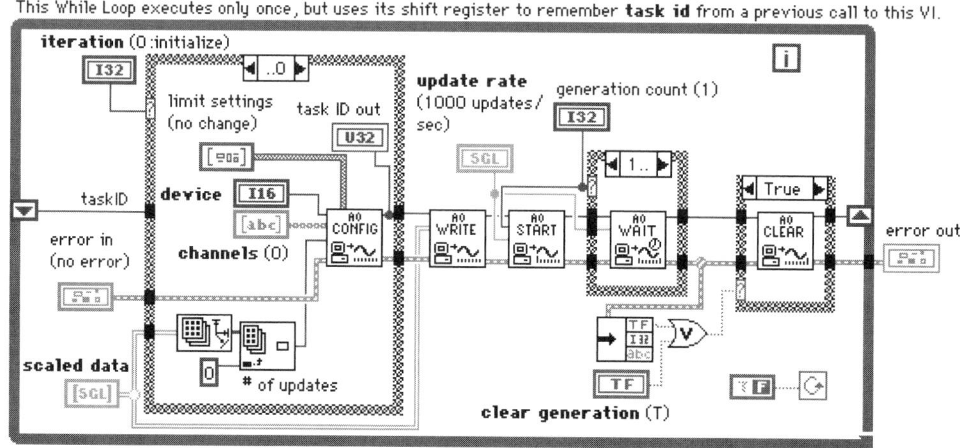

Figure 9.2
Diagram window of **AO Waveform Gen.vi**.

You should recognize that both **AO Write.vi** and **AO Start.vi** will repeat in Figure 9.2, whereas only **AO Write.vi** will do so in Figure 9.3 (see frame 1 of the **Case** structure) if you put **AO Waveform Gen.vi** and **AO Continuous Gen.vi** in a **While Loop** to generate waveforms continuously. This section will elaborate on this with the input **number of buffer iterations** of **AO Start.vi**. Notice that in Figure 9.2, one (1) is wired to **number of buffer iterations**

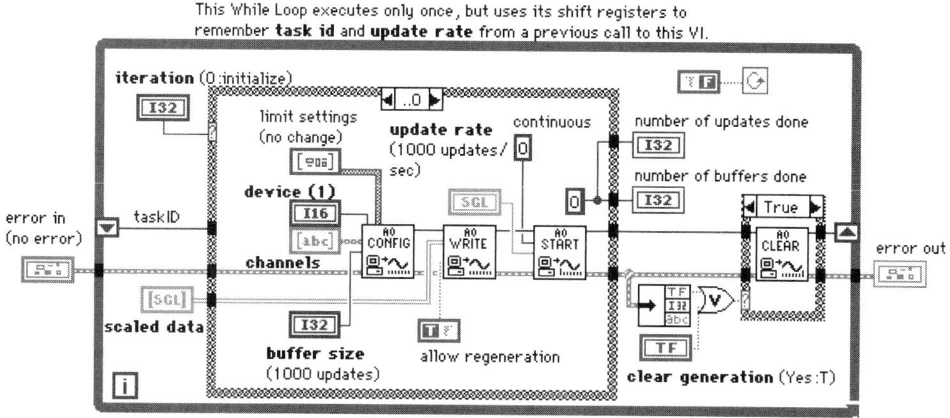

Figure 9.3
Diagram window of **AO Continuous Gen.vi**.

whereas zero (0) is wired in Figure 9.3. Input 0 in Figure 9.3 implies that the waveform generation will be continuous using the circular buffer. Therefore, you will only need to write to the buffer repeatedly using **AO Write.vi** in order to update analog output channels continuously. Also, immediate waveform change during its generation can be achieved by writing different waveform data to the circular buffer with **AO Write.vi**.

However, in Figure 9.2, where 1 is set for the input **number of buffer iteration** of **AO Start.vi**, a single buffer is used instead of a continuous circular buffer so that the board needs to be initiated (not configured) again every time LabVIEW fills up the single buffer. If you want to generate waveforms continuously using **AO Waveform Gen.vi**, set the input **number of buffer iterations** to 0 without putting **AO Waveform Gen.vi** in a **While Loop**. If you use the method in Figure 9.2 to generate continuous waveforms, however, you will not be able to modify the waveforms immediately and will have to call **AO Clear.vi** to stop the waveform generation and restart the VI with a new waveform. As a rule of thumb, to generate a continuous waveform while changing it during its generation, or if the length of the waveform data is too long for the buffer in the memory so that more than a simple single buffer is needed, use **AO Continuous Gen.vi** in a **While Loop** using the circular buffer. All of the three continuous waveform generation examples, **Continuous Generation.vi**, **Generate Continuous Sinewave.vi**, and **Function Generator.vi**, use either **AO Continuous Gen.vi**, which is the case in **Continouous Generation.vi**, or the same structure as in **AO Continuous Gen.vi** (repeating only **AO Write.vi** in a **While Loop** with input 0 for the input **number of buffer iteration** of **AO Start.vi**). Oherwise, **AO Waveform Gen.vi** will do the different types of continuous waveform generation depending on the value of the input **number of buffer iteration** of **AO Start.vi**, which is the control **generation count** in **AO Waveform Gen.vi**. The three continuous waveform generation examples can be found in the analog output directory under the daq example folder.

Unlike analog input data acquisition, continuous generation while changing the waveform during the generation is quite important, and therefore the following examples are listed for each case. As was pointed out in the earlier chapters about data acquisition, learning which examples can be used for which case is the goal in the chapters about data acquisition. Using the various techniques presented in Chapter 8, you can easily customize the following examples for your application instead of starting with a new VI. All of the examples discussed in this section are assumed to be in the following directory: LabVIEW directory >> **examples** >> **daq** >> **anlogout** >> **anlogout.llb**.

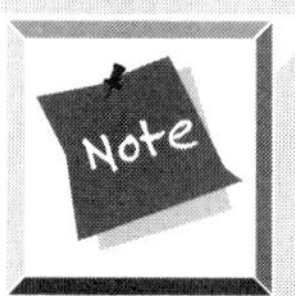

Both **AO Waveform Gen.vi** and **AO Continous Gen.vi** can generate continuous waveforms. **AO Waveform Gen.vi** can generate an infinite number of repeated waveforms by setting 0 for the control **generation count**, which is the input to **number of buffer iteration** of **AO Start.vi**. If you set **generation count** to a positive integer number n, it will generate n number of repeated waveforms and stop the generation. Note that **AO Waveform Gen.vi** generates continuous waveforms with 0 or positive integer value of **generation count** *without* being used in a **While Loop** even though it has the input **iteration** available. (You need to apply an input to the input **clear generation(T)**. See Problem 9.2 for more details.) On the other hand, **AO Continuous Gen.vi** is designed to be used in a **While Loop** to generate continuous waveforms especially when the waveform needs to be changed during its generation. Also, all of the three continuous waveform generation examples, **Continuous Generation.vi**, **Generate Continuous Sinewave.vi**, and **Function Generator.vi**, use the same structure (**AO Continuous Gen.vi** in a **While Loop**).

1. *Continous generation while not changing the waveform*
   Continuous Generation.vi
2. *Continous generation while changing the waveform*
   Generate Continuous Sinewave.vi
   Function Generator.vi

Note that the format of the waveform data for the input **scaled data** of both **AO Waveform Gen.vi** and **AO Continuous Gen.vi** is a 2-D array, and each *column* will be the data for each output channel. Therefore, you may have to use **Transpose 2D Array.vi** from the subpalette **Functions >> Array** if the data for each channel are stored in each row of the input 2-D array.

9.3 High-Level Analog Output VIs

In the subpalette **Analog Output**, there are four high-level VIs in the top row: **AO Generate Waveform.vi**, **AO Generate Waveforms.vi**, **AO Update Channel.vi**, and **AO Update Channels.vi**. The first two are for generating a waveform such as an AC signal from a single or multiple analog output channels. (Singular Waveform in the name of the VI implies that it is for a single channel, and plural Waveforms, for multiple channels.) Similarly, the last two are for single and multiple channels to perform single-point update, which is more like a DC generation. (If you can change the DC value fast by calling the VI quickly and repeatedly, the output could look like an AC signal, but this is not a realistic solution to generating an AC signal.) These VIs are ready to be used with the least amount of modification required. If you would like to use them in a loop, such as a **While Loop** or a **For Loop**, you may apply the same technique described in Section 8.3 in Chapter 8 or use the utility VIs that were discussed in the previous section.

9.4 Intermediate-Level Analog Output VIs

This section will discuss intermediate-level VIs and their functionality using an example in the following directory: LabVIEW directory >> **examples >> daq >> anlogout >> anlogout.llb**.

There are five intermediate VIs shown in the second row in Figure 9.1, and they give us much more control over the waveform generation than the high-level VIs do. Recall that the high-level VIs are ready to be used with limited control options. Consider Figure 9.4, which shows a part of the diagram window of the example **Generate Continuous Sinewave.vi**.

The example in Figure 9.4 generates a sine wave continuously, and this figure will be used to discuss the five intermediate analog output VIs. It again shows the three stages of data acquisition: Configuration, Read or Write, and Reset. The first stage is performed by **AO Config.vi**, where you can set the buffer size for the buffered waveform generation, output range, and other input parameters, such as output channels and device number. The second stage Write (or Update) is achieved by **AO Write.vi** and **AO Start.vi**. In this second stage, **AO Write.vi** first writes the waveform data to

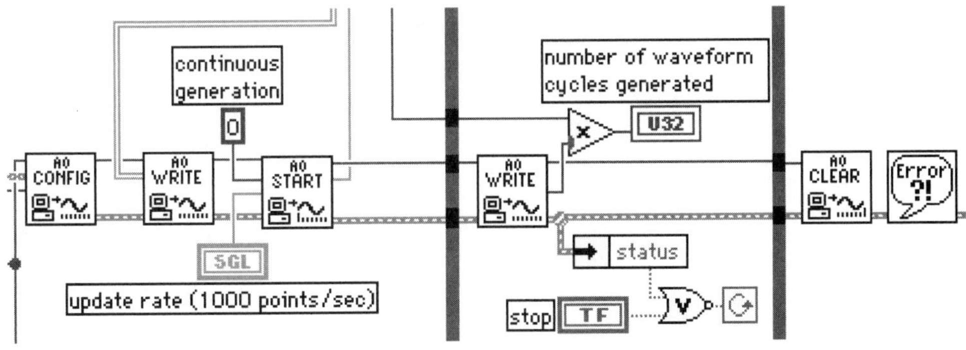

Figure 9.4
Part of the diagram window of **Generate Continuous Sinewave.vi**.

a buffer in the memory, and **AO Start.vi** initiates the board to generate a waveform on the output channel(s) specified at **AO Config.vi**. Note that **AO Write.vi** takes a 2-D array as its waveform data, and each *column* corresponds to the data to be written to each analog output channel. Therefore, you may need to use **Transpose 2D Array.vi**, which is available in the **Functions >> Array** subpalette, to transpose a row-oriented 2-D data array. Also, unlike analog input data acquisition, where **AI Start.vi** comes before **AI Read.vi**, **AO Write.vi** is called before **AO Start.vi** in analog output data acquisition.

The following two inputs of **AO Start.vi** need further discussion: **number of buffer iteration** (default is 1) and **update rate** (default is 1000 updates per second). **AO Start.vi** starts the waveform generation on a board at a rate specified in **update rate**. The input **number of buffer iteration** controls how many times the board will repeat the waveform generation before it stops, and it receives the input value from the control **generation count** in the case of the analog output utility VI **AO Waveform Gen.vi**. (See Figure 9.2.)

In Figure 9.4, there is a **While Loop** containing **AO Write.vi.** If you want to generate a waveform continuously without changing it during the generation, having two **AO Write.vis** (one in the **While Loop** and the other outside the **While Loop**) is redundant. You can customize the VI such that setting zero (0) to **number of buffer iteration** of **AO Start.vi** without the second **AO Write.vi** in the **While Loop** suffices to perform the continuous waveform generation. A modified version of Figure 9.4 is shown in Figure 9.5.

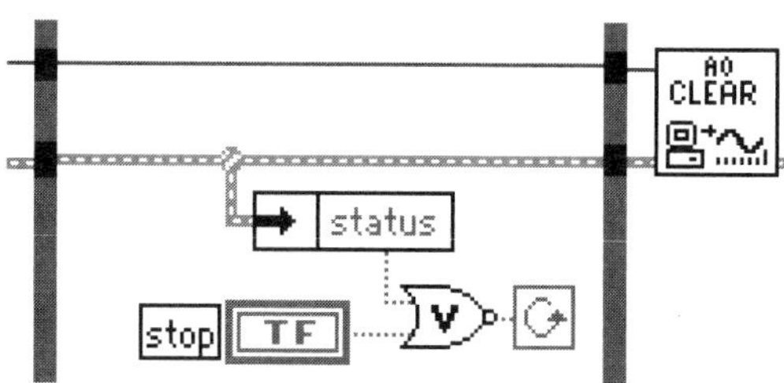

Figure 9.5
A modified version of **Generate Continuous Sinewave.vi**. **AO Write.vi** in the **While Loop** is deleted since it is redundant to have it in the loop to generate the same waveform continuously. The **While Loop** is still there in order to hold calling **AO Clear.vi**. The loop will stop and terminate the waveform generation when either the **stop** button is pressed or there is an error during the process.

However, if you want to change the waveform spontaneously during the waveform generation, new waveform data need to be written to the buffer again; therefore, one more **AO Write.vi** needs to be in the **While Loop** to continuously update the waveform data.

AO Wait.vi is an intermediate analog output VI that releases the CPU time to allow other tasks to be performed while the board is generating waveforms. Note that this VI must be used only for a finite number of repeated waveform generation. In other words, if the input **number of buffer iteration** of **AO Start.vi** is zero (0), implying continuous waveform generation, **AO Wait.vi** must not be used at all because it will never allow other processes to be performed. (**AO Wait.vi** will wait until the infinite number of waveform generation, which will never end, is completed.) This is illustrated in the **Case** structure in the middle of Figure 9.2 and is shown again for convenience in Figure 9.6.

In Figure 9.2, the input **generation count** is also wired to the input **number of buffer iteration** of **AO Start.vi**. As shown in Figure 9.6, if the waveform data are to be repeated a positive number of times (nonzero integer value), **AO Waveform Gen.vi** will call **AO Wait.vi** in frame 1 of the **Case** structure after calling **AO Start.vi**. If the **number of buffer iteration** is zero (0), implying continuous waveform generation, **AO Wait.vi** is bypassed (frame 0 of the **Case** structure).

AO Clear.vi is identical with **AI Clear.vi** in the analog input data acquisition and resets all of the settings on the board.

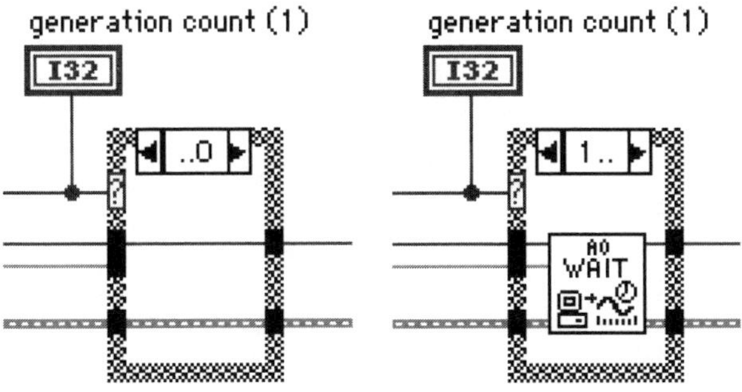

Figure 9.6
The second **Case** structure in the diagram window of **AO Waveform Gen.vi**, which is an analog output utility VI. This is to show the use of **AO Wait.vi**, which should be used to generate a continuous waveform only for a finite number of times.

9.5 Single/Circular-Buffered Waveform Generation

Single-buffered analog output in LabVIEW refers to a single waveform generation; for example, write an array of waveform data to a buffer in the memory and generate a waveform once by sending the data to an analog output channel or channels. One way to achieve it is to call the analog output utility VI **AO Waveform Gen.vi** once with one (1) set for the input **generation count**. Recall that the two high-level VIs, **AO Generate Waveform.vi** and **AO Generate Waveforms.vi**, also use **AO Waveform Gen.vi** to generate waveforms. On the other hand, circular-buffered analog output refers to continuous waveform generation where immediate waveform change may be necessary, or the waveform data are too long to fit in the buffer in the memory. Circular-buffered waveform generation can be realized by calling **AO Waveform Gen.vi** *once* with zero (0) for the input **generation count** if spontaneous waveform change is not necessary. Other good examples of circular-buffered waveform generation are **Generate Continuous Sinewave.vi**, **Function Generator.vi**, and **Continuous Generation.vi**. Note that **Generate Continuous Sinewave.vi** and **Function Generator.vi** can change the waveform during its generation by updating the buffer with new waveform data in the **While Loop** using **AO Write.vi**. (A portion of the diagram window of **Generate Continuous Sinewave.vi** was shown in Figure 9.4.)

9.6 Gain and Channel String for Analog Output

Similar to the input **input limits** of **AI Config.vi**, **AO Config.vi** provides a cluster input terminal where you can specify different output limits for different output channels. Consider the example shown in Figure 9.7.

How to set different limits to different channels is the same as that with **AI Config.vi**. In Figure 9.7, each element in all three array inputs corresponds to each other. Output channels 0 through 2 will generate –4 V DC, where its high and low limits are defined as 10 V and –10 V. Unlike the input cluster of **AI Config.vi**, you have one additional input **reference source** as the third element in the cluster. This input has three options: **no change**, **internal**, and **external**. The first two options depend on the data acquisition board that you will be using to perform analog output (either DC, AC, or arbitrary waveform generation). The first option, **no change**, means that there is no change compared to the default settings. The second option, **internal**, allows you to change the maximum and the minimum limit of out-

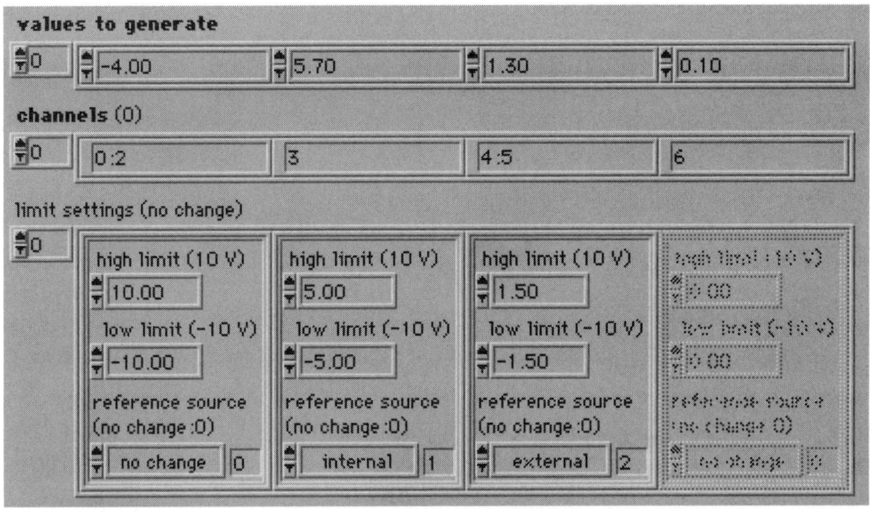

Figure 9.7
Channel string and cluster array inputs of **AO Config.vi** for output limits. The input terminal of **AO Config.vi** is labeled **limit settings**.

put signal using the settings that the board supports. The last option, **external**, allows you to supply your own output limits to the board.

Having those three selections in mind, consider Figure 9.7 again. It illustrates DC generation using multiple output channels. The output channels 0 through 2 generate –4 V DC using the default limit settings of the board that you are using. (The default settings are shown in parentheses as ± 10 V.) The low limit will be the same value as the high limit with the opposite polarity if the polarity setting in the driver software is set to bipolar. If it is set to unipolar, the low limit will be zero. Also, note that if the board is not software configurable but has jumper switches on the board to configure input settings, your settings in LabVIEW will be overwritten by the hardware settings. The driver software is what allows LabVIEW to communicate with the boards and is provided by the vendor of the boards. It comes in the form of Extensions on Macintosh platforms and DLLs on PC platforms.

Hardware settings using jumper switches on the board will always overwrite the software settings set in either the driver software or LabVIEW. If the board is software configurable, LabVIEW settings will overwrite those in the driver software.

According to Figure 9.7, channel 3 will generate 5 V DC using *one* of the output limit settings available on the board, and it is assumed that the board in Figure 9.7 supports ± 5 V. (Since the desired value for channel 3 is out of the limit, the output will be 5 V instead of 5.7 V.) The third channel string, 4:5, indicates that channels 4 and 5 will generate 1.3 V, which falls in the range of ± 1.5 V, and these limits are provided externally by the user. Since the number of cluster elements in the array **limit settings**, which is three, is less than the number of channel string elements in the array **channels**, which is four, the fourth channel string will use the settings in the third cluster element, which uses an external reference source with ± 1.5 V range.

Usually the number of output channels is two with only one external reference input pin, which is usually labeled EXTREF. In that case, you would not need more than one cluster element in the array **limit settings** and both of the two analog output channels will share the single external reference source if you choose *external* for **reference source**. However, if the board supplies you with more than two analog output channels, it usually comes with multiple external reference sources, which are labeled EXTREF0, EXTREF1, and so forth. Then you can apply different limit settings to different channels

Chapter 8 discussed the relationship between the input gains, which are set by input limits, and the resolution of the board, which directly influences the accuracy in the representation of input signals. It concluded that you must choose the limits that are the closest to the input signal range for the maximum accuracy in the representation of input signals. Similarly, for analog output data acquisition, you must choose the closest limit settings to the range of the output signals to be generated in order to achieve the maximum utilization of the resolution of the board for digital-to-analog (D/A) conversion. If you set a value that exceeds the high limit, or the low limit for the input **values to generate** in Figure 9.7, the output will saturate at each limit.

9.7 Triggered AO and Simultaneous AI/AO

This section will discuss triggered analog output and simultaneous analog input and output together due to the following reasons. Some data acquisition boards have a separate clock for analog input and analog output so that they can run at a separate rate independently. Those advanced types of boards usually provide both analog and digital triggering capability. In that case, you can perform triggered analog output independent of analog input, and both analog and digital triggering can be applied. However, not all of the boards have a separate clock for analog input and analog output, and they share the same clock for both operations, instead. Those types of boards usually provide only digital triggering capability. Since they use the same clock, you should use **AI Start.vi** where you can specify the trigger type and the clock source since **AO Start.vi** does not have an input terminal for the trigger type specification. However, you specify the trigger type with **AI Start.vi** and the clock source type with **AO Start.vi**. (See item 4 in the following list of examples.) This implies that **AI Start.vi** must be used to specify the trigger type for triggered analog output data acquisition with a board that does not have a separate clock for analog input and analog output. Therefore, this section will discuss simultaneous AI/AO with triggered AO together. Triggering signal connections on the boards without a separate clock varies depending on the type of boards, so you may refer to the help information of the VI examples that are listed regarding the details about signal connection. All of the triggered analog output examples can be found in LabVIEW directory >> **examples** >> **daq** >> **anlogout** >> **anlogout.llb**, and all of the simultaneous analog input and output examples can be found in LabVIEW directory >> **examples** >> **daq** >> **anlog_io** >> **anlog_io.llb**. The following are examples, with key VIs and features in each case.

1. *Digital hardware triggering with boards that have separate clocks for AI and AO*
   (a) Examples: **Generate N Updates-E-series D-Trig.vi** (for finite generation), **Cont Generation-E-series D-Trig.vi** (for continuous generation)
   (b) Parameters of **AO Trigger and Gate Config.vi**: **trigger or gate source**, **trigger or gate condition**, **trigger or gate source specification**
   (c) Parameters of **AO Start.vi**: **number of buffer iterations** (only for continuous generation with value 0)
   (d) Usage of **AO Wait.vi** (only for finite generation)
   (e) Trigger signal connection: PFI0/TRIG1 or other PFI pins
2. *Digital hardware triggering with boards that have separate clocks for AI and AO for simultaneous AI/AO*
   (a) Examples: **Simul AI/AO Buffered Trigger (E-series MIO).vi**
   (b) Parameters of **AI Start.vi**: **edge or slope**, **trigger type**, **analog chan (-) & level**
   (c) Parameters of **AO Trigger and Gate Config.vi**: **trigger or gate source**, **trigger or gate condition**, **trigger or gate source specification**
   (d) Trigger signal connection: PFI0 (default) or other PFI lines specified in trigger source specification input of **AO Trigger and Gate Config.vi**

3. *Analog hardware triggering with boards that have separate clocks for AI and AO*
   (a) Examples: **Generate N Updates-E-series A-Trig.vi** (for finite generation), **Cont Generation-E-series A-Trig.vi** (for continuous generation)
   (b) Use **AI Group Config.vi** between **AO Config.vi** and **AO Write.vi**.
   (c) Use **AI Trigger Config.vi** between **AO Write.vi** and **AO Trigger and Gate Config.vi**, and the parameters of **AI Trigger Config.vi** are **additional trigger specification**, **level**, **trigger type**, **mode**, **trigger source**, and **trigger or pause condition**.
   (d) Parameters of **AO Trigger and Gate Config.vi**: **trigger or gate source**, **trigger or gate condition**
   (e) Usage of **AO Wait.vi** (only for finite generation)
   (f) Trigger signal connection: PFI0
4. *Digital hardware triggering with boards that share the same clock for AI and AO (Must be done as simultaneous AI/AO and only digital triggering is available.)*
   (a) Examples: **Simul AI/AO Buffered Trigger (legacy MIO).vi, Simul AI/AO Buffered Trigger (Lab/1200).vi**
   (b) Parameters of **AI Start.vi**: **trigger type**, **number of scans to acquire** (0 for continuous acquisition)
   (c) Parameters of **AO Start.vi**: **clock source**, **number of buffer iterations** (0 for continuous acquisition)
   (d) Trigger signal connection: START TRIG*, EXTTRIG. Extra connection is required depending on the number of analog input channels. Also, signal connection varies depending on the type of data acquisition boards. Refer to the online help for each example VI for more details about signal connection.

9.8 Waveform Generation with External Clock

In a manner similar to analog input with an external clock, you can provide an external clock for analog output, too. Due to the similarity of the concepts as well as signal connection and VI usage, this section will conclude by listing examples, key VIs, and parameters.

1. *External Update Clock*
   (a) Examples: **Generate N Update-ExtUpdateClk.vi** (for finite generation), **Cont Generation-ExtClk.vi** (for continuous generation)
   (b) Parameters of **AO Start.vi**: **clock source**, **number of buffer iterations** (for continuous generation with value 0)

(c) Usage of **AO Wait.vi** (only for finite generation)
(d) TTL-level clock signal connection: PFI5/UPDATE*, OUT2, or EXTUPDATE* depending on the type of data acquisition boards. Refer to the manual of the board of interest.

PROBLEMS

9.1 Open the example **Function Generator.vi** from the following directory:

`LabVIEW directory` **>> examples >> daq >> anlogout >> anlogout.llb**

(a) Open another example **Cont Acq&Chart (buffered).vi** from LabVIEW directory **>> examples >> daq >> anlogin >> anlogin.llb**. Have those two examples open next to each other. Connect the pins of your data acquisition board properly so that the analog output generated by **Function Generator.vi** is fed directly to an analog input channel specified in **Cont Acq&Chart (buffered).vi**. Run the analog output example followed by the analog input example. This setup will allow you to see the waveform generated by **Function Generator.vi** on the chart in **Cont Acq&Chart (buffered).vi**. Try different waveforms and different input values while running the two examples. Can you change the input(s) without restarting **Function Generator.vi**?
(b) You should have realized in (a) that **Function Generator.vi** is capable of changing the waveform generated in real time; for example, you can change the values of controls **signal type**, **amplitude**, **offset**, **update rate**, and **phase.** Explain how it can reflect the change(s) without stopping and restarting the VI.

9.2 Using **AO Waveform Gen.vi**, design a VI **P09_02.vi** that can generate a 2-Hz sine wave either continuously or a finite number of times. Once you complete **P09_02.vi**, open the example **Cont Acq&Chart (buffered).vi** from LabVIEW directory **>> examples >> daq >> anlogin >> anlogin.llb** next to it, and run both to view the waveform you are generating. (Proper wiring will be needed. For the signal connection, consult with your manual for the data acquisition board.) In order to complete this problem, the following steps may help:
(a) Generate a sine waveform of two cycles using 1000 samples using **Sine Pattern.vi** from **Functions >> Analysis >> Signal Generation**. If you use 1000 samples per second as the update rate, this waveform will become a 2-Hz sine wave.
(b) The output of **Sine Pattern.vi** is 1-D; therefore, convert it to a 2-D array using **Build Array**. This will create a 2-D array where the first row (not column) is the generated data. Therefore, you need to pass it through **Transpose 2D Array** to make it column oriented.

(c) To generate a 2-Hz sine wave for *n* times, you don't need to use the **While Loop**. A single call to **AO Waveform Gen.vi** with a proper value to the input **generation count(1)** suffices to achieve such a task.
(d) To generate a 2-Hz sine wave continuously, you can do it with or without the **While Loop**. Again, proper input settings for **generation count(1)** and **clear generation(T)** will allow you to achieve such a task.

9.3 Using **AO Continuous Gen.vi**, design a VI **P09_03.vi** that can generate a 2-Hz sine wave continuously. Once you complete **P09_03.vi**, open the example **Cont Acq&Chart (buffered).vi** from LabVIEW directory **>> examples >> daq >> anlogin >> anlogin.llb** next to it, and run both to view the waveform you are generating. (Proper wiring will be needed. For the signal connection, consult with your manual for the data acquisition board.) This problem is similar to Problem 9.2 except that **AO Continuous Gen.vi** is designed to generate a continuous waveform in a **While Loop**. Signal generation part for a 2-Hz sine wave in Problem 9.2 can be used again. Remember to transpose the 2-D array created by **Build Array**.

9.4 Consider the analog output settings in Figure P9.4 to answer the following questions. Assume that your data acquisition board has eight analog output channels with a default range ± 10 V and eight individual external reference input connections. Also, assume that the board supports the range ± 5 V.
(a) What DC values do you expect to see at analog output channels 0 through 7?
(b) Explain the settings in **limit settings (no change)** for each channel.
(c) In order to create the settings in Figure P9.4, you need to provide an external power source. What voltage does it have to be?

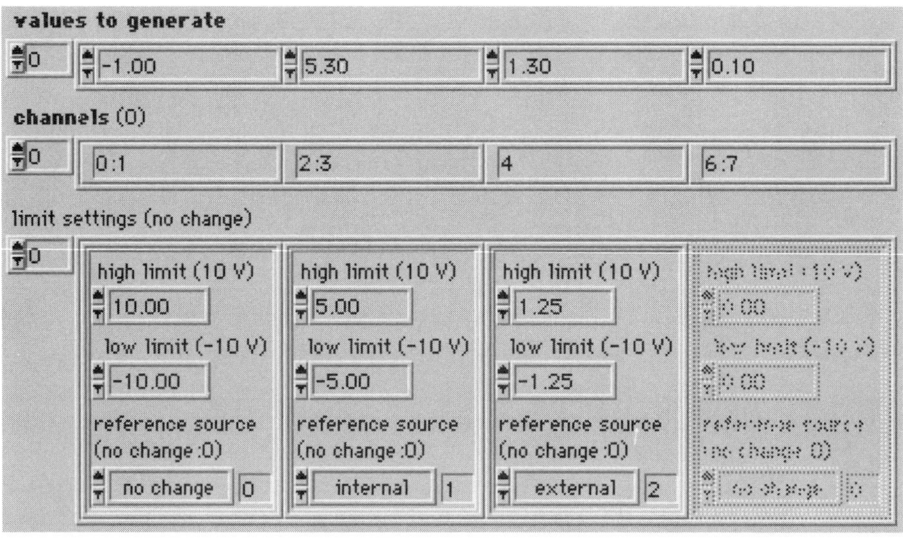

Figure P9.4

Data Acquisition: Digital I/O and Counters

10

10.1 Relays and Switches (Digital I/O)

In some applications, you may need to control switching devices based on your observations; for example, if the temperature drops below a threshold, you may want to turn on the furnace, or if the voltage exceeds a certain level, you may want to turn on the alarm. This type of data acquisition can be monitored and controlled by LabVIEW and is said to be digital input and output. Mostly, devices called relays control such on/off switching mechanisms. The relays are electronic switches that connect or disconnect circuits based on an externally supplied voltage, and the voltage level varies depending on the type of relays. Without loss of generality, assume 0 V and 5 V for the off/on voltages for the relays.

If you want to control a relay using LabVIEW and a data acquisition board, you need to generate either 0 V or 5 V from a digital channel, which is called a digital line. If you want to control more than a single relay, multiple binary states (0 V or 5 V) must be generated from multiple digital lines. Such a group of multiple digital lines is said to be a digital port. The number of digital lines in a digital port is often called port width. The port width can be manually changed in LabVIEW but is usually 4 or 8. This implies that you can generate a binary pattern from a port that consists of multiple digital lines. If the port width is 4, you will be able to generate a word of binary pattern. If it is 8, a byte (8-bit binary pattern) can be generated. The direction of digital lines or ports indicates whether they are used as input or output.

The expressions digital *line*, digital *port*, *port width*, and *direction* were just introduced using an example of relays. In some situations, you may wish to control a digital device based on a sequence of binary messages. This brings up an issue of digital communication, which is commonly referred to as digital *handshaking*. In digital handshaking, two (or more) digital machines (such as a computer and a scanner) communicate with each other by indicating incidents when one needs new data, and when one receives the data, using designated digital lines. The first incident is called REQ, as in request, and the latter ACK, as in acknowledge. As was the case with the analog input and output data acquisition, you can use the circular buffer for continuous digital input and output data acquisition, including handshaking. In case of a finite number of digital input or output data acquisition processes, a simple buffer can be used, and **DIO Wait.vi**, which is similar to **AO Wait.vi**, needs to be implemented. (Recall that you used **AO Wait.vi** for a finite number of waveform generations in Chapter 9.)

The three stages of digital data acquisition are again identical: Configuration, Read or Write, and Reset. Configuration is performed by **DIO Config.vi**, and Read or Write by **DIO Read.vi** or **DIO Write.vi**, and Reset by **DIO Clear.vi**. **DIO Wait.vi** is similar to **AO Wait.vi**, so you should use **DIO Wait.vi** for only a finite number of digital output data acquisitions. The organization of VIs in the **Digital I/O** subpalette is similar, too, except that there is no utility subpalette (Figure 10.1). The four VIs in the first row in Figure 10.1 are the high-level VIs, and they are ready to be used for digital input or output using a port or individual digital lines. However, **DIO Single Read/Write.vi** in the last row can be treated as the utility VI, which provides you with more control than high-level VIs, as well as the input **iteration** for repeating acquisition. The VIs in the second row are the intermediate-level ones. (**DIO Clear.vi** is located in the last row even though it is the intermediate-level VI.)

The general order of VI call is also similar. In the analog input data acquisition, the order was **AI Config.vi-AI Start.vi-AI Read.vi-AI Clear.vi**, and in the analog output data acquisition, it was **AO Config.vi-AO Write.vi-AO Start.vi-AO Wait.vi-AO Clear.vi**. For the digital input data acquisition, it is **DIO Config.vi-DIO Start.vi-DIO Read.vi-DIO Clear.vi**, and **DIO Config .vi-DIO Write.vi-DIO Start.vi-DIO Wait.vi-DIO Clear.vi** for the digital output data acquisition.

10.1.1 Immediate Digital Input and Output VIs

Immediate digital input (DI) and digital output (DO) do not use any buffers in the memory to store the data. All four high-level VIs in the first row in Figure 10.1 can be used for the immediate DI and DO applications. These VIs will read/write from/to a single digital line or a digital port of port

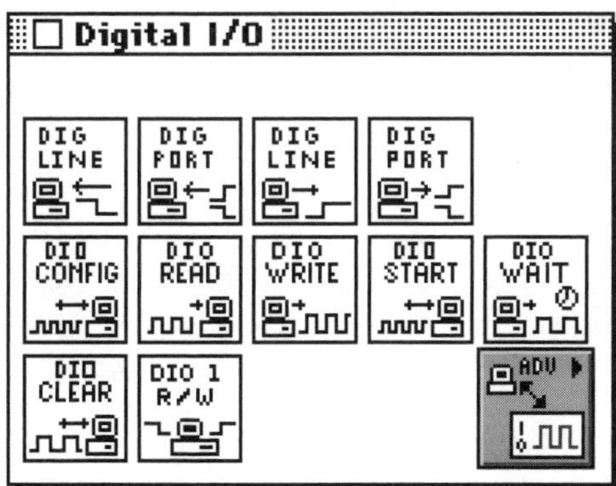

Figure 10.1 **Digital I/O** subpalette.

width 4 or 8 depending on the type of boards. Since they all have a terminal **iteration**, you can use them in a **While Loop** for repeated DI or DO. However, these VIs do not provide triggering or handshaking capability. In order to perform triggered digital data acquisition or handshaking, a buffered handshaking technique should be used. The next subsection will discuss different handshaking methods using various types of buffers.

10.1.2 Nonbuffered and Simple/Circular-Buffered Handshaking

Digital handshaking is based on a series of TTL pulses generated by a timing source. If the source is internal, handshaking is performed based on the onboard clock. If it is external, a third-party device provides TTL pulses to perform handshaking. When the handshaking is performed, only a single digital value (either 0 V or 5 V) is generated (DO) or read (DI) at a time. If multiple ports are grouped together, the data will be interleaved; for example, if you have ports 0 and 1 as a group with the same direction, the data structure will be <data from port 0> <data from port 1> <data from port 0> <data from port 1> . . . where the time elapses from left to right.

Signal connection for handshaking also varies depending on the boards; for example, some use only REQest (REQ) and ACKnowledge (ACK) lines, and some use STroBe (STB), Input Buffer Full (IBF), ACK, and Output Buffer Full (OBF) lines. Therefore, users should refer to the manuals of the boards for the complete and accurate signal connection information.

The difference between nonbuffered and buffered handshaking is whether the communication is intended to be continuous or not. In other words, if there are more than a few countable handshaking TTL pulses, buffered handshaking must be used. If the communication is intended for a short period of time using few TTL pulses, nonbuffered handshaking may suffice.

As for the data acquisition boards, there are generally two types of data acquisition boards that support handshaking. One type supports all of the functionality, such as triggered handshaking, eligibility for both internal and external handshaking timing source, and simple- and circular-buffered handshaking capability. The other type supports only external handshaking timing source and simple-buffered handshaking. For the sake of convenience in discussion, the first type (which supports all of the functionality) will be abbreviated as type A, and the other as type B in this subsection. Also, all of the examples in this subsection can be found in the following directory unless otherwise stated: LabVIEW directory >> **examples** >> **daq** >> **digital** >> **digio.llb**.

First, nonbuffered handshaking can be performed using **DIO Single Read/Write.vi** with the following input settings in a **While Loop**:

1. *Key parameters for nonbuffered handshaking with* ***DIO Single Read/ Write.vi***
   (a) **group direction**: 1 for input, and 2 for output
   (b) **operation code**: 1 for nonbuffered handshaking
   (c) **number to read** for group direction 1, and **updates to write** for group direction 2. For group direction 1 (input), the total number of data read will be the multiplication of the number of ports in the group and **number to read**. For group direction 2 (output), the total number of data written will be the multiplication of the number of ports in the group and **updates to write**. In either case, the data from each port will be interleaved, as mentioned previously in this subsection.

In order to perform simple-buffered handshaking using data acquisition boards of types A and B, the VIs that need to be called are as follows. (Key parameters are also indicated.)

1. *Digital input using simple-buffered handshaking with boards of type A*
   (a) General order of VI call: **DIO Config.vi-DIO Start.vi-DIO Read.vi-DIO Wait.vi-DIO Clear.vi**
   (b) Key parameters of **DIO Config.vi**: **handshaking mode parameters** (Triggering mode can be chosen using this input.)
   (c) Key parameters of **DIO Start.vi**: **handshake source**, **clock frequency**. If **handshake source** is 1 for internal, **clock frequency** must be specified. For an external source, select I/O connector or RTSI connection, and refer to the manual of the board being used for the specific signal connection.
   (d) Key parameters of **DIO Wait.vi**: **direction** (1 for input)
2. *Digital input using simple-buffered handshaking with boards of type B*
   (a) General order of VI call: **DIO Config.vi-Digital Buffer Control.vi-DIO Read.vi-DIO Wait.vi-DIO Clear.vi**
   (b) Key parameters of **DIO Config.vi**: **handshaking mode parameters** *not* used
   (c) Key parameters of **Digital Buffer Control.vi**: **control code** (0 for start)
   (d) Key parameters of **DIO Wait.vi**: **direction** (1 for input)
3. *Digital output using simple-buffered handshaking with boards of type A*
   (a) General order of VI call: **DIO Config.vi-DIO Write.vi-DIO Start.vi-DIO Wait.vi-DIO Clear.vi**
   (b) Key parameters of **DIO Config.vi**: **handshaking mode parameters** (Triggering mode can be chosen using this input.)

   (c) Key parameters of **DIO Start.vi**: **handshake source**, **clock frequency**. If **handshake source** is 1 for internal, **clock frequency** must be specified. For an external source, select I/O connector or RTSI connection, and refer to the manual of the board being used for the specific signal connection.
   (d) Key parameters of **DIO Wait.vi**: **direction** (2 for output)
4. *Digital output using simple-buffered handshaking with boards of type B*
   (a) General order of VI call: **DIO Config.vi-DIO Write.vi-Digital Buffer Control.vi-DIO Wait.vi-DIO Clear.vi**
   (b) Key parameters of **DIO Config.vi**: **handshaking mode parameters** *not* used.
   (c) Key parameters of **Digital Buffer Control.vi**: **control code** (0 for start)
   (d) Key parameters of **DIO Wait.vi**: **direction** (2 for output)

A good example of simple-buffered handshaking is **Digital Buffered Handshaking.vi**, and its diagram shows similar VI calls as mentioned previously.

As was the case with the analog input and analog output data acquisition regarding the usage of the circular buffer, continuous DI and DO must use the circular buffer, and a good example of circular-buffered handshaking is **Digital DblBuffered Pattern Generation.vi**. This example illustrates both directions (digital input and digital output) using the circular buffer. Note that only the boards of type A support circular-buffered handshaking. Therefore, when you perform circular-buffered handshaking using the boards of type A, the internal clock option can be used in addition to the external clock option, since type A has an on-board clock available for handshaking. However, make sure to specify the input **clock frequency** when using **internal** for the input **handshake source** of **DIO Start.vi**. Also, the input **number of scans/updates to acquire or generate** of **DIO Start.vi** should be set to zero for a continuous operation.

10.2 Data Acquisition with Counters

Figure 10.2 shows the VIs for data acquisition using counters. The first row contains the high-level VIs, the first subpalette in the second row contains the intermediate-level VIs, and the second subpalette in the second row has the advanced-level VIs.

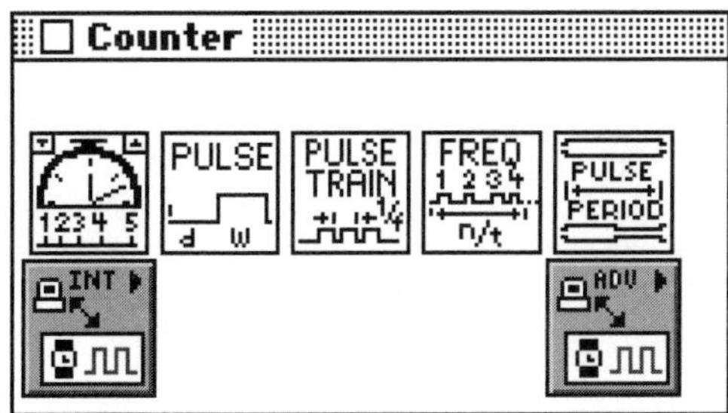

Figure 10.2 **Counter** subpalette.

Applications with counters are fairly simple, and as long as you understand the fundamental concept behind each different task with counter VIs, you merely need to know which example(s) to use for your case. Therefore, this section will first discuss the basic concept of counter operation and then provide a list of examples for each different application.

First, definitions of a few terms are introduced in order to make this discussion easy. There are three terminals on each on-board counter: GATE, SOURCE, and OUT. GATE and SOURCE are the two inputs where TTL signals are fed, and OUT is the output where a single TTL pulse or a sequence of TTL pulses (pulse train) is generated. In the case of a single TTL pulse generation, you can make it either toggle (from low to high or high to low) or complete a cycle of a single TTL pulse. Also, you can use a series of counters with OUT from a previous stage wired to SOURCE or GATE of the next stage to extend the level of operation. For example, you can extend the range of count by having the next additional counter count the carry out TTL pulse from the current counter. This would resemble a counter with double digits instead of a single digit.

The duty cycle of a TTL pulse is a percentile measure that indicates the ratio between the high state (5 V) and the low state (0 V) for a given TTL pulse. The high and the low state are also called phase 1 and phase 2, respectively, in LabVIEW. Therefore, if you have a TTL pulse of 50% duty cycle, the ratio between phase 1 and phase 2 would be 1:1. When you generate a TTL pulse in LabVIEW, the duty cycle can manually be controlled.

One of the important measures of different counters is the resolution. The resolution of a counter indicates how many bits the counter uses to represent a number (count), and the more bits available, the higher count you can achieve (higher resolution). Conventional resolution on the market is usually either 16 bit or 24 bit (sometimes 12 bit). Suppose that you have a counter with 2-bit resolution. Then you can easily conclude that the maximum count would be 3 since you can count 0, 1, 2, and 3. Extending this

Table 10.1 *Comparison of the four most popular counter chips that LabVIEW supports: DAQ-STC, NI-TIO-ASIC, Am9513, and 8253/54.*

	Edge Detection	Resolution	Count Direction	Internal Timebase (Hz)	Miscellanies
DAQ-STC	R[1], F[2]	24 bit	I[3], D[4]	100K, 20M	
NI-TIOI-ASIC	R[1], F[2]	32 bit	I[3], D[4]	80M, 125M	
Am9513	R[1], F[2]	16 bit	I[3], D[4]	100, 1K, 10K, 100K, 1M	
8253/54	R[1]	16 bit	D[4]	1M, 2M	Only reverse-logic TTL pulse can be generated.

[1]Rising edge.
[2]Falling edge.
[3]Increment.
[4]Decrement.

simple example, you can come up with $2^n - 1$ for the maximum count with a n-bit resolution counter.

LabVIEW supports many types of counter chips, and the four most popular ones are DAQ-STC, NI-TIO-ASIC, Am9513, and 8253/54. Consider Table 10.1 for the complete information about those four types. Table 10.1 shows the resolution, count direction, and available internal timebase for each type of counter chips. For all of the four types of counter chips, SOURCE can use both internal or external TTL timebase signals.

10.2.1 Basic Concept behind Different Counter Operations

The next few subsections will introduce different types of counter operations, such as pulse generation (either continuous or finite), width measurement of a single TTL pulse, measurement of period/frequency of a TTL pulse train, and event counting. Event counting refers to an operation where the counter increments or decrements when it recognizes the high state of the input TTL pulse train present at the SOURCE terminal. Note that the signal to SOURCE can be either an internal timebase or an external TTL signal, as mentioned previously.

Even though the counter operations just listed are called by different names, the basic concept behind them is fairly simple and similar: You measure the attribute of an unknown TTL pulse or a pulse train using a known TTL pulse or a pulse train provided either internally or externally. Consider the two TTL pulse trains A and B in Figure 10.3. Suppose that you want to measure the pulse width of Pulse A while the state is high. It can easily be calculated if you know either the frequency or the period of Pulse B, assuming that the duty cycle of Pulse B is 50% and that the start time of Pulse A and B is synchronized. (If the pulse width of Pulse B is much smaller than that of Pulse A, such a calculation would be valid.) If you count how many high states of Pulse B are present while the state of Pulse A is high in Figure 10.3, the pulse width of Pulse A would be twice the period of Pulse B since you have two high states of Pulse B present while Pulse A is high. Once you know the pulse width of Pulse A in time, you can calculate the frequency of Pulse A simply by inverting its period in seconds. This simple illustration explains the basic concept behind pulse width measurement, period and frequency measurement, and elapsed-time measurement. You merely need to be aware of the signal connection and which example(s) you need to use after the fundamental concept is understood.

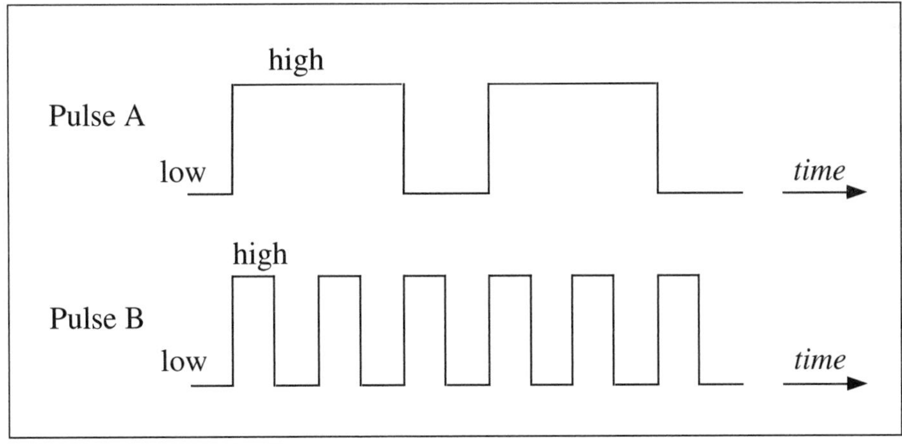

Figure 10.3
Two TTL pulse trains to illustrate an example of counter operations: pulse width measurement.

Since the operations with counter chips and their basic concepts are fairly simple, each different operation will be explored simply by showing the VIs to be used and the appropriate LabVIEW examples for each case. For details about signal connections, online help for each example should suffice. The directories of each example for the rest of this chapter are assumed to be one of the following:

LabVIEW directory >> **examples >> daq >> counter >> daq-stc.llb**
LabVIEW directory >> **examples >> daq >> counter >> am9513.llb**
LabVIEW directory >> **examples >> daq >> counter >> 8253.llb**

It will be understood that examples for DAQ-STC counter chips are located in the first directory, those for AM9513 counter chips in the second, and so forth. Note that if you use intermediate-level VIs for DAQ-STC or AM9513 counter chips, you should use **Counter Stop.vi** in order to stop the counter without resetting the entire data acquisition board. To stop 8253/54 counter chips, choose **reset** for the input **control code** of **ICTR Control.vi**. Also, you should be aware of the fact that 8253/54 counter chips have the SOURCE of counter 0 internally connected to the internal timebase.

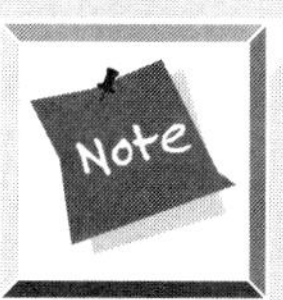

*In order to stop the counter chips DAQ-STC or AM9513, use **Counter Stop.vi**. To stop 8253/54 counter chip, use **ICTR Control.vi** with **reset** for the input **control code**. Also, 8253/54 counter chips have the SOURCE of counter 0 internally connected to the internal timebase.*

10.2.2 Single TTL Pulse Generation

DAQ-STC and AM9513

1. Easy VIs
   VIs: **Generate Delayed Pulse.vi**
   Examples: (1) **Delayed Pulse-Easy (DAQ-STC).vi**
   (2) **Delayed Pulse-Easy (9513).vi**
2. Intermediate VIs
   VIs: **Delayed Pulse Generator Config.vi**
   Counter Start.vi
   Counter Stop.vi

Examples: (1) **Delayed Pulse-Int (DAQ-STC).vi**
(2) **Delayed Pulse-Int (9513).vi**

8253/54

VIs: **ICTR Control.vi**
timebase generator (8253).vi
Examples: **Delayed Pulse (8253).vi**

10.2.3 TTL Pulse Train Generation (Continuous)

DAQ-STC and AM9513

1. Easy VIs
   VIs: **Generate Pulse Train.vi**
   Examples: (1) **Cont Pulse Train-Easy (DAQ-STC).vi**
   (2) **Cont Pulse Train-Easy (9513).vi**
2. Intermediate VIs
   VIs: **Continuous Pulse Generator Config.vi**
   Counter Start.vi
   Counter Stop.vi
   Examples: (1) **Cont Pulse Train-Int (DAQ-STC).vi**
   (2) **Cont Pulse Train-Int (9513).vi**

8253/54

VIs: **ICTR Control.vi**
generate pulse train (8253).vi which uses **timebase generator (8253).vi**
Examples: **Cont Pulse Train (8253).vi**

10.2.4 TTL Pulse Train Generation (Finite)

DAQ-STC

1. Easy VIs
   VIs: **Generate Pulse Train.vi**
   Wait+ (ms).vi
   Counter Stop.vi

Examples: **Finite Pulse Train-Easy (DAQ-STC).vi**

2. Intermediate VIs

   VIs: **Continuous Pulse Generator Config.vi**
   Delayed Pulse Generator Config.vi
   Adjacent Counters.vi
   Counter Start.vi
   Wait+ (ms).vi
   Counter Stop.vi

   Examples: **Finite Pulse Train-Int (DAQ-STC).vi**

3. Generating a finite TTL pulse train of frequency = internal timebase/n, where $n = 1, 2, \ldots, 16$. This TTL pulse train can be generated even if none of the counters is available, and its duty cycle is 50%.

   VIs: **CTR Control.vi**

   Examples: **Generate Pulse Train on FREQ_OUT.vi**

4. Generating a finite TTL pulse train using *internal* wiring of OUT to GATE of the next counter chip.

   VIs: **CTR Mode Config.vi**

   Examples: **Finite Pulse Train-Adv (DAQ-STC).vi**

AM9513

1. Easy VIs

   VIs: same as DAQ-STC

   Examples: **Finite Pulse Train-Easy (9513).vi**

2. Intermediate VIs

   VIs: same as DAQ-STC

   Examples: **Finite Pulse Train-Int (9513).vi**

3. Generating a finite TTL pulse train of frequency = internal timebase/n, where $n = 1, 2, \ldots, 16$. This TTL pulse train can be generated even if none of the counters is available, and its duty cycle is 50%.

   VIs: same as DAQ-STC

   Examples: **Generate Pulse Train on FOUT.vi**

8253/54

VIs: **ICTR Control.vi**

Examples: **Finite Pulse Train (8253).vi**

All three counter chips need to be used.

You need an inverter to provide a positive polarity pulse to the GATE of the next counter.

10.2.5 Event and Elapsed-Time Counting

DAQ-STC and AM9513

1. Easy VIs
   VIs: **Count Events or Time.vi**
   Examples: (1) **Count Events-Easy (DAQ-STC).vi**
   (2) **Count Events-Easy (9513).vi**
   (3) **Count Time-Easy (DAQ-STC).vi**
   (4) **Count Time-Easy (9513).vi**
2. Intermediate VIs
   VIs: **Event or Time Counter Config.vi**
   Counter Start.vi
   Counter Read.vi
   Counter Stop.vi
   Examples: (1) **Count Events-Int (DAQ-STC).vi**
   (2) **Count Events-Int (9513).vi**
   (3) **Count Time-Int (DAQ-STC).vi**
   (4) **Count Time-Int (9513).vi**

8253/54

VIs: **ICTR Control.vi**
Examples: (1) **Count Events (8253).vi**
(2) **Count Time (8253).vi**

10.2.6 Pulse Width Measurement

The basic concept of pulse width measurement was illustrated in Figure 10.3. As the assumption (which was made in the example associated with Figure 10.3 in Section 10.2.1) implicates, the faster the timebase, the more accurate the measurement will be. For example, the smaller the pulse width of Pulse B gets in Figure 10.3, the error in the pulse width measurement of

Pulse A would clearly become smaller. Make sure that the pulse width to be measured does not exceed the maximum time duration the counter can measure for a given timebase. For example, if the timebase (Pulse B in Figure 10.3) is 100 KHz with 24-bit resolution, the maximum time duration that the counter can measure would be $2^{24} \times (100000)^{-1}$ seconds, which is approximately 167.77 seconds. Also, make sure that the pulse whose width is to be measured transits from low to high after the counter starts.

Examples for each different case follow and are provided in the next few subsections. The directories of each example were previously mentioned in Section 10.2.1.

DAQ-STC

1. Easy VIs
   VIs: **Measure Pulse Width or Period.vi**
   Examples: **Measure Pulse-Easy (DAQ-STC).vi**
2. Intermediate VIs
   VIs: **Pulse Width or Period Meas Config.vi**
   Counter Start.vi
   Counter Read.vi
   Counter Stop.vi
   Examples: N/A
3. Simple-buffered pulse width measurement
   VIs: **CTR Group Config.vi**
   CTR Buffer Config.vi
   CTR Mode Config.vi
   CTR Control.vi
   CTR Buffer Read.vi
   Examples: **Meas Buffered Pulse-Period (DAQ-STC).vi**
4. Circular buffer is not supported.

AM9513

1. Easy VIs
   VIs: same as DAQ-STC
   Examples: **Measure Pulse-Easy (9513).vi**

2. Intermediate VIs
   VIs: same as DAQ-STC
   Examples: N/A

8253/54

VIs: **ICTR Control.vi**
get timebase (8253).vi
timebase generator (8253).vi

Examples: (1) **Measure Short Pulse Width (8253).vi**
(2) **Measure Long Pulse Width (8253).vi**
In order to measure a low-state pulse width (the pulse width while it stays at low state 0 V), use an inverter between the pulse source and the GATE of counter 0.

10.2.7 Period and Frequency Measurement (High-Frequency Signals)

DAQ-STC and AM9513

1. Easy VIs
   VIs: **Measure Frequency.vi**
   Examples: (1) **Measure Frequency-Easy (DAQ-STC).vi**
   (2) **Measure Frequency-Easy (9513).vi**
2. Intermediate VIs
   VIs: **Event or Time Counter Config.vi**
   Adjacent Counters.vi
   Delayed Pulse Generator Config.vi
   Counter Start.vi
   CTR Control.vi
   Counter Read.vi
   Counter Stop.vi
   Examples: N/A

8253/54

VIs: **ICTR Control.vi**
get timebase (8253).vi
Wait+ (ms).vi

Examples: **Measure Frequency > 1 kHz (8253).vi**

10.2.8 Period and Frequency Measurement (Low-Frequency Signals)

DAQ-STC and AM9513

1. Easy VIs

   VIs: **Measure Pulse Width or Period.vi**

   Examples: (1) **Measure Period-Easy (DAQ-STC).vi**
   (2) **Measure Period-Easy (9513).vi**

2. Intermediate VIs

   VIs: **Pulse Width or Period Meas Config.vi**
   Counter Start.vi
   Counter Read.vi
   Counter Stop.vi

   Examples: N/A

8253/54

VIs: **ICTR Control.vi**
timebase generator (8253).vi

Examples: **Measure Frequency < 1 KHz (8253).vi**

PROBLEMS

10.1 Open either one of the following examples based on the type of counter chip that your board has:

```
Cont Pulse Train-Easy (DAQ-STC).vi
Cont Pulse Train-Easy (9513).vi
Cont Pulse Train (8253).vi
```

Each of those examples can be found in one of the following directories:

`LabVIEW directory >>` **`examples`** `>>` **`daq`** `>>` **`counter`** `>>` **`daq-stc.llb`**

`LabVIEW directory >>` **`examples`** `>>` **`daq`** `>>` **`counter`** `>>` **`am9513.llb`**

`LabVIEW directory >>` **`examples`** `>>` **`daq`** `>>` **`counter`** `>>` **`8253.llb`**

Set each parameter properly so that the example generates a TTL pulse train of 10 Hz with duty cycle 0.7 (70%) from counter 0. Wire the output OUT of counter 0 to the analog input channel 0 of your board. Now, open another example **Cont Acq&Chart (buffered).vi** from LabVIEW directory >> **examples** >> **daq** >> **anlogin** >> **anlogin.llb**. Set each parameter properly to show the input signal at channel 0 at scan rate 1 KHz. Run both examples, and see if you observe a 10-Hz TTL pulse train in **Cont Acq&Chart (buffered).vi**. Change the duty cycle and observe the change in the shape of the pulse train.

10.2 The example VI that was used in Problem 10.1 will be used again in this problem except for **Cont Acq&Chart (buffered).vi**. Additionally, open another one of the following examples based on the type of counter chip that your board has in the directories listed in Problem 10.1:

`Count Events-Easy (DAQ-STC).vi`

`Count Events-Easy (9513).vi`

`Count Events (8253).vi`

Using the example in Problem 10.1, generate a TTL pulse train of 1 Hz from counter 0. In the example you just opened for this problem, set each parameter properly so that it will count the number of high states in the input TTL pulse train using counter 1. Run the example in Problem 10.1 first followed by one of the examples just opened. Describe what you observe. Change the frequency of the TTL pulse train, and comment on the change you observe.

10.3 Open the example **Digital 1 Port.vi** from the following directory:

`LabVIEW directory >>` **`examples`** `>>` **`daq`** `>>` **`digital`** `>>` **`digio.llb`**

Save the VI as **P10_03.vi** first, and answer the following questions:

(a) Explain briefly what the example VI does.

(b) Modify the VI such that it generates a binary pattern corresponding to integers 0 through 15 repeatedly.

11 File Input and Output

The last few chapters have discussed different types of data acquisition. This chapter will complete the discussion of data acquisition by presenting you how to store the acquired data safely.

The process of saving data to or retrieving data from a file is often called file input and output (file I/O), and the importance of it cannot be overemphasized. File I/O is also closely related to string manipulation, a topic treated separately in Chapter 12.

The subpalette **File I/O** is shown in Figure 11.1. The organization of VIs and functions is similar to other data acquisition subpalettes: High-level VIs are in the first row, intermediate-level VIs and functions are in the second row, and advanced VIs and functions are at the bottom.

11.1 High-Level File I/O VIs

Figure 11.1 shows that the **File I/O** subpalette has five high-level VIs in the first row: **Write To Spreadsheet File.vi**, **Read From Spreadsheet File.vi**, **Write Characters To File.vi**, **Read Characters From File.vi**, and **Read Lines**

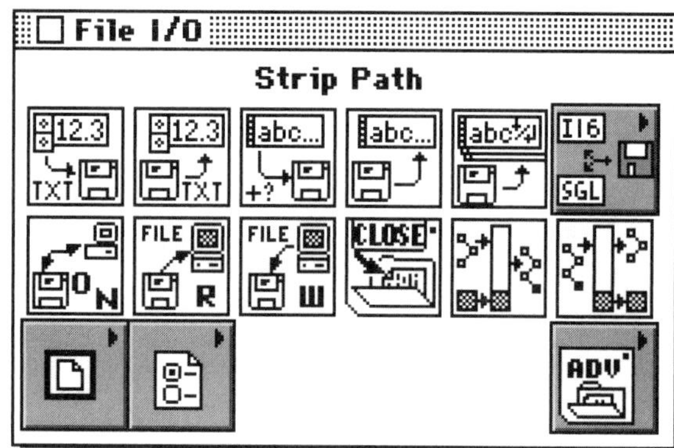

Figure 11.1 **File I/O** subpalette.

From File.vi. These VIs open an existing file or create a new one, read or write data, and close the file. In other words, they perform the complete file I/O procedure. The three-step procedure is similar to the three stages in data acquisition. Each individual step will be studied in detail in the next section, where intermediate-level VIs will be discussed.

High-level VI functionality is self-explanatory. The first two can be used to write or read an array of numeric data to or from a file specified in the input **file path**. The data written by **Write To Spreadsheet File.vi** will be in spreadsheet format. This implies that the data written will be in American Standard Code for Information Interchange (ASCII) format so that you can open the file and view the data using any third-party spreadsheet application. If the written data are binary, they will not be legible using a spreadsheet application. Reading and writing a binary data file are discussed in detail in a later section.

The next two high-level VIs, **Write Characters To File.vi** and **Read Characters From File.vi**, can write or read characters to or from a specified file. The last one, **Read Lines From File.vi**, reads a specified number of lines from a file starting at a specified offset and returns the data in string format. Therefore, the first five high-level VIs deal with reading or writing character data from or to a file. This requires that you specify the file in the input **file path** and the data will be read or stored.

The last one in the first row in Figure 11.1 is the subpalette **Binary File VIs**. This contains four VIs (Figure 11.2) for reading or writing binary data from or to a file. The first two are **Read From I16 File.vi** and **Write To I16 File.vi**, and they read or write data in 16-bit integer type binary format. If the data are float type, you should use the last two VIs, **Read From SGL File.vi** and **Write To SGL File.vi**. Like the other high-level VIs, these VIs can be used with simple input specification.

Before proceeding to the intermediate-level VIs and functions, this section will further discuss two types of data: binary and ASCII. In binary, each number occupies the same number of bits or bytes: 4 bytes (32 bits) for single precision (SGL) and 2 bytes (16 bits) for integer (I16) type. However, in ASCII, *each* digit (character) requires one byte so that multiple-digit numbers will

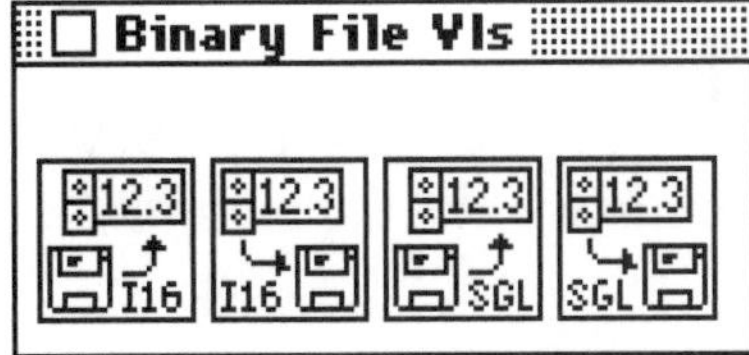

Figure 11.2 **Binary File VIs** subpalette in the **File I/O** subpalette.

require multiple bytes. For example, the numbers 123 and 4 both require 16 bits (2 bytes) if you save them as I16 binary type, whereas 3 bytes are necessary for 123 and 1 byte for 4 if you save them as ASCII type. Therefore, each type has an advantage and a disadvantage: ASCII files can be viewed and understood using spreadsheet applications, but the number of bits for each datum varies. Therefore, the size of ASCII files is likely to be much bigger than that of binary files. (Try saving the same data in two different formats using high-level VIs and observe the size of each file.)

Binary files cannot be viewed and understood using spreadsheet applications, but the number of bits for each datum is fixed, and the size of binary files is usually much smaller than ASCII files. Therefore, the choice of binary or ASCII type can be based on priorities as to file size, legibility, and capability of finding arbitrary elements in data. (With binary data, you can locate any arbitrary element in the data just by knowing the offset of the element from the beginning of the binary file. This is because the number of bits for each binary datum is fixed.)

11.2 Intermediate-Level File I/O VIs and Functions

The middle row in Figure 11.1 has one intermediate-level VI and five LabVIEW functions. As was briefly mentioned earlier, there are three steps in file I/O procedure: File Open, Read or Write, and File Close. If you want to read or write continuously, the second step needs to be embedded in a **While Loop**. Another similarity lies in the parameter **refnum**, whose role is the same as **task ID** in data acquisition. The **task ID** is what identifies different data acquisition boards in the computer, and so is the **refnum** in file I/O operations. Therefore, all of the intermediate-level VI and functions use the **refnum** to identify files. This section will now look at each one.

Open/Create/Replace File.vi This VI opens, creates, or replaces a file specified in the input **file path**. You can select the different options of **function (open:0)** in the input to create a new file, or open or replace a preexisting file. The output **file size (bytes)** returns the size of the file in bytes.

Read File This function reads data from a file regardless of data type—either binary or ASCII. You specify the type of data using the input **byte stream type**. If a string is wired to this input, **Read File** will view the data as ASCII type. If you wire floating-type data (for ex-

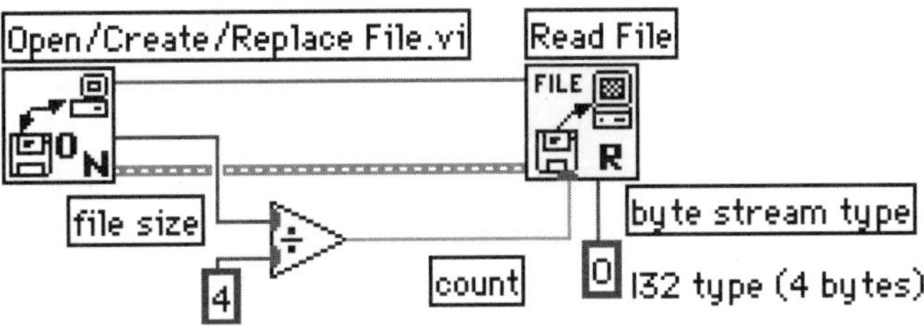

Figure 11.3
Reading all of the binary data of I32 type using **Read File** and **Open/Create/Replace File.vi**. This example assumes that the type of binary data is I32 (4 bytes). Therefore, you need to divide the output **file size** by the size of each binary datum (4 bytes), and the result is fed to the input **count** of **Read File**. Note that **file size** is in byte, and a numeric constant 0 of type I32 is wired to the input **byte stream type** to specify the input type.

ample, an array of SGL), it will interpret the data as floating. Note that the input **count** is not in terms of bytes but the number of data to be read. This input can be specified for either binary or ASCII files using the output **file size (bytes)** of **Open/Create/Replace File.vi**, as shown in Figures 11.3 and 11.4.

There are three important parameters of **Read File**: **byte stream type**, **convert eol (F)**, and **line mode (F)**. **byte stream type** and the other two are mutually exclusive. In other words, you can use either **byte stream type** alone, or **convert eol** and **line mode** without using **byte stream type**, but not all three simultaneously. You specify the type of data that **Read File** will read in with **byte stream type**. For example, Figure 11.3 shows that the data type will be I32 so that each data element will be 4 bytes. If you use **convert eol** and **line mode**, instead, the following

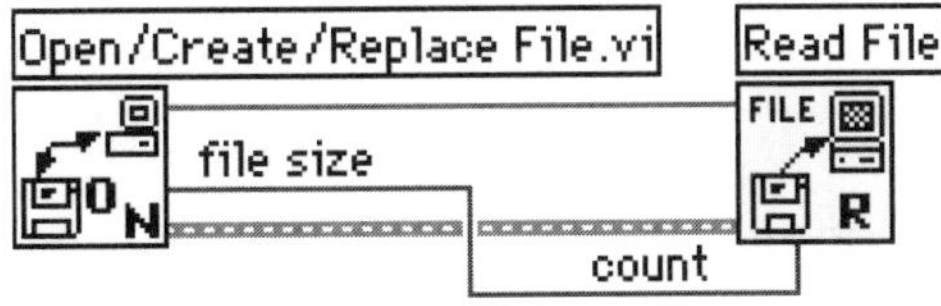

Figure 11.4
Reading all of the ASCII data using **Read File** and **Open/Create/Replace File.vi**. Since each ASCII character is one byte, you can directly connect the output **file size** to the input **count** of **Read File**. Also, note that you don't have to specify the data type using **byte stream type** of **Read File**.

will happen: If **line mode** is TRUE, **Read File** will continue to read until the end of file is reached. If you specify **count** with **line mode** set to TRUE, **Read File** will continue to read until either the end of the file is reached or the number of data specified in **count** has been read, whichever happens first. The input **convert eol** determines if you want to convert the end of line (EOL) character to line feed, which is the EOL character in LabVIEW. You are given the ability to **convert eol** since the EOL characters are different on different platforms: for example, they are carriage return and line feed on Windows, carriage return on Macintoshes, and line feed on UNIX.

As for the type of data that **Read File** returns, it will be string type if you specify only the input **count**. (If you do not specify the data type with **byte stream type**, **Read File** will assume that the data type is string; therefore, **Read File** will return string data.) If you specify only the data type with **byte stream type**, the data returned will be the same type as **byte stream type**. If you specify both **byte stream type** and **count**, the output will be an array of the same type as **byte stream type**. The results from different settings are summarized in Table 11.1.

Write File The two most important parameters of this function are **header** and **data**. Similar to the input **byte stream type** of **Read File**, if string data are wired to the input **data**, the function will write them as ASCII data. If data of floating type are wired to **data**, they will be written as binary data. Since it is not easy to view binary files, you may want to include header information about the binary files when you create them. In later discussion of reading and writing binary data, you will see that it is not critical to have this header information, but it can be helpful when interpreting and retrieving data. The input

Table 11.1 Comparison of the data returned by ***Read File*** *with different input settings.*

byte stream type	***count***	***data returned***
unspecified	specified	string
specified	unspecified	same type as **byte stream type**
specified	specified	array of the same type as **byte stream type**

header allows you to create the header information automatically if the Boolean input is set to TRUE.

Close File This function closes a file whose reference number is specified in the input **refnum**.

Build Path and **Strip Path** These two functions perform opposite operations. **Build Path** takes two inputs, **base path** and **name or relative path**. It appends the string in **name or relative path** to **base path** and returns the resulting concatenated string as a path. **Strip Path** does exactly the opposite task: Given a path, it extracts the last subdirectory in the path, or file name if there is no subdirectory in the path.

11.3 Additional File I/O Subpalettes

In Figure 11.1, there are three subpalettes in the bottom row: **File Constants**, **Configuration File VIs**, and **Advanced File Functions**. The first, **File Constants**, contains all of the constants for file I/O applications, and the third one, **Advanced File Functions**, contains the low-level functions used in the high-level VIs. The second one, **Configuration File VIs**, is a new subpalette that was not available in LabVIEW 4.x or earlier versions. LabVIEW 5.0 has introduced this new subpalette, which contains VIs that allow you to control the configuration files, such as the **.ini** file on PC platforms or the configuration file on Macintosh platforms. Actually, the file that gets created by those VIs is platform independent since it is simply an ASCII text file.

The next few sections will discuss different cases of file I/O: reading from or writing to a binary file once or continuously, reading from or writing to an ASCII file once or continuously, and reading from or writing to a file that contains both binary and ASCII data. The last case would be useful if you would like to take advantage of both binary and ASCII data. In other words, you may wish to save the header information in ASCII format followed by binary data. This will allow you to view the header information using word processing applications while saving disk space by saving data in binary format since binary files are much smaller than ASCII files. Also, you can access binary data randomly since the size of each binary datum is uniform. (See Section 11.2 for more details.) Each case study will be presented with an example in the next few sections.

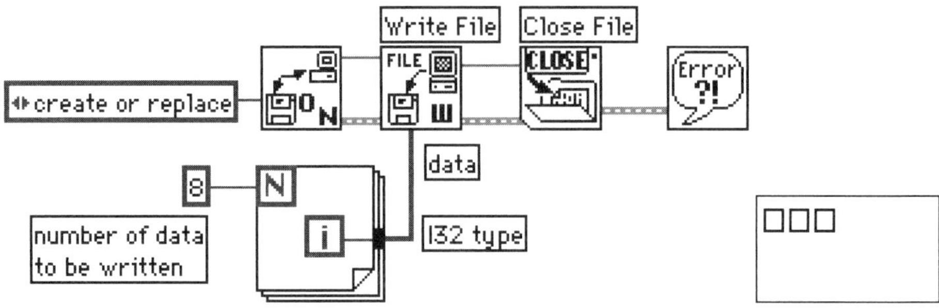

Figure 11.5
Writing 1-D data in binary format, and the view of the file created by this example.

11.4 Writing 1-D Data in Binary Format

Suppose that you have a 1-D array of data you want to save in binary format. The example in Figure 11.5 can perform such a task. In Figure 11.5, you are writing an array of eight elements in I32 (4 bytes) binary format. The data are wired to the input **data** of **Write File**, and the view of the resulting binary file in a word processing application is also shown. As you can see, the eight numbers (0, 1, 2, . . . , 7) in 4-byte binary format are illegible.

11.5 Reading 1-D Binary Data

Figure 11.6 shows how to read binary data from a file created by the example in Figure 11.5. Recall that the output **file size** of **Open/Create/Replace File.vi** is in byte. Since each datum is 4 bytes because it is type I32, you must divide **file size** by 4 to compute the total number of data and wire the result to the input **count** of **Read File**. Note that since the input **byte stream type** is wired, the inputs **convert eof** and **line mode** are disabled. (See Section 11.2.)

If you do not wire a numeric constant of type I32 to **byte stream type** of **Read File**, this example will return the data in string type, not in an array of I32 type. Since you specified **count**, **Read File** returns the data in an array. (See Table 11.1.)

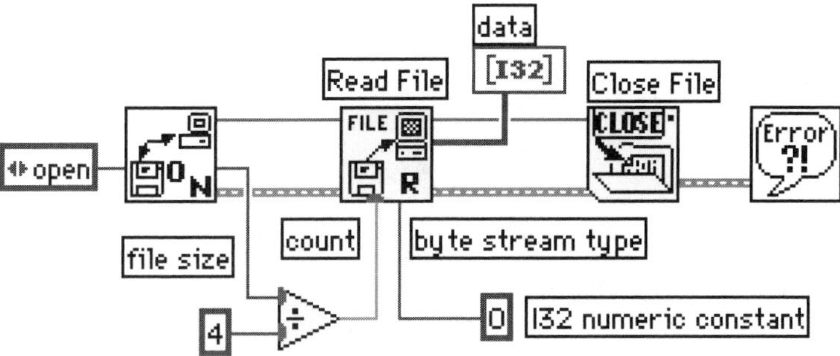

Figure 11.6
Reading 1-D binary data.

11.6 Writing 1-D Data in Binary Format Continuously

If you want to save data in binary format while acquiring them, you can use the example illustrated in Figure 11.7. In this example, you are saving the index as a 1-D data with two elements every one second until either the **QUIT** button is pressed or an error occurs from **Write File**. If you need to process the data as you acquire them, proper analysis VIs can be included in the **While Loop**. Writing a single data sample in binary format continuously is just a special case of 1-D array data. To read the 1-D binary data that was continuously saved using Figure 11.7, the examples in Figure 11.6 or Figure 11.8 in the next section can be used.

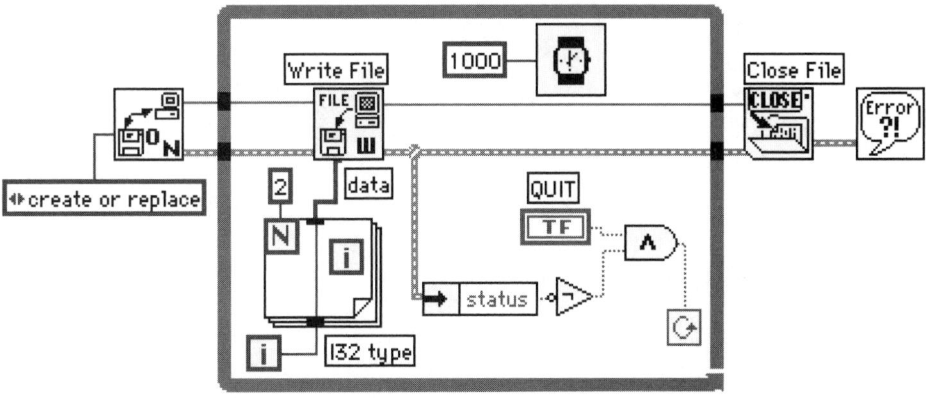

Figure 11.7
Writing 1-D data in binary format continuously.

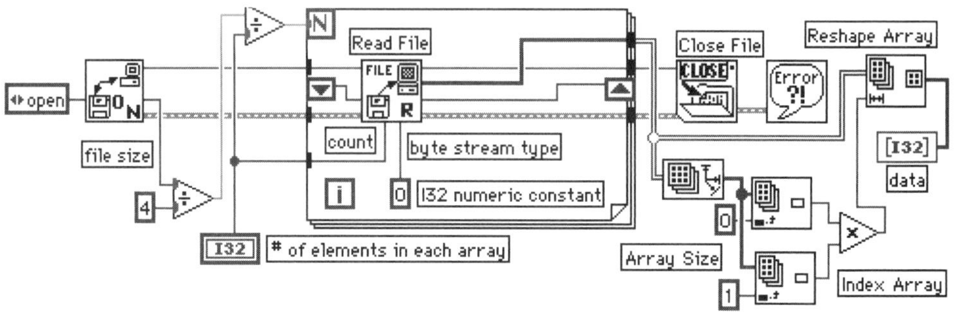

Figure 11.8
Reading 1-D binary data continuously. Note the usage of the input **pos offset (0)** and the output **offset** of **Read File**. The offset values are in bytes; therefore, the value of **offset** will increment by 8 since you are reading two I32 type (4 bytes) binary samples at each iteration.

11.7 Reading 1-D Binary Data Continuously

In some applications, you may want to read a small portion of binary data at a time and process them individually since the size of data is too big to be read in and stored in LabVIEW by single reading. Such a case can be realized by using the example in Figure 11.8. (Figure 11.8 assumes that the data to be read were saved using the example in Figure 11.7: two-element 1-D array data at each iteration.) The input **pos offset (0)** and the output **offset** of **Read File** are now utilized in order to start the reading at the correct location in the binary input file. In this example, the output of the **For Loop** is a 2-D array; therefore, reshaping it into a 1-D array is done after the **For Loop**. Furthermore, the value of offset is in bytes, so the value of the output **offset** of **Read File** will increment by 8 since you are reading two 4-byte (I32 type) data samples at each iteration.

11.8 Writing 2-D Data in Binary Format

If you are saving data acquired from multiple channels, the resulting data will be a 2-D array, and each column of the 2-D array will correspond to the data from each channel specified in the channel string input. If you plan to save the 2-D array data in binary, you must exercise caution. **Write File** does

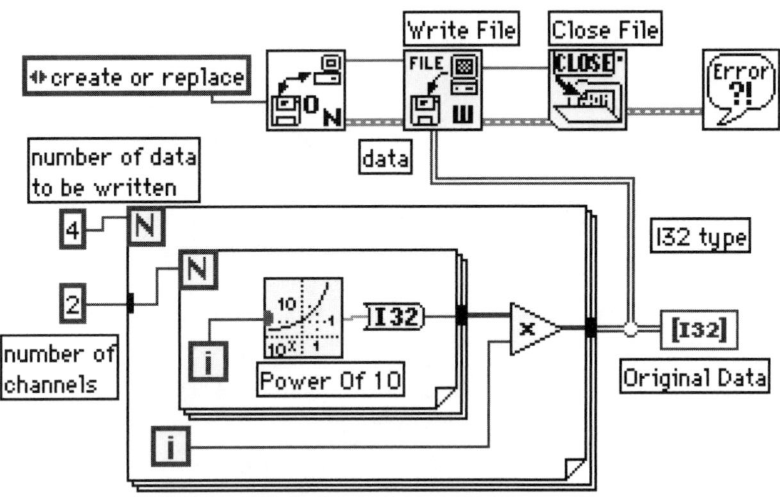

Figure 11.9
Writing 2-D data in binary format.

not care about the dimension of input data; it will treat a 2-D data array as a 1-D array. Therefore, you must do extra work when you *read* the data in order to interpret them correctly. The next section will discuss how to read 2-D binary data.

In Figure 11.9, you are saving a 2-D array of I32 type numbers in binary format, and the 2-D data array is labeled **Original Data** in the figure. This is also shown in Figure 11.10, where the data from the first channel are 0, 1, 2, and 3, and those from the second channel, 0, 10, 20, and 30. The function **Write File** saves the 2-D data array as a 1-D array while interleaving the data from each channel, and **Data saved** shows the resulting 1-D array in Figure 11.10. Therefore, you must be careful when you *read* the data.

11.9 Reading 2-D Binary Data

Since the 2-D data array is saved as a 1-D array as shown in Figure 11.10, an extra step is required to reconstruct the original 2-D data array after reading the binary data file. The reading part is the same as Figure 11.6, which shows how to read 1-D binary data.

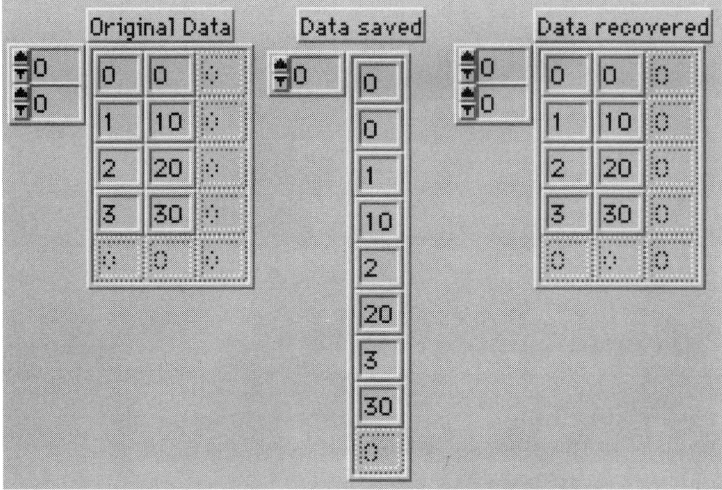

Figure 11.10
View of original data, binary data saved in a file, and the data reconstructed.

Once you bring in the data in 1-D format, you need to reshape the array based on the *a priori* information: the number of input channels. As shown in Figure 11.11, you will use **Array Size** and **Reshape Array** to reshape the 1-D array to a 2-D array. **Data recovered** shows the resulting 2-D array in Figure 11.10. As you can see, the original 2-D data array has been correctly reconstructed from the 1-D array. However, you may include the information about the number of input channels when you create the binary file. Even though this information is called header information, you may save it either at the beginning or at the end of the file.

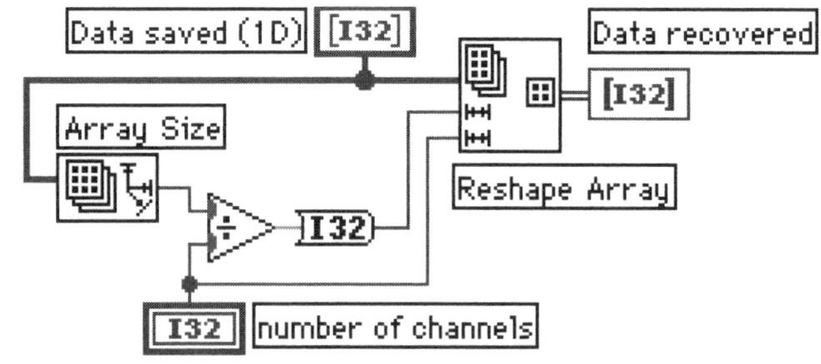

Figure 11.11 Extra steps to read 2-D binary data using Figure 11.6.

11.10 Writing/Reading 2-D Binary Data Continuously

Continuously writing 2-D data in binary format can easily be realized by using the example in Figure 11.9 except for having **Write File** in a **While Loop**. Previously, Figure 11.10 showed that 2-D data arrays were stored as 1-D data arrays. You have also seen how you can continuously read 1-D binary data in Figure 11.8. Therefore, continuously reading 2-D binary data can be done by reading the stored data in 1-D array format continuously using the example in Figure 11.8 while reshaping the 1-D array into a 2-D array using the extra steps in Figure 11.11. Therefore, the detailed VI design will be left to you as an exercise. (See Problem 11.3.)

11.11 Writing Data in ASCII Format

If you prefer to save data in ASCII format so they can be viewed in spreadsheet or word processing software, the example in Figure 11.12 can be used. The view of the resulting file in such third-party software is also shown. If the data are numeric instead of string, you can easily convert the numeric data into string type before feeding the data to **Write File**. Use the functions in **Functions** palette >> **String >> Additional String to Number Functions**. The resulting numeric data in string format can then be wired to the input **data** of **Write File**. The subsequent steps are identical with those illustrated in Figure 11.12.

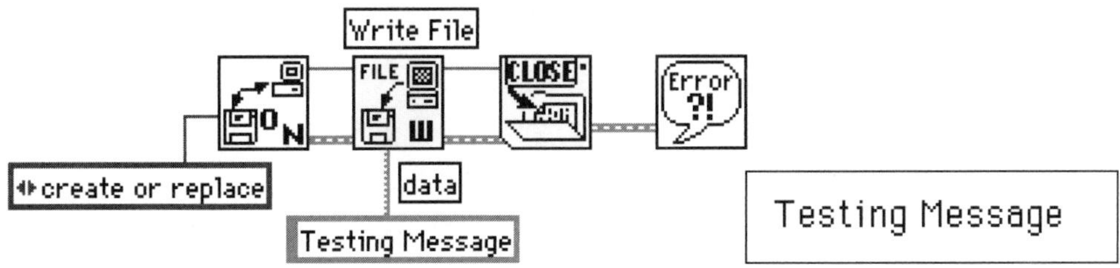

Figure 11.12
Writing data in ASCII format and the view of the file created by this example.

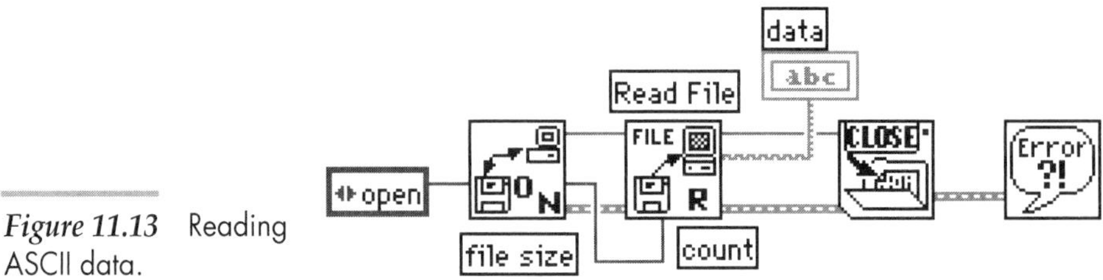

Figure 11.13 Reading ASCII data.

11.12 Reading ASCII Data

Figure 11.13 allows you to read the ASCII data saved by Figure 11.12. Again, if your data are numeric ASCII string type, you can convert them into numeric numbers using the functions in the subpalette **Functions** palette >> **String >> Additional String to Number Functions**. This raises a need for further discussion of string functions, which will be explored in detail in Chapter 12.

11.13 Writing Data in ASCII Format Continuously

In Figure 11.14, you have a **For Loop**, instead of a **While Loop**, to repeat **Write File** for a finite number of times. You have multiple string data elements to be written in an array, and you repeatedly call **Write File** to write them continuously. If the data are numeric type, they can be converted into string format be-

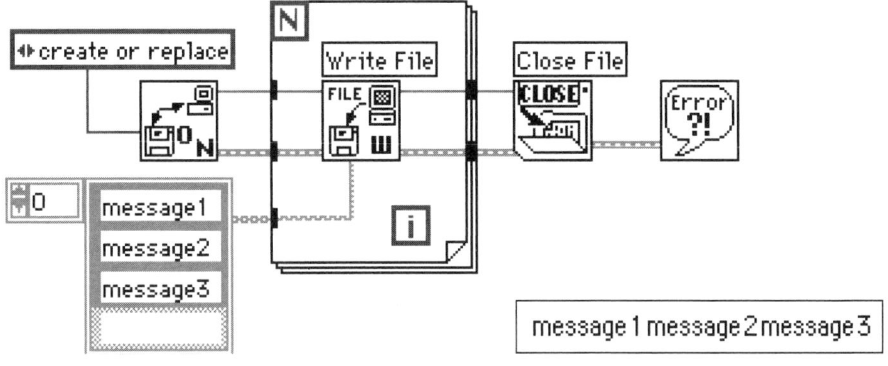

Figure 11.14
Writing data in ASCII format continuously.

fore you wire them to **Write File**. (See Section 11.11.) Figure 11.14 shows the view of the file created by this example, and you can view and understand the data. In actual applications, the 1-D string array should be placed in the loop and replaced by a data source that continuously generates string data. Also, the **For Loop** may be replaced by a **While Loop** to repeat indefinitely.

11.14 Writing Data in Both ASCII and Binary Format

You may want to use this method in order to use ASCII format for the header information in a binary data file. This will allow you to view the data file and read the header information using word processing or spreadsheet software, even though the data portion will be unreadable.

Figure 11.15 shows an example of achieving such a task. You have three different types of inputs in this example: **String input**, **I16 Input**, and **DBL Input**. In other words, you are not only mixing ASCII and binary type, but the type of binary data is also different (I16 and DBL). Note that accurate information about the data structure of the binary file will be the key to the success of correct data recovery. (This is one of the disadvantages of binary files: If you do not have the accurate information about the data structure, data recovery may not be done correctly.)

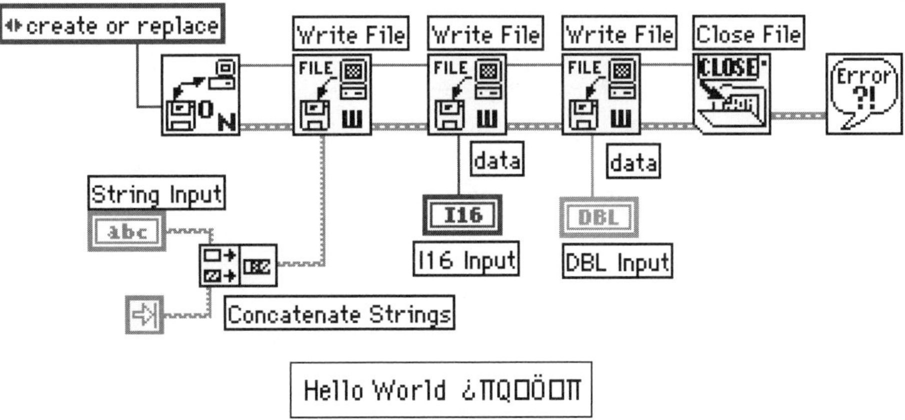

Figure 11.15
Writing data in both ASCII and binary format, and the view of the file created by this example.

The view of the file created by this example is also shown in the figure. The example uses `Hello World` for **String Input** so that it can be viewed in a third-party application as mentioned earlier. Note that the rest of the data are not readable since they are in binary format. The trick used in this example is as follows: Choose an ASCII character (which is assumed to be never used in the header) and append it to the header. Then use it to find the boundary between the header and the binary data. The TAB character is used for that purpose in this example.

11.15 Reading a Mixture of ASCII and Binary Data

Based on the trick used in Figure 11.15, first find the TAB character to extract the ASCII header information by using **Match Pattern** from **Functions** palette >> **String** (Figure 11.16). The output, **offset past match** of **Match Pattern**, returns the offset value where the match occurred plus 1, and it will be used for the input **count** of **Read File**. Actually, the first **Read File** returns both the header information and the TAB character because **offset past match** starts at zero. For example, if you compare *abc* with *b*, a match occurs on the second character *b*, and **offset past match** will return 2, the offset of the letter *c*. (The offset of character *a* is zero.) Therefore, if you read two characters from the beginning with **Read File**, it will return *ab*, instead of *a* only. The letter *b* corresponds to the TAB character in Figure 11.16.

However, the TAB character will show up as a tab space that does not affect the content of the header, so it can be safely ignored. If you prefer to remove the TAB key in the display of the header, it can be deleted using one of the string functions. (String functions are treated in detail in Chapter 12.)

After reading the ASCII header, the second **Read File** reads the I16 type binary data starting at the offset where the first **Read File** stopped reading. The same procedure will read the DBL binary data. Note that the input **count** of the last two **Read File**s is not specified; therefore, the data returned are a single value of the same type as **byte stream type.** (See Table 11.1.) If the **count** were specified, the data returned would be an array. For reference, the size of each data type in LabVIEW is as follows:

I8, **U8**: 1 byte, **I16**, **U16**: 2 bytes, **I32**, **U32**, **SGL**: 4 bytes

DBL, **CSG**: 8 bytes, **CDB**: 16 bytes

EXT: 10 bytes on PCs, 12 bytes on Macs, 16 bytes on Suns

CXT: 20 bytes on PCs, 24 bytes on Macs, 32 bytes on Suns

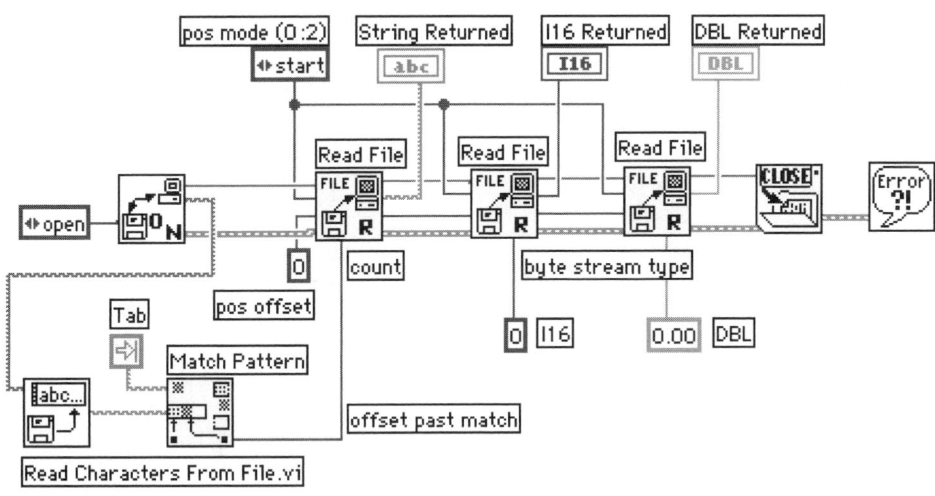

Figure 11.16
Reading a mixture of ASCII and binary data.

11.16 Two Easy Ways to Save Data

LabVIEW provides a couple of quick and easy ways to save your data using a technique called datalogging. This is especially helpful if you have a complicated data structure containing numerous data elements of different types or if you do not want to deal with file I/O details.

Method 1

The VI steps are the same as outlined in the earlier file I/O discussion. But datalogging will allow you to deal with complicated data structure without knowing the details of it. An example is shown in Figure 11.17. Note that use of this method requires that you specify the datalog file type when opening or creating a file.

Recall that you used **Open/Create/Replace File.vi** to open or create files in the previous sections, but it cannot be used with datalogging since it does not provide the input terminal **datalog type**, where you specify the datalog file type. Therefore, you need to use either of the two low-level functions **New File** or **Open File** to create or open datalog files, since they provide the

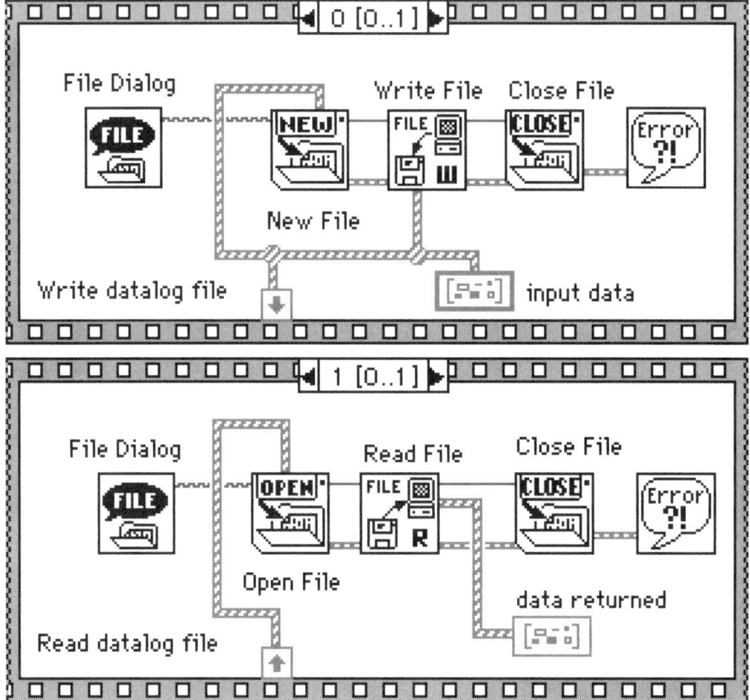

Figure 11.17
Creating and reading a datalog file. Frame 0 creates a datalog file, and frame 1 reads it, and displays the data. In frame 0, a cluster control **input data** is wired to the input **datalog type** of **New File** to inform LabVIEW of the data structure in the input. In frame 1, **input data** from frame 0 is wired to the input **datalog type** of **Open File**. In fact, a template that has the same data structure as the input is good enough for the input **datalog type** of **Open File**. But, for the sake of convenience, the control **input data** is used in this example. The same is true with **New File**.

input **datalog type**. Consequently, as shown in Figure 11.17, **Open/Create/Replace File.vi** will seem to have been replaced by the following two functions: **File Dialog** and **New File**, or **File Dialog** and **Open File**. The example in Figure 11.17 creates a datalog file in the first frame and reads the data in the second frame and displays them. The input data cluster and the data read are shown in Figure 11.18.

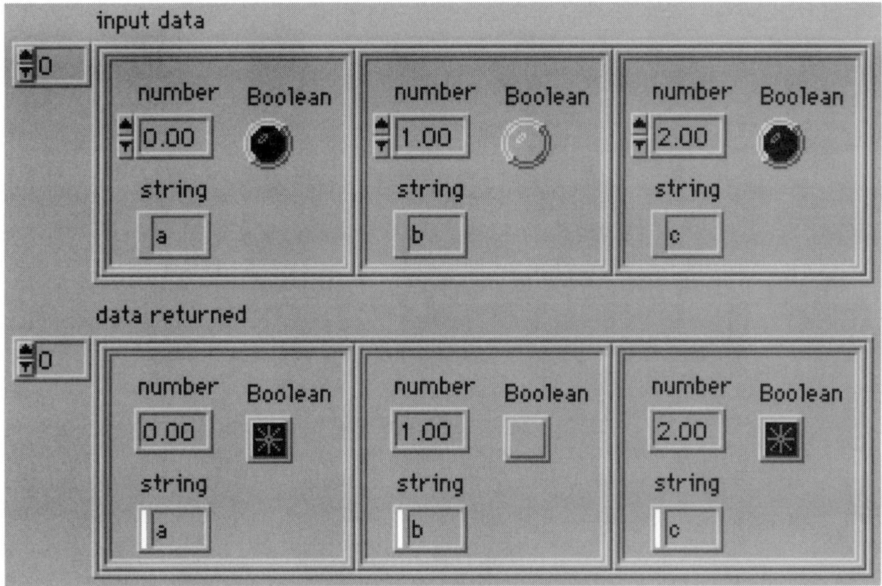

Figure 11.18
Input data and the data read using the example in Figure 11.17.

Method 2

The second method may be quite interesting to you since you do not need to use any file I/O VI or function at all. Suppose that you have a VI that acquires data, processes them, and displays the result. A simple example of such a VI is shown in Figure 11.19. In the example, five random numbers are generated (acquired), their mean is computed, and a Boolean LED indicates if the mean is greater than 0.4.

Now you wish to save all of the data at each execution of the VI without using any file I/O VI or function. Follow these steps to do so:

Step 1: Go to the pull-down menu **Operate** and select **Data Logging >> Log . . . .** This will prompt you to enter a name of the datalog file that is about to be created. Enter a name, and acknowledge the dialog message.

Step 2: Go to the pull-down menu **Operate** again, and select **Log at Completion**. A check mark will appear after you select the option.

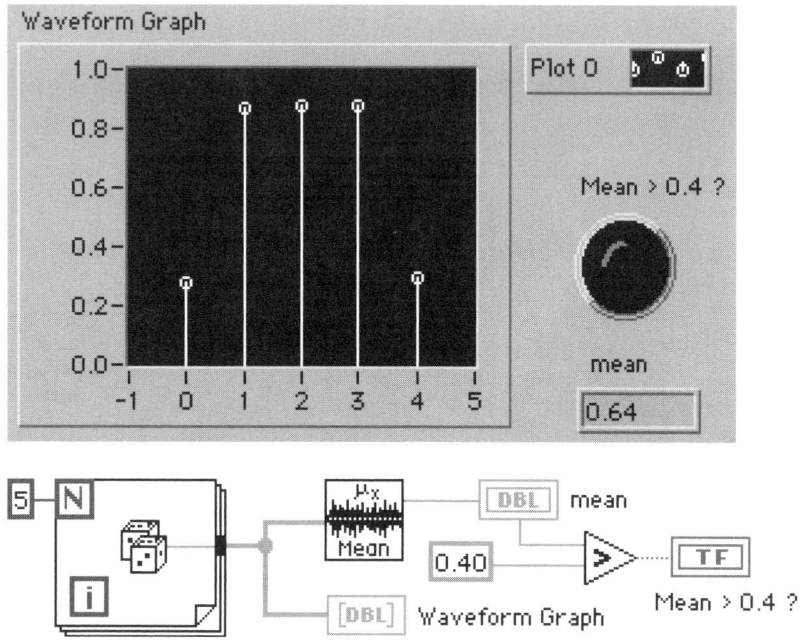

Figure 11.19
An example VI where data are acquired, analyzed, and displayed. All of the data shown in this VI can be saved to a datalog file using method 2 without using any file I/O VI or function.

Step 3: Run the VI several times. Upon the completion of each execution, all of the data acquired or generated in the VI will be saved in the file created in Step 1. Close the VI.

Step 4: To retrieve the data, the datalog file created in Step 1 must be available. If you moved it to a different location, you must link the VI to the datalog file again. To do this, reopen the VI.

Step 5: Go to the pull-down menu **Operate**, and select **Data Logging >> Change Log File Binding . . . .** This will prompt you to locate the datalog file. Select the file that was created in Step 1.

Step 6: Select **Operate >> Data Logging >> Retrieve . . . .** This will change the front panel of the VI, as shown in Figure 11.20. The range of screen shot is shown as [0..9], indicating that there are 10 sets of data available. Different screen shots can be selected by clicking the up/down arrows. To exit this mode, click **OK**.

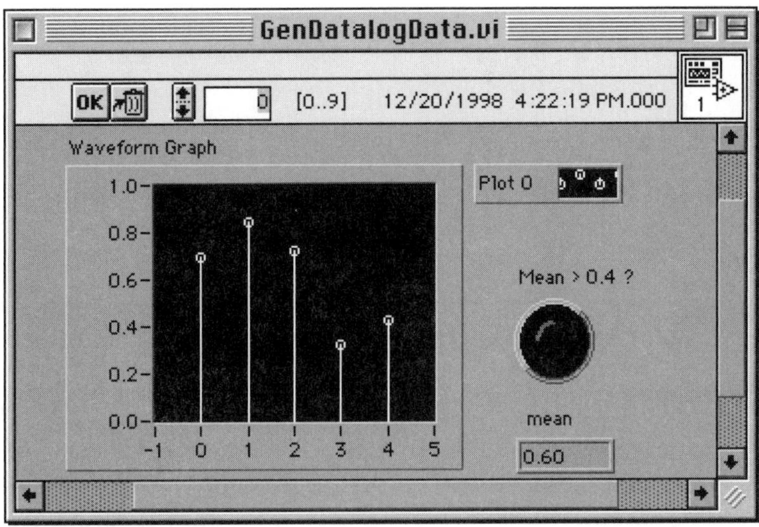

Figure 11.20
Front panel of the VI in Figure 11.19 in datalog data retrieval mode.

Those six steps are simple enough to follow, but if you want programmatically to retrieve the data from this VI you will need to follow these steps:

Step 1: Go to the diagram window of the VI where you want to import the data from a datalog file programmatically. Bring up the **Functions** palette, and choose **Select a VI . . . .** Then select the VI that was used to acquire and analyze the data, and save them in a datalog file. This will place an icon of the VI in the diagram window.

Step 2: Right click on the VI that was just placed in the diagram window and select **Enable Database Access**. This will create a halo around the VI, as shown in Figure 11.21.

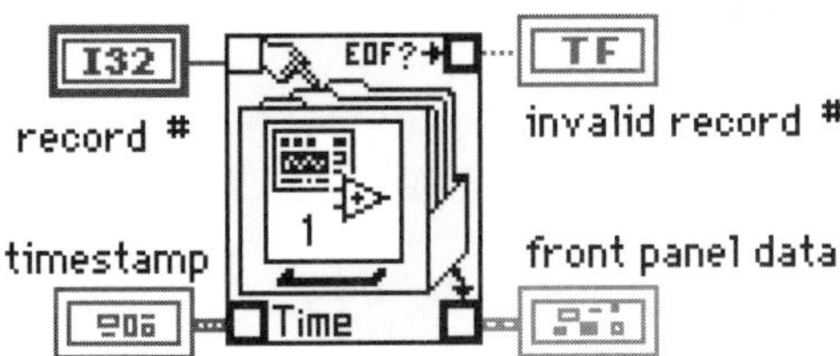

Figure 11.21 A halo encompasses the VI that was used for data acquisition when you select **Enable Database Access** from its pop-up menu. There are three indicators and one control to retrieve the data from the VI.

Step 3: Create a control to the terminal **record #**, and three indicators to **timestamp**, **Invalid record #**, and **front panel data**. If the VI surrounded by the halo has lost the link to its datalog file, there will be no data to be retrieved. Therefore, you need to go back to the VI that was used to acquire and analyze data, link it to the datalog file you want to retrieve, and save the VI. Then the data will be available using the method shown in Figure 11.21.

Note that the data type of **time stamp** and **front panel data** is cluster. Therefore, you will need to use **Unbundle By Name** or **Unbundle** to extract any specific type of data. The advantage of method 1 and method 2 is their convenience. The disadvantage is more overhead, especially in method 2. In other words, if you have a complicated data structure with an enormous amount of data, it can slow down your VI significantly. Therefore, either method should be used with care.

PROBLEMS

11.1 Use the high-level file I/O VI **Write Characters To File.vi** to complete the following problems.

(a) Create a VI **P11_01a.vi**, and place **Write Character To File.vi** in the diagram. Create two controls respectively to the two inputs of **Write Characters To File.vi**: **file path (dialog if empty)** and **character string**. Enter some characters in **character string**, and run the VI. Explain what **P11_01a.vi** does if you specify the file path and if not. In which case do you get a prompt to specify the name of the file to be written?

(b) Save **P11_01a.vi** as **P11_01b.vi** first, and go to the diagram window of **P11_01b.vi**, where you have **Write Characters To File.vi** is located. Double click on it to open its front panel, and save it as **WCTF_11_01b.vi** in your own directory, not in the LabVIEW directory. (See Figure P11.1.) Go to the diagram window of **WCTF_11_01b.vi** using the shortcut **Ctrl-E**. You will see that a series of sub VIs are wired. Among them, double click on **Open/Create/Replace File.vi** to open its front panel. When its front panel is open, save it as **OCRF_11_01b.vi** in your own directory again, not in the LabVIEW directory. This technique must be exercised every time you modify any LabVIEW VIs or examples: saving it under a different name first.

Run **P11_01b.vi** with the input **file path (dialog if empty)** left blank. Describe what happens. Now, run it again after manually entering the

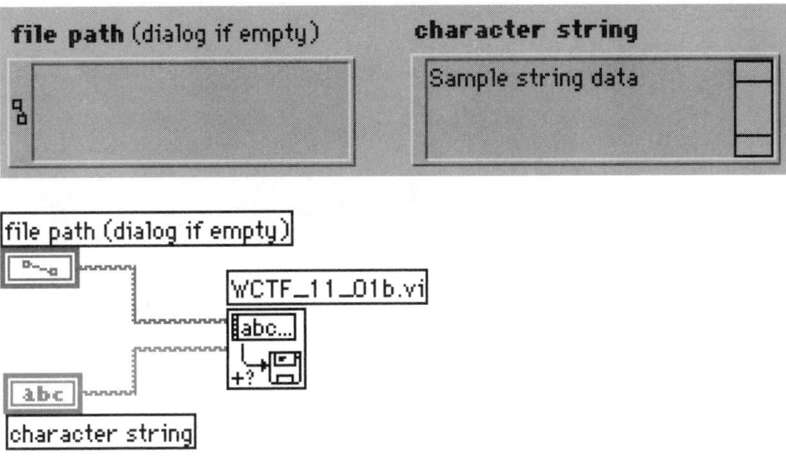

Figure P11.1
Front panel and diagram window of **P11_01b.vi**.

file name that was created in (a) in **file path (dialog if empty)**. In other words, run **P11_01b.vi** to write to the *same* file that you just created in (a). Describe what happens then.

(c) Now, modify **OCRF_11_01b.vi** to make **P11_01b.vi** perform the following task: If you do not specify the input **file path (dialog if empty)**, then **P11_01b.vi** prompts you to enter the file name to be written. If the file path is specified, especially with a pre-existing file name, in **file path (dialog if empty)**, **P11_01b.vi** does not prompt you for the file path input unlike the case in (b) above. Note that such a functionality is very useful in the following situation: You create a new file and write your data to it first. Beyond that point, you can keep saving the latest data to the same file repeatedly, without getting any prompt message. This feature is similar to the functionality of **SAVE** in commercial applications. Save the modified **P11_01b.vi** as **P11_01c.vi**.

11.2 Create a VI **P11_02.vi** that writes a 1-D array of three random numbers every 2 seconds to a file **bin1Ddata.dat** continuously in binary format, until a button **QUIT** is pressed. When the continuous write is complete, **P11_02.vi** performs continuous reading such that a three-element 1-D random binary data array is being read at a time. After the data read is complete, **P11_02.vi** generates another 1-D array in which each element is the average of three-element 1-D random data array. Assume that the random numbers are DBL type.

11.3 Section 11.10 briefly discusses how to write and read 2-D binary data continuously. This problem will design such a VI **P11_03.vi** that writes a 2-D

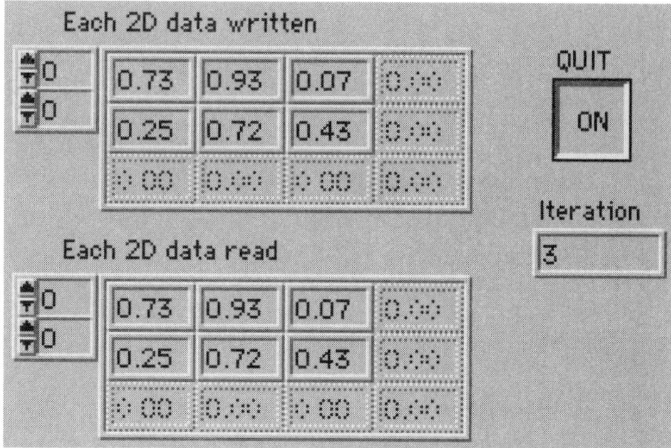

Figure P11.3

random number array at every 1 second continuously to a file **bin2Ddata.dat** while displaying the current 2-D data array. The data write stops when you press the button **QUIT**, and the data read will immediately follow from the same file while displaying each 2-D array for 1 second. The front panel of a possible solution is shown in Figure P11.3. (You are acquiring two data samples from three input channels at every 1 second.)

11.4 (a) What is the advantage and the disadvantage of ASCII and binary files?

(b) Build the example in Figure 11.15, and save it as **P11_04b.vi**. After saving the file, view it using a text editor. Explain your observation.

(c) Build a VI **P11_04c.vi**, which reads the file you created in (b) using the example in Figure 11.16. Did you get the original data back? Explain your observation.

12 String Manipulation

In the **Functions** palette, the subpalette **String** contains functions for string-related tasks. String manipulation is the core part of file input and output applications and is also very important in GPIB, VISA, and serial communication. Despite the important role of those functions, their functionality is quite simple and self-explanatory. For example, the function **String Length** returns the length of input string, **Concatenate Strings** combines its input strings and generates a concatenated version of them, **String Subset** returns a portion of input string, and so forth. Therefore, not all of the functions will be discussed except those that need extra attention in their usage.

First, string functions are colored yellow, implying that they are at the lowest level. Therefore, any further modification is not allowed. Second, the most important task of string functions may be the data conversion between numeric data and ASCII data. The data conversion must be well understood, and you should get to feel comfortable with it, especially between numeric and ASCII format. The subpalette **Conversion** contains the functions for various types of conversion, such as file path to strings, strings to file path, etc. The other subpalette, **Additional String to Number Functions**, contains the functions for the conversion between ASCII data and numeric data. Figure 12.1 shows the subpalette **Additional String to Number Functions**.

All of the functions in the top row except **Format & Append** can be used to convert numeric type data to string type data. The first three convert integers to strings, and the other three convert floating numbers to strings. All of the functions in the bottom row except **Format & Strip** are for the opposite task: converting string type data to numeric type data. The first three in the bottom row convert strings to integers, and the fourth one converts strings

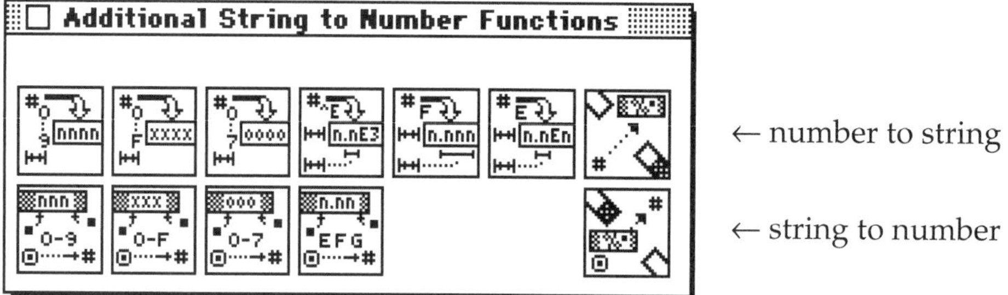

Figure 12.1
The subpalette **Additional String to Number Functions** of the subpalette **String**. This contains all of the functions for data conversion between numeric type and string type.

to floating numbers. The last functions in each row, **Format & Append** and **Format & Strip**, are discussed separately in the next section.

12.1 Format & Append and Format & Strip

Format & Append and **Format & Strip** are a pair for a similar task in the opposite way. **Format & Append** converts the input number to a string and appends it to the input string based on a user-specified format and has three inputs: **string ("")**, **number (0)**, and **format string**. On the other hand, **Format & Strip** identifies a number in string format at the *beginning* of the input string and converts it to a number based on the format that you specify. For example, if the input string is `3.5high`, **Format & Strip** can extract the numeric string `3.5` and converts it to a number 3.5, not string `3.5`. If, however, the input string is `a3.5high` with an extra character `a`, **Format & Strip** cannot extract the numeric string `3.5` without preprocessing the input string since the numeric string `3.5` is not at the beginning of the input string. **Format & Strip** has three inputs and two outputs: **string**, **format string**, and **default (0 dbl)** are the inputs, and **number** and **output string** are the outputs.

Figure 12.2 shows an example of the usage of the input **format string** of **Format & Append** in the following format:

```
%08.3f
```

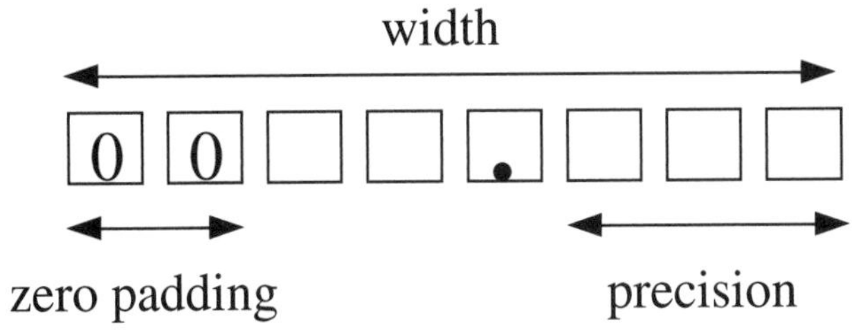

Figure 12.2
Number representation %08.3f in the input **format string** of **Format & Append**. Note that precision can only be specified when formatting data *into* strings; therefore, precision parameter is only meaningful for the functions **Format & Append** and **Format Into String**. The function **Format Into String** is discussed in Section 12.2.

The format denotes that you will use eight digits to represent the input number in string format where the last three digits are for precision and the floating point goes to the fourth digit from the right. Note that you have a zero (`0`) before the `8` in this example, and that will return a string (numeric string) with zeros inserted before the actual number. Figure 12.2 shows two zeros, assuming that the actual number will appear in the two digits above the floating point. For example, if you use `%08.3f` to display 1.4, there will be three zeros before 1 so that the output string will be 0001.400 with two zeros at the end. Different numeric string outputs for different input formats are shown in Table 12.1. The character `f` stands for floating-point number representation or fractional number representation. Other options, such as `d` (decimal), `o` (octal), `e` (engineering), or `x` (hexadecimal), are also available. The dash (`-`) controls the justification of the numeric string output, and its appearance in the input **format string** will place the numeric string output portion immediately after the string output.

In Table 12.1, note that the dash (`-`) and the zero (`0`) have no effect on the engineering representation (case 7 and 8). Also, if the decimal format (`d`) is chosen, the number in the **output string** becomes rounded up (case 4, 5, and 6). In case 1, you are using 7 digits to represent 1.75 with the last 3 digits for precision; however, due to the dash (-) sign, 1.750 follows immediately

Table 12.1 *Comparison of* ***output string*** *of* ***Format & Append*** *for different settings of* ***format string****. The symbol • indicates a white space.*

	string	**number**	**format string**	**output string**
case 1	`test`	`1.75`	`%-7.3f`	test1.750••
case 2	`test`	`1.75`	`%07.3f`	test001.750
case 3	`test`	`1.75`	`%7.3f`	test••1.750
case 4	`test`	`1.75`	`%-7.3d`	test2••••••
case 5	`test`	`1.75`	`%07.3d`	test0000002
case 6	`test`	`1.75`	`%7.3d`	test••••••2
case 7	`test`	`1.75`	`%-7.3e`	test1.750E+0
case 8	`test`	`1.75`	`%07.3e`	test1.750E+0
case 9	`test`	`1.75`	`%7.3e`	test1.750E+0

*Table 12.2 Effect of the input **default** on the output **number** of **Format & Strip**.*

	string	default	format string	number	output string
case 1	`1.75test`	unwired	`%f`	1.75	test
case 2	`1.75test`	unwired	`%d`	1	.75test
case 3	`1.75test`	0 (type I16)	`%f`	2 (type I16)	test
case 4	`1.75test`	0 (type I16)	`%d`	1 (type I16)	.75test

after the string input `test`. The number 1.750 uses 5 digits (1 digit for floating point (.)); therefore, two white spaces were put at the end.

The format options available for the input **format string** of **Format & Strip** are `%d` (decimal), `%o` (octal), `%x` (hexadecimal), `%e` (engineering), and `%f` (floating point or fractional). The effect of different input **default** is shown in Table 12.2.

It is interesting to note that if you have the input **default** wired, the type of the output **number** is controlled by the type of the input **default**. In Table 12.2, case 3 shows that a numeric constant zero of type I16 is wired to **default**, and `%f` is wired to **format string**. Then the output **number** returns a number 2 of type I16 after rounding up the number 1.75, and the remaining string becomes "test". However, if `%d` is wired to **format string**, it returns 1 and the remaining string will be ".75test", including the floating point (.).

12.2 Format Into String and Scan From String

Format Into String is similar to **Format & Append** except that **Format Into String** can have multiple inputs. Another important difference is that **Format Into String** provides you with an input dialog box for the input **Format String**. The dialog box can be brought up by double clicking on **Format Into String** or selecting **Edit Format String** from the pop-up menu of **Format Into String**. Examples are shown in Table 12.3, where there are three inputs. (LabVIEW labels them **Input 1**, **Input 2**, and **Input 3**.)

The first two inputs are DBL and I16 numeric values and the last input is a string. The table shows a different combination of input formats for the input **Format String** and its corresponding output strings. Observe the difference

Table 12.3 Examples of different outputs of ***Format Into String****. The symbol • indicates a white space.*

	Input String	three inputs	Format String	Output String
case 1	A	1.23 4 !	%f%d%s	A1.2300004!
case 2	A	same as case 1	•%f•%d•%s	A•1.230000•4•!
case 3	A	same as case 1	•%6.3f•%2d•%2s	A••1.230••4••!
case 4	A	same as case 1	•%6.3f•%2d•B•%2s	A••1.230••4•B••!

in specifying the number of digits and its effect on the output string. You can also include strings in the input **Format String**, and case 4 shows an example with a single character `B`. The letter `B` is inserted in the input **Format String**, and **Output String** consequently displays it. As shown in cases 1 and 2, eight digits are used in the representation of `%f`, and the method in Figure 12.2 can be used to customize the number of digits.

The counterpart of **Format Into String** is **Scan From String**, which is similar to **Format & Strip**, with multiple output capability. Like the case of **Format & Strip**, **Scan From String** has the input **Default Value** for each output. Examples of **Scan From String** for different input format settings are shown in Table 12.4.

Table 12.4 illustrates many interesting aspects. Case 4 shows the effect of the **Default Value**, and only the second one (`0` of type I16) is wired for the second output. Note the effect of the number of digits on the output (`%s%1f` vs. `%s%f` in case 3). Also, each data element in the **Input String** must be separated by a white space; otherwise, each data element is not correctly recovered. (The outputs in cases 1 and 2 are incorrect.) In case 4, there are two effects to be noticed. First, the data type of the second output is determined by **Default Value**. Therefore, the second output is I16 since zero (0) in **Default Value** is I16. Second, the number of digits has been specified (`1`); therefore, the second output returns only one digit of I16 type (1 in `12.23`). In case 5, the second output returns two digits of I16 type (12 in `12.23`). The remaining portion of the second data is returned in **Remaining String**. Case 7 shows the effect of the wrong number of digits for the first string data element `abc`. Case 8 shows a correct usage of **Scan From String** with three

Table 12.4 Examples of different outputs of ***Scan From String****. The symbol • indicates a white space. The two lines in* ***Input String*** *and* ***Remaining String*** *are continuous without any space between.*

	Input String	**format string**	**Default Value**	outputs	**Remaining String**
case 1	`abc1.23`	`%s%f`	all unwired	abc1.23 0.00	
case 2	`abc,1.23`	`%s%f`	all unwired	abc,1.23 0.00	
case 3	`abc•1.23`	`%s%f`	all unwired	abc 1.23	
case 4	`abc•12.23`	`%s%1f`	unwired `0` (Type I16)	abc 1	2.23
case 5	`abc•12.23`	`%s%2f`	unwired `0` (Type I16)	abc 12	.23
case 6	`abc,•1.23`	`%s%f`	all unwired	abc, 1.23	
case 7	`abc•1.23`	`%2s%f`	all unwired	ab 0.00	c•1.23
case 8	`abc•3.72+` `4.56i•12345.67`	`%s%f%f`	all unwired	abc 3.72+4.56i 12345.67	
case 9	`abc•3.72+` `4.56i•12345.67`	`%s%f%f`	all unwired	abc 3.72 4.56	i•12345.67
case 10	`abc•3.72+` `4.56i•12345.67`	`%2s%f%f`	all unwired	ab 0.00+0.00i 0.00	c•3.72+4.56i• 12345.67
case 11	`abc•3.72+` `4.56i•12345.67`	`%s%1f%f`	all unwired	abc 3.00+0.00i 0.72	+4.56i•12345.67
case 12	`abc•3.72+` `4.56i•12345.67`	`%s%f%2f`	all unwired	abc 3.72+4.56i 12.00	345.67

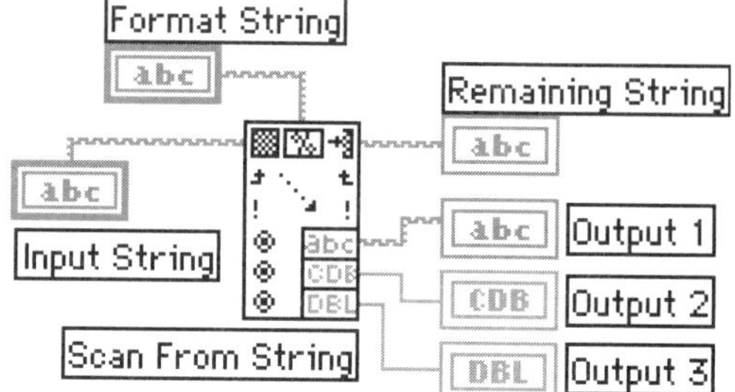

Figure 12.3 The number of output terminals and their data type must correctly be set when using **Scan From String**.

different data elements: string, complex number, and floating number. Note that the output indicator type has been accordingly set to string, CDB (complex double), and DBL (Figure 12.3). Case 9 shows the effect of the wrong data type for the output. In this case, the data type of the second output is DBL. Then the second complex data element, `3.72+4.56i`, is understood as a floating number; therefore, the second output returns only the real part `3.72` and the imaginary part is returned in the third output. The remaining data are returned in the **Remaining String**. Cases 10, 11, and 12 all show the effect of the wrong number of digits in the **format string**. Note that unlike **Format Into String**, as shown in Table 12.3, any white space in the **format string** has no effect on the output of **Scan From String**.

When you create indicators for the output of **Scan From String**, the number of output terminals and their data type must be set correctly. In Figure 12.3, it is assumed that the **Input String** contains three data elements, such as case 8 through case 12 in Table 12.4. Accordingly, you must have three outputs (**Output 1**, **Output 2**, and **Output 3**), and their types are properly set to string, complex double, and double precision, respectively.

12.3 Conversion between Array and Spreadsheet String

If you have an array of numeric data and want to convert it to a numeric string, or vice versa, you can use either **Array To Spreadsheet String** or **Spreadsheet String To Array**. As was noted in Figure 12.2, *precision* in the input format is meaningful only when formatting numeric data into strings using **Format & Append** or **Format Into String**. (See Table 12.1 or Table 12.3.) On the other hand, when converting numeric string data to their original numeric format, only the *data type* such as `f`, `d`, `o`, `e`, `x`, and the number of digits to be read will be meaningful. (See Table 12.4.)

Precision in Figure 12.2 can be controlled only when converting numeric data into strings. Therefore, ***Format & Append****,* ***Format Into String****, and* ***Array To Spreadsheet String*** *can specify the precision. However, the counterparts of those functions can only control the number of digits to be read and cannot control the precision. These functions are* ***Format & Strip****,* ***Scan From String****, and* ***Spreadsheet String To Array****.*

The function **Array To Spreadsheet String** takes an array of numeric data of *any* dimension—try three-dimensional numeric input array—and generates a string that contains the numeric data using the format specified in **format string**. The counterpart of **Array To Spreadsheet String** is **Spreadsheet String To Array,** and it takes a numeric string in an array *format* (not an actual array) as input and converts it back to its original numeric array using the type specified in **format string**. Note that **Spreadsheet String To Array** can take both numeric string data and character string data in an array *format*. Examples of each case are shown in Figures 12.4 and 12.5, respectively.

In Figure 12.4, note that you have not specified the input **array type** since its default is 2-D DBL. As was mentioned earlier, you cannot specify the precision in the input **format string** since you are not converting numeric data into strings but can only control the number of digits to be read. Therefore, inputs such as `%f` or `%3f` will work whereas inputs such as `%4.1f` will not.

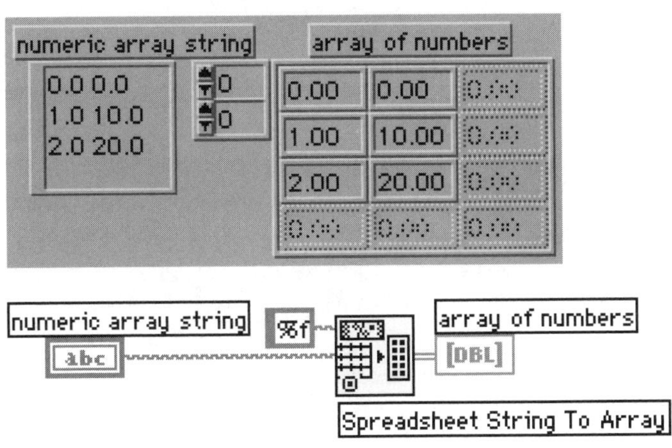

Figure 12.4
Example of converting a string input of numeric 2-D array data using **Spreadsheet String To Array**.

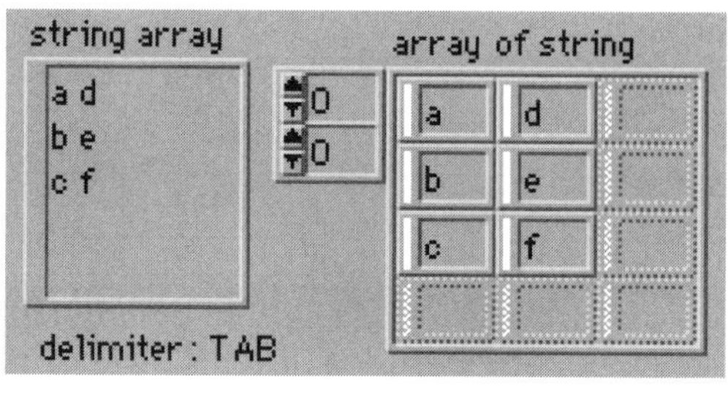

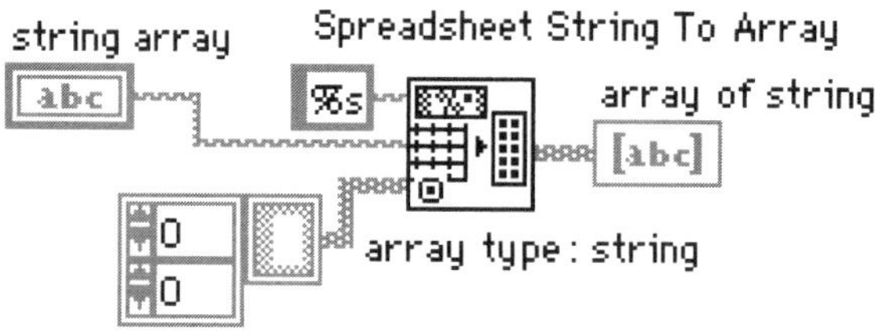

Figure 12.5
Example of converting a string input of string 2-D array data using **Spreadsheet String To Array**.

Figure 12.5 is an example of an input **string array** containing character string data in an array format. To convert it into an array of character string, you need to specify the type of array using the input **array type**. Note that in Figure 12.5, you have a 2-D array of string type wired to **array type**. As a result, the output is a 2-D array of string data.

12.4 Conversion of Binary String to Numeric Data

Chapter 11 discussed reading and writing binary and ASCII files and how you can have files with binary data saved in ASCII format. Examples of how to generate such files are illustrated again in Figures 12.6–12.8.

The example in Figure 12.6 saves a 1-D array of data in binary format. If you read the file in Figure 12.6 using the method in Figure 12.8, you will end up with an ASCII file containing binary data. Now, how would you convert the binary data back to numbers that you can understand without using the method in Figure 12.7? If the binary data are type U8 (unsigned 8 bit), you can use **String To Byte Array** found in **Functions** palette >> **String >> Conversion**. That function converts the input string into an array of unsigned

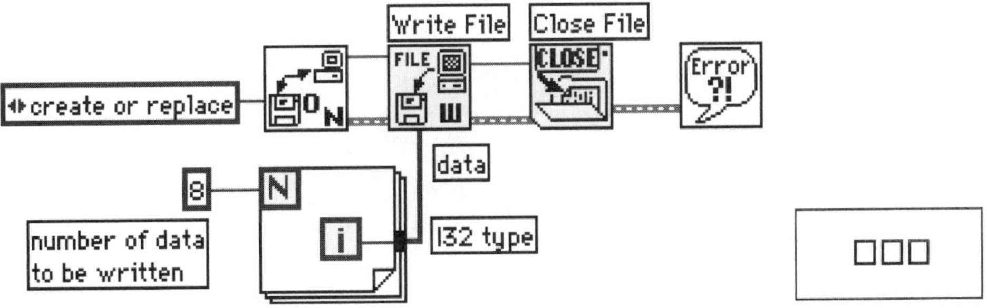

Figure 12.6
Writing 1-D data in binary format, and the view of the file created by this example.

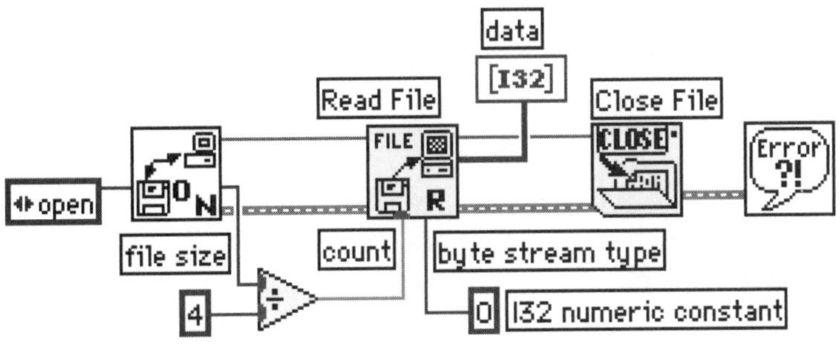

Figure 12.7
Reading 1-D binary data.

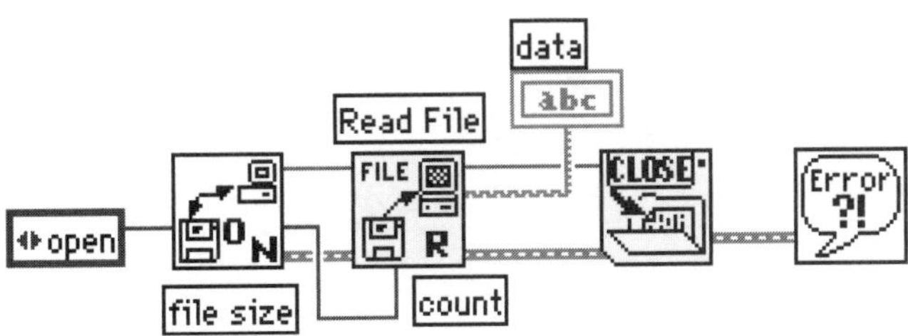

Figure 12.8
Reading ASCII data.

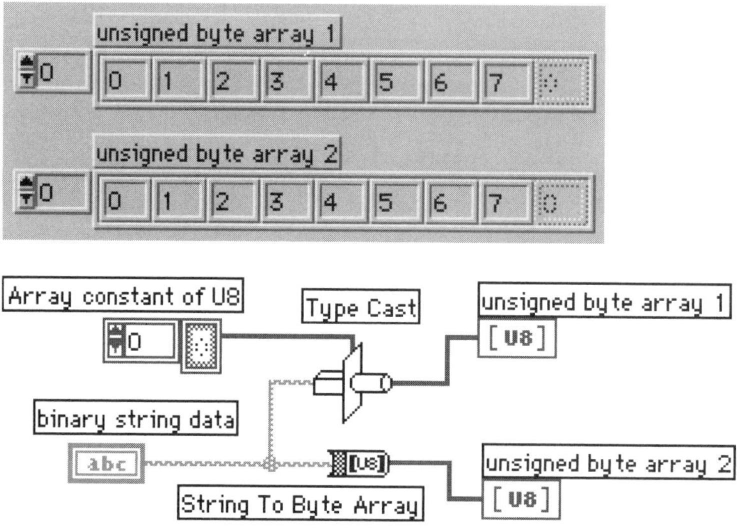

Figure 12.9
Two functions to recover U8 binary string data: **String To Byte Array** and **Type Case**.

bytes. Therefore, if you have a U8 binary file created by the method in Figure 12.6 and have read the file using the methods in Figure 12.8, feed the read data to **String To Byte Array** to recover the original data. There is also an alternative method of converting the binary string data, and it involves using **Type Cast** to convert the binary string data to their original format. It is important to note that this alternative method is not limited to U8 case but can be applied to any type of data conversion except for *arrays of strings* or *2-D arrays of numeric data*.

Figure 12.9 shows an example of using **String To Byte Array** and **Type Cast**. (**Type Cast** can be found in **Functions** palette >> **Advanced >> Data Manipulation**.) The binary data contained in the string control **binary string data** are 0, 1, 2, 3, 4, 5, 6, and 7 of U8 type. The indicators **unsigned byte array 1** and **unsigned byte array 2** in Figure 12.9 show that both methods

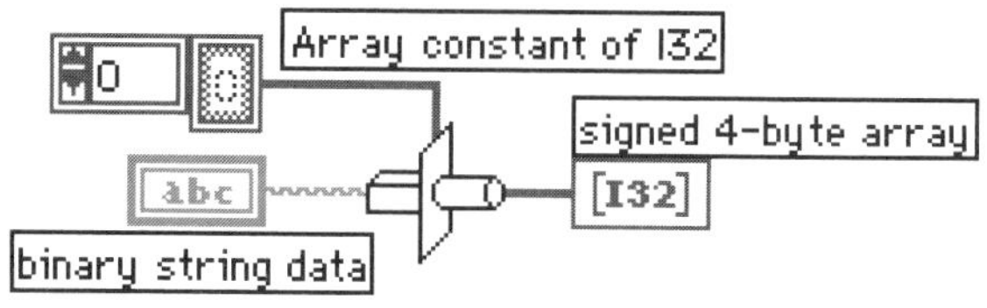

Figure 12.10 Conversion of binary string data of type I32 using **Type Cast**.

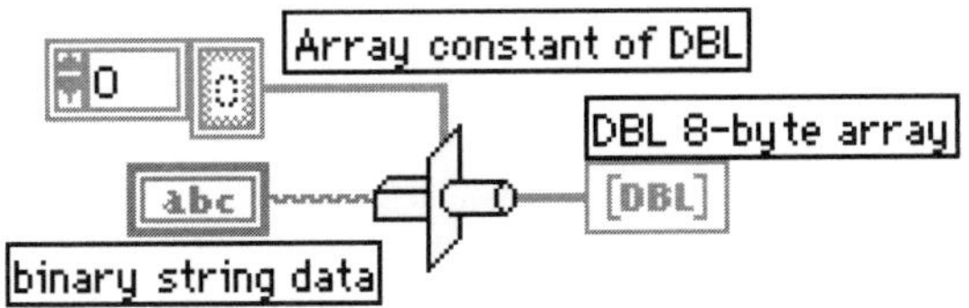

Figure 12.11 Conversion of binary string data of type DBL using **Type Cast**.

recovered the data correctly. Since **Type Cast** can apply to *any* type of data conversion except for arrays of strings or 2-D arrays of numeric data, here are two more examples where **binary string data** contains binary data of type I32 and DBL. Figures 12.10 and 12.11 show examples of converting binary string data of I32 and DBL using **Type Cast**. These techniques are useful when you want to convert data returned by GPIB or serial instruments since they often send data as binary strings.

Type Cast *can take any type of binary data or 1-D arrays of binary data as its input. However, it cannot take **arrays of strings** or **2-D arrays of numeric data** as its input.*

PROBLEMS

12.1 Write a VI **P12_01.vi** that performs the following tasks:
The user can enter a password, and **P12_01.vi** compares it with a password dictionary, which currently has five entries. If there is a match, a message "**Access Granted.**" is displayed on the screen. Otherwise, "**Access Denied.**" is displayed.

12.2 Suppose that you want to save the following data in ASCII format to a file:

0	sweep0	0
1	sweep1	10
2	sweep2	20
3	sweep3	30
4	sweep4	40

Each column is separated by a tab character. Design a VI **P12_02.vi** that can save such a file. Hint: You may use **Format Into String** and the method described in Chapter 11 to write data in ASCII format continuously.

12.3 Suppose that you have the following string data:

```
Pump 1 ON
Pump 2 ON
Pump 3 ON
```

Each line is assumed to be separated by carriage return. Now, given the preceding string data, you want to create the following new string data:

```
Current Status: Pump 2 ON
```

In other words, you want to pick the second line of the given string data and append it to **Current Status**. Use **Pick Line & Append** to accomplish this task, and save the VI as **P12_03.vi**.

12.4 Save **P12_03.vi** as **P12_04.vi**. Then repeat Problem 12.3 using **Index & Append**. In other words, the preceding string data consisting of three lines is now given as a 1-D string array.

12.5 Design a VI that acquires a random number between 0 and 1 every 1 second. If the value is above 0.7, you want to display **Current Status: TOO HIGH**. Otherwise, **Current Status: normal** is displayed. The VI will continue to operate until a Boolean button **QUIT** is pressed. Write the VI using **Select & Append**, and save it as **P12_05.vi**. In this problem, you are *not* allowed to use the **Case** structure!

12.6 Suppose that you were given the following string data consisting of the protocol and the data from your scanner:

```
IMGDAT:     1001001000111100
```

Assume that you had the list of command protocols that your scanner transmitted as follows:

```
STAT:
ONOFF:
IMGDAT:
```

Now, write a VI that compares the protocol from the scanner with your list and returns only the data portion without the protocol and the delimiter using **Index & Strip**; save it as **P12_06.vi**. Assume that the protocol and the data from the scanner are separated by a TAB character, and the VI just needs to return the binary data in string format.

12.7 Repeat Problem 12.6 assuming that there are only two protocols from the scanner: IMGDAT and ONOFF. Also, use **Select & Strip** instead of **Index & Strip** to extract the binary string data without the TAB delimiter. Save the VI as **P12_07.vi**.

12.8 Design a stop watch that displays the elapsed time in minutes and seconds. Display 1/10 of second, too. Use **Tick Count (ms)**, which can be found in **Functions >> Time & Dialog**. Save the VI as **P12_08.vi**. The front panel of a sample solution is shown in Figure P12.8.

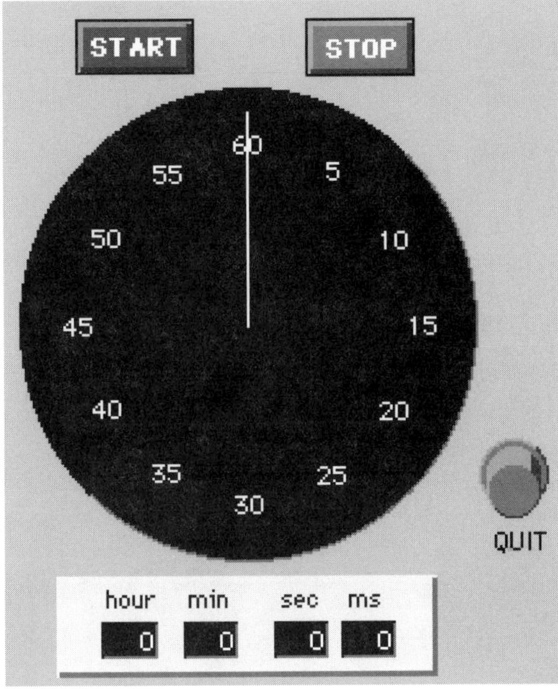

Figure P12.8

Instrument Control

13

This chapter will discuss how you can use LabVIEW to control instruments using different protocols, such as GPIB, VISA, or serial commands. It will first present a brief background about each one and follow up with example VIs and tips on troubleshooting communication problems with your instruments. The principle of instrument control in LabVIEW is much simpler due to the following rationale: Since the types of instruments vary from vendor to vendor, there should be a universal way to control them. Therefore, the simplification of procedure naturally comes as a solution, and the principle becomes simpler. This chapter will begin with GPIB communication.

13.1 GPIB (IEEE-488)

GPIB stands for general purpose interface bus and was introduced by Hewlett Packard during the 1960s and 1970s. Shortly after its appearance, it became IEEE (institute of electronics and electrical engineers) standard protocol. Due to this standardization, IEEE-488 and GPIB are often understood to be one entity, but they actually are not. The idea behind GPIB was simultaneously to control multiple instruments using a computer. In order to accomplish such a task, a new type of GPIB was designed. GPIB uses addresses between 0 and 30 to communicate with different instruments, including the bus of the computer. GPIB boards in PCs usually use address 0, and external instruments use one address between 1 and 30. Note that GPIB uses bytes as the length of each data segment to communicate with instruments and employs special characters to signify the end of each data transmission. The end of data transmission can also be detected by counting the number of bytes of data received in lieu of looking for any special character in the string data stream.

Since each data pattern is 1-byte long, ASCII characters are most often used as the data format. This implies that all of the string functions will play an important role in instrument control since all of the commands and data will be ASCII characters. Note that ASCII data will need to be converted to numeric data before they can be processed.

Some instruments return data as binary strings, but such a case can easily be handled by converting binary strings to their original numeric format, using the methods presented in Chapters 11 and 12. Also, some instruments may use more than 1 byte as the length of each data segment and then transmit the lower byte first. This means that the arrival order of the data is important and the received data must be properly reordered before being processed. This can be done using the functions in **Functions** palette >> **Advanced >> Data Manipulation**.

The method for controlling instruments using LabVIEW can be summarized in five steps: (1) Install the driver software that will connect LabVIEW with the GPIB board that is to be installed on a computer. (2) Shut down the computer and install the GPIB board. (3) Configure the hardware using the driver software to make sure that the board is recognized by the operating system. (4) Connect instruments to the GPIB board through an appropriate cable. (5) Check the link between the GPIB board and the instruments by sending a simple command to them in LabVIEW. Once all of the steps go through without any error, the system is ready to initiate the communication with the instruments through proper string commands. Refer to the manuals of the instruments for the complete set of available commands.

As for the installation of the GPIB board and driver software, it is recommended that the board be installed *after* the driver software is installed regardless of the platforms. A proper installation procedure can eliminate most of the simple problems in GPIB communication. Sometimes you may need to check with the basic input/output system (BIOS) or the Extensions depending on your platform. You need to make sure that the GPIB board supports the settings in the BIOS and that the version of the BIOS is up to date. (The latest version of BIOS is usually available from the vendor of your PC.) On Macs, make sure that the driver Extensions does not conflict with preexisting ones. If you have a conflict, first find out which is not compatible with the driver Extensions and remove, update, or change the order of the Extensions loading sequence. For more details about Extensions conflict on Macs, refer to the manual of your Macintosh.

13.2 RS-232, RS-449, RS-422, and RS-423

Serial devices transmit and receive data bits in a sequential order; therefore, they have a much lower data transmission rate in comparison to parallel communications. Yet serial communication is still widely used due to its simplicity and ready-to-run nature. For instance, almost all computers on the market have at least one serial port already available so that you do not have to purchase any extra hardware for serial communication. The application of serial communication is also broad, ranging from modem communication to complicated cockpit display in aircraft simulators.

Serial data transmission rates are often measured in bits per second (bps) or baud rate. The bps indicates how many data bits can be transmitted per second, whereas baud rate represents how many states can be transmitted per second. For example, if two states of data must be represented, 1 bit would suffice to denote all of the states (either high or low). In that case, bps and baud rate are equal. However, if you use four states to represent data, you would need 2 bits, and bps would be twice the baud rate. Therefore, bps and baud rate will be the same only for a two-state data representation system.

The binary pattern for the serial communication consists of the following bits: start bit, data bit, parity bit, and stop bit. Some of those parameters are shown in the front panel of **Serial Port Init.vi**, which can be found in the subpalette **Functions >> Instrument I/O >> Serial**.

Serial communication uses reverse logic meaning that low state is represented by a binary bit 1, and high state is represented by a binary bit 0. This is especially noteworthy when *parity bit* is concerned. The purpose of the parity bit is to provide an error-checking mechanism. Usually, the parity bit is divided into two types: even parity and odd parity. Even parity adds a binary bit either 1 or 0 to the data to make the total number of 1's in the data even. Odd parity is the opposite case of even parity in terms of the total number of 1's. Reverse-logic logistics in serial communication must be taken into account when the input parameter **parity** is set, as in Figure 13.1.

Start bit indicates when *data bit* starts; for example, data bit starts when start bit is high (binary bit 0). *Stop bit* is the counterpart of start bit, and it indicates when data bit ends using either 1, 1.5, or 2 bits.

Since the serial communication can be a two-way communication, two serial devices (talker and listener) can continuously send and receive data. This is known as handshaking, and there are two types of serial handshaking communication: software handshaking (SWHS) and hardware handshaking (HWHS). SWHS interleaves two control characters (XON and XOFF) at the beginning and at the end of message bits to indicate the start and the end of message bits. On the other hand, HWHS monitors the voltage level on handshaking lines to distinguish the beginning of the message from the end of the message. (These parameters are also shown in Figure 13.1.)

Examples of different types of serial port communications are RS-232 and RS-449. RS-449 has two types depending on the modulation scheme: RS-422 and RS-423. Since different types use different cables with different pin assignments, proper cable conversion would be necessary to ensure successful serial communication. A later section will present an important method for testing serial communication between a computer and a serial device.

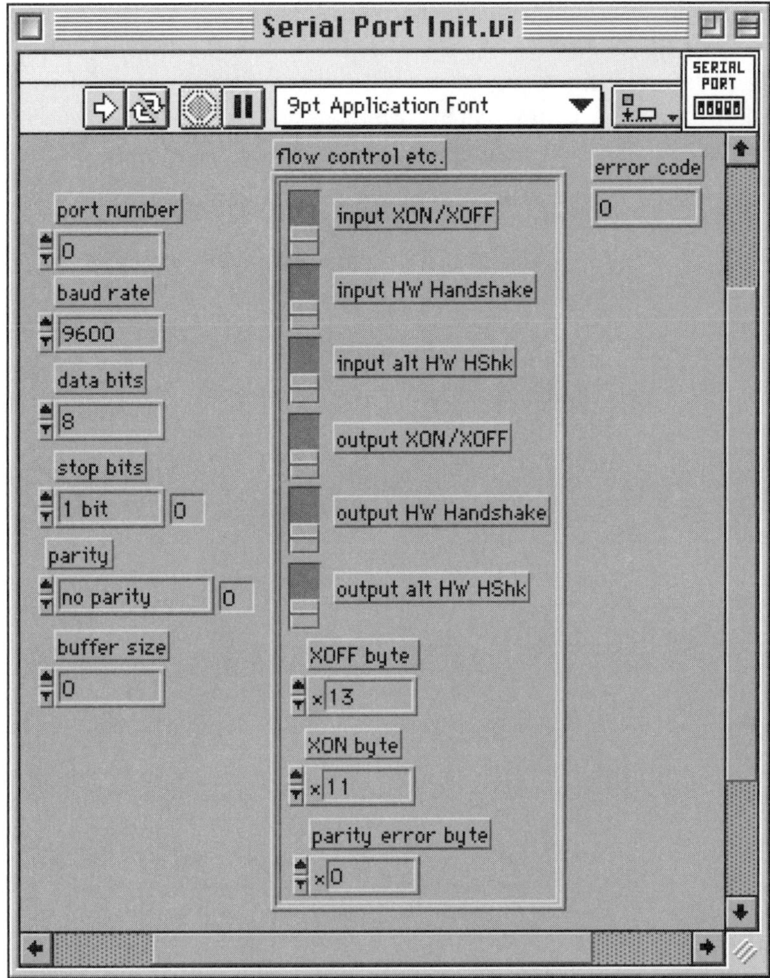

Figure 13.1
Front panel of **Serial Port Init.vi** showing different parameters in serial communication protocols.

13.3 VISA

Virtual instrument software architecture (VISA) is a global set of protocols that can control GPIB, serial, and other devices, including VXI. It is a set of protocols agreed upon by more than 40 major instrument companies so that almost all of the instruments can be controlled by LabVIEW's VISA command functions.

The steps of communication using VISA functions are similar to those for data acquisition: **VISA Open**, **VISA Read** or **VISA Write**, and **VISA Close**. LabVIEW identifies which session is active by the parameter **VISA session** generated by **VISA Open. VISA session** is similar to the **task ID** of data acquisition VIs or the **refnum** of file I/O functions.

Setting up a communication path with an instrument or instruments using VISA protocols is quite straightforward. A set of VIs designated for a specific instrument is called a driver, and more than 600 instrument drivers have already been written. These drivers are available on the second LabVIEW CD, at the National Instruments Internet site, or from instrument vendors. Once you obtain the driver for your instrument(s), test the communication channel using the VI **Getting Started.vi**, which can be found in the driver **.llb** files. Once the communication path is verified, the rest is LabVIEW programming itself.

LabVIEW 5.x also provides the Instrument Wizard to guide you through the steps that are necessary to establish the communication path between a computer and an instrument or instruments. Instrument Wizard is similar to DAQ Wizard, which was first introduced in LabVIEW 4.1. The purpose of the two wizard programs is to get beginners started in a short period of time. However, once you acquire the basics of data acquisition and instrument control, minimize your dependence on the wizard programs and use LabVIEW examples to build your own applications. Such a recommendation is made because the experience of building your own application by yourself is very important in the long run in G-programming. Given this suggestion, those two wizard programs will not be discussed in this book.

13.4 Functions for GPIB

There are two subpalettes, **GPIB** and **GPIB 488.2**, in **Functions >> Instrument I/O**. The subpalette **GPIB 488.2** contains the functions that have evolved from the ones in the subpalette **GPIB**, providing more functionality. The most commonly used GPIB functions are **GPIB Read**, **GPIB Write**, and **GPIB Status** in **Function >> Instrument I/O >> GPIB**.

GPIB Read This function reads bytes of string data, and the number of bytes to be read is specified in the input **byte count**.

GPIB Write This function writes string data to a GPIB instrument whose address is specified in the input **address string**.

GPIB Status This function returns the status of the GPIB controller.

Using these three commonly used GPIB functions, you can build an example that performs a basic GPIB communication by first sending a command to an instrument, then listening and reading the data that come from the instrument, and finally checking the status. The sequence of function calls are **GPIB Write-GPIB Read-GPIB Status**. Note that **GPIB Write** and **GPIB Read** send and read data in string format. Therefore, string functions will need to be used to construct a complete string command from pieces of words, or to convert ASCII or binary string data into numeric data. Refer to Chapter 12 for a complete discussion of string manipulation.

A good LabVIEW example for checking the communication with GPIB instruments is **LabVIEW <-> GPIB.vi**, and it can be found in the following directory:

```
LabVIEW directory >> examples >> instr >> smplgpib.llb
```

This example allows you to write a string command and read string data so that the communication setup can be verified.

13.5 VIs for Serial Communications

There are five serial VIs in **Functions >> Instrument I/O >> Serial**, but this section will only cover the four that are most commonly used.

Bytes At Serial Port.vi This VI returns how many bytes are available in the output **byte count** at the serial port specified in the input **port number**.

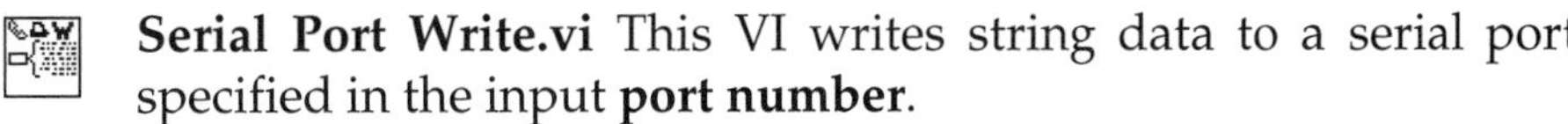

Serial Port Write.vi This VI writes string data to a serial port specified in the input **port number**.

Serial Port Read.vi This VI reads bytes of string data and returns them in the output **string read**. The number of bytes to be read is specified in the input **requested byte count**.

Serial Port Init.vi This VI initializes a serial port specified in the input **port number** using various parameters. See Figure 13.1 for available parameters.

The steps for serial communication are similar to those for GPIB communication. First, initialize a serial port with the parameters shown in Figure 13.1. Next, send string data and check how many bytes are returned and available at the serial port. Finally, read the entire data based on the byte count.

13.6 Functions and VIs for VISA

This section will examine five low-level functions for VISA protocols since they are most commonly used. Then it will discuss the five high-level VIs. First, consider the low-level function **VISA Open**.

VISA Open This function initiates communication with an instrument specified in the input **resource name** and **VISA session (for class)**. This returns a **VISA session** that indicates the current session.

VISA Write This function writes or sends string data specified in the input **write buffer** to an instrument identified by the **VISA session** that was previously generated by **VISA Open**.

VISA Read This function reads bytes of string data and returns them in the output **read buffer**. The number of bytes to be read is specified in the input **byte count**.

VISA Close This function closes the session specified in the input **VISA session**.

VISA Clear This function terminates the communication with an instrument identified by **VISA session**. Note that **VISA Close** only closes the session and does not clear the communication setup.

There are five high-level VIs for VISA in **Functions >> Instrument I/O >> VISA**, and they are shown in Figure 13.2.

Figure 13.2 High-level VIs for VISA communication.

 **Easy VISA Find Resources.vi** This VI finds all of the resources available for VISA communication.

 **Easy VISA Read.vi** This VI reads string data from an instrument specified in the input **resource name**.

 **Easy VISA Write.vi** This VI writes string data to an instrument specified in the input **resource name**.

 **Easy VISA Write & Read.vi** This VI first writes string data and then reads data returned by the instrument specified in the input **resource name**.

 **Easy VISA Serial Write & Read.vi** This VI is similar to **Easy VISA Write & Read.vi**, but is for instruments using serial communication.

13.7 Testing Serial Communication

If you have a communication problem with serial devices, the following simple test will help you trace the problem source. The idea behind this test, the so called self-loop test, is to write some simple string data to a serial port of your computer and read them back into LabVIEW. If the string data come back to LabVIEW, then the problem is likely from the serial devices, not from the computer.

In order to administer this test, you will need a serial communication cable. Open the example **LabVIEW <-> Serial.vi** from the following path: LabVIEW directory >> **examples >> instr >> smplserl.llb**. The front panel of **LabVIEW <-> Serial.vi** is shown in Figure 13.3. Make sure that the cable is securely connected to one of your serial ports, and short (connect) the two pins for RxD (reception of data) and TxD (transmission of data). Their pin numbers are usually 2 and 3 for any type of serial cables (DB-9 or DB-25). There are already input strings `Sample String` in the input **Serial Write**. Set the input **Action** to **Write**, check the input **Port Number**, and run the VI. When the VI stops, set the input **Action** to **Read** this time, and run the VI

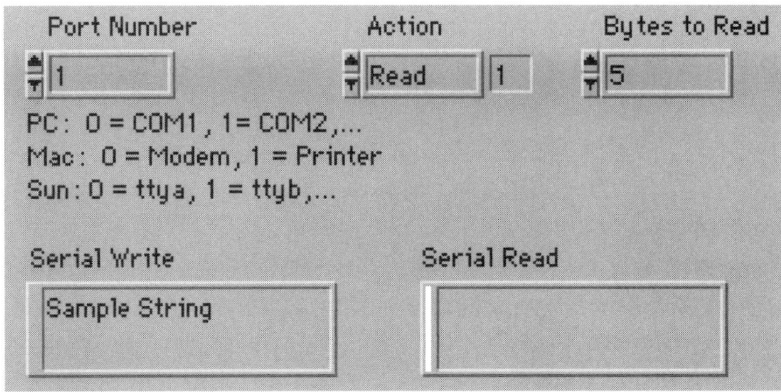

Figure 13.3
Front panel of **LabVIEW <-> Serial.vi** for testing serial devices.

again. If the string data `Sample String` comes back in the string indicator **Serial Read**, the test has been successful. Therefore, you should check the serial device to locate the problem source because it is neither the computer nor the cable. If the test fails, however, you should examine the BIOS to check the settings on the computer. Also, the LabVIEW preference file **labview.ini**, found in the LabVIEW directory on PC platforms, should be examined to see if the numbers in the input **Port Number** and the serial ports of the computer match. Discussion of **labview.ini** file is presented in the next section.

13.8 Parallel Port for Serial Communication

It is possible to use parallel ports for serial communication only on PC platforms. In order to do so, **labview.ini** must be modified. If your operating system is Windows 3.x, you can have only up to eight serial ports, but the maximum number goes up to 256 if your operating system is Windows 95 or later or Windows NT.

Open **labview.ini** in LabVIEW directory using any text editor, and add the following:

```
serialDevices = "COM1;COM2;COM3;LPT1;LPT2"
```

This will map COM1 to port 0, COM2 to port 1, COM3 to port 2, LPT1 to port 3, and LPT2 to port 4. Therefore, by adding this line to **labview.ini**, you can assign a port number for serial communication to a parallel port of your computer.

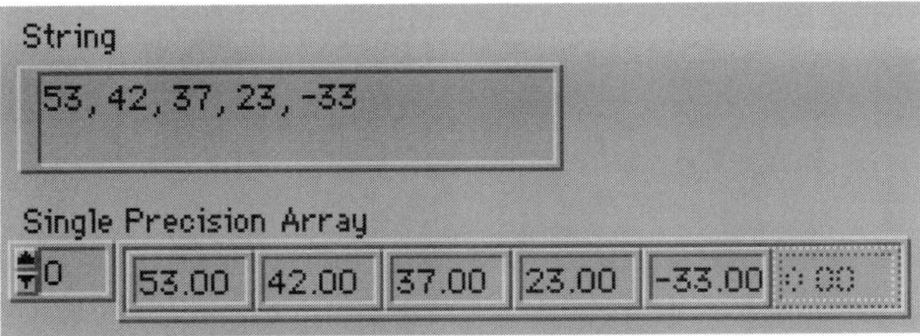

Figure 13.4
A partial view of the front panel of **Extract Numbers.vi** in LabVIEW directory >> **examples** >> **general** >> **strings.llb**. This VI extracts numeric strings separated by a comma and returns them in a numeric array. The separator can be any character, and it can be changed in the diagram window.

13.9 String Data from Instruments

Chapter 12 discussed cases where you had string data either in ASCII or binary format and you wanted to convert them to their original numeric format. Since instruments return binary or ASCII string data, the methods in Chapter 12 can be applied directly to convert the acquired string data to numbers. Since Chapter 12 covers the conversion of binary and ASCII string data to numeric data in detail, this section will introduce two additional examples of converting ASCII string data to numeric data. Those examples are **Extract Numbers.vi** and **Parse String.vi**, and they are located in the following directory: LabVIEW directory >> **examples >> general >> strings.llb**.

The partial views of the front panels of the two examples are shown in Figures 13.4 and 13.5. In Figure 13.4, the control **String** contains the data returned by an instrument, and each data string is separated by a comma. When you run the VI, it will return an array of numeric data, as shown in Figure 13.4. In fact, the data separator, which is the comma in this case, can

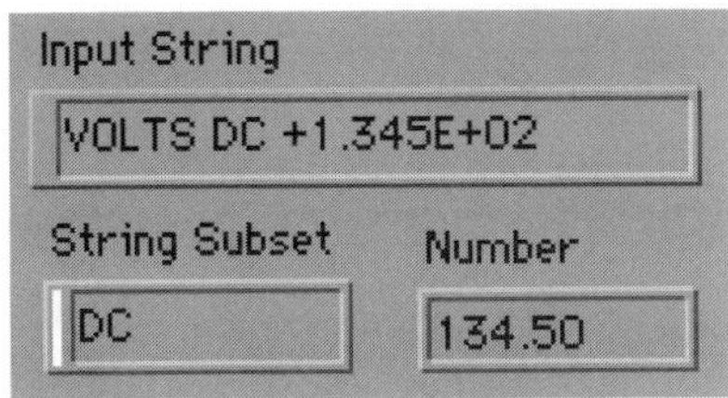

Figure 13.5 A partial view of the front panel of **Parse String.vi** in LabVIEW directory >> **examples** >> **general** >> **strings.llb**. This VI parses the string in **Input String** into string data and numeric data.

be any character, and different separators can be set in the diagram window. The second example in Figure 13.5 is for a case where your instrument returns a mixture of numeric string and text string. It can extract numeric data embedded in any part of the input text data.

There are many other interesting examples, such as **Parse Postfix Expression.vi** and **Parse Arithmetic Expression.vi**, related to the conversion of string data to numeric data in LabVIEW. These can also be found in LabVIEW directory >> **examples >> general >> strings.llb** and can give you more insight into the different usage of string functions. Therefore, you are encouraged to examine the other examples and experiment with them to better understand string manipulation.

PROBLEMS

13.1 Suppose that your instrument returned the following string data:

```
35, 24, 57, 77, -12, -44, 21, 26, 53, 32, -28
```

Note that each numeric string datum is separated by a comma. Design a VI that converts such string data into a numeric 1-D array, and save it as **P13_01.vi**.

13.2 Suppose that the following were the sequence of string data returned by your encoder:

```
UP•+43•-1.23E+01
DN•+22•+2.31E-03
DN•-11•+1.95E-01
```

The symbol • represents a white space. Design a VI **P13_02.vi** that returns three 1-D arrays where each array consists of the data in each field. (The first

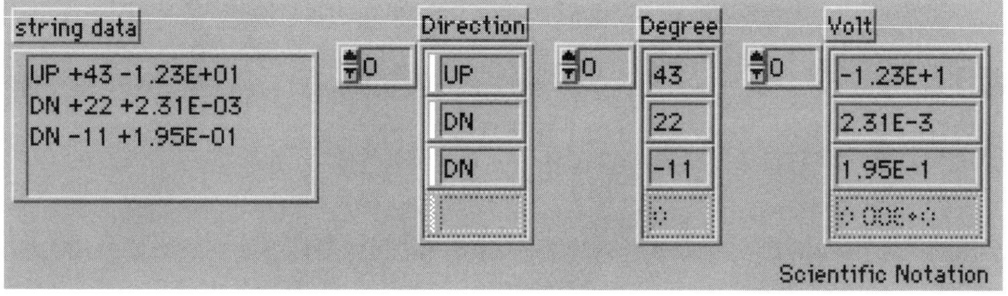

Figure P13.2

field is **Direction**, the second field is **Degree**, and the last is **Volt.**) The front panel of a sample solution is given in Figure P13.2. Hint: **Match Pattern** uses the sign + as a special character used in its option setting. Also, note that each row consists of 17 characters, including the end-of-line character at the end of each row.

13.3 Suppose that your power supply could generate DC –3.1 V or AC 2.15 V peak-to-peak by the following string commands:

```
GEN•DC•-3.10•V
GEN•AC•+2.15•V
```

The symbol • in the commands indicates a white space. Design **P13_03.vi** that automatically generates a string command based on the selection of AC or DC, and the amplitude value. In other words, **P13_03.vi** will generate a proper string command based on the user input for AC/DC and the signal amplitude. Note that you must include the positive sign (+) in the string command for positive amplitude values. The front panel of a sample solution is shown in Figure P13.3. Hint: You may use either **Pick Line & Append**, **Index & Append**, or **Select & Append**.

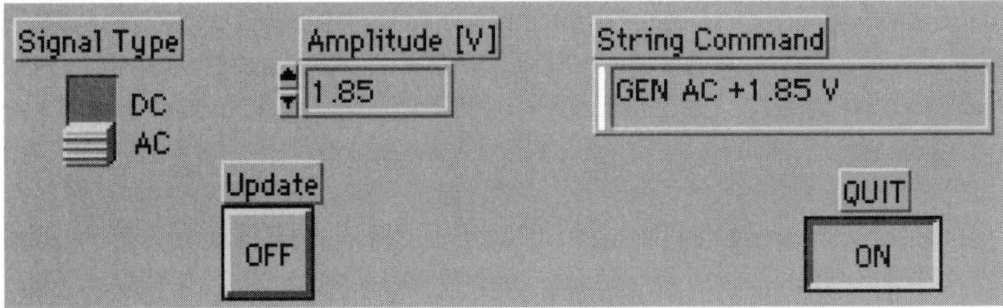

Figure P13.3

Data Analysis

14

The ultimate goal of data acquisition is to analyze data and extract meaningful information based on the result of the analysis. Since there are so many different types of information that you might need to extract from data, the types of data analysis methods vary widely. This chapter will categorize the realm of data analysis and discuss it.

One category is stochastic or deterministic data analysis. Stochastic data analysis is also called random data analysis since it is about randomly varying data. Therefore, you will need to use some type of probabilistic measure. The counterpart of stochastic analysis is deterministic data analysis. In such a case, the analysis may not involve any probabilistic measure since the data are not random. For example, a signal that can be represented by an equation of nonrandom variables is deterministic since you can predict the behavior of the signal accurately as long as you know the parameters in the equation. A sine wave can be an example: If you know the frequency, the amplitude, and the phase shift, you can completely predict the signal at any given time incident. The other category is time and frequency data analysis. Data analysis in the time domain focuses on the characteristics of the data (signal) in time, such as waveform recovery, data fitting, impulse or step response analysis of a linear system, and so forth. On the other hand, data analysis in the frequency domain studies the behavior of the data in terms of frequency, such as Fourier analysis, spectrum analysis, filtering, or frequency response analysis of a linear system. This naturally raises the need for a good understanding of the two systems, linear and nonlinear systems, since the techniques you will study in this chapter mostly can only be applied to a linear system. This chapter begins with a discussion of linear and nonlinear system analysis.

14.1 Linear and Nonlinear System Analysis

14.1.1 Characteristics of a System

A system is an entity that describes certain phenomena, and system analysts always try to model it mathematically so that they can apply all of the various known mathematical tools to analyze it. A TV set can be a system where you enter multiple inputs (remote control signal, power from AC outlet, etc.) to obtain multiple outputs (images on the screen, sound, other information display, etc.). Anything that you encounter in daily life can be represented

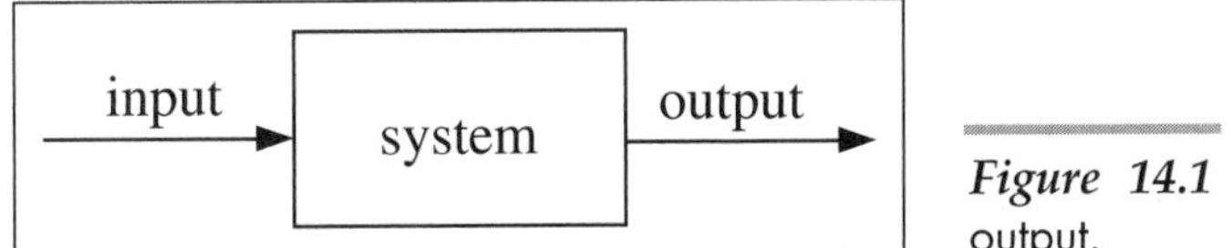

Figure 14.1 System with input and output.

as a system, and there are generally five benchmarks to describe a system: memory, dependence on time, linearity, causality, and stability.

If a system depends only on the input at the current time incident, it is called a memoryless system. Suppose that the input, the output, and the system in Figure 14.1 are all defined in continuous time, and $x(t)$, $y(t)$, and $h(t)$ are used to represent each of them, respectively. The following are some examples of memory and memoryless systems:

a continuous system with memory: $y(t) = x(t - 1)$
a continuous system without memory: $y(t) = x(t)$

If the output does not depend on the time delay occurring in the input, the system is called time-invariant. This can be restated as follows: A continuous time invariant (TI) system satisfies

$$y(t) = \Im[x(t)] \Rightarrow y(t - t_o) = \Im[x(t - t_o)]$$

where $\Im[\bullet]$ indicates a function (or a system) of an argument $\bullet$.

As for the linearity, one tends to couple it with a function of a straight line, which is partially correct. A linear system is defined as follows: A continuous linear system satisfies

$$y_1(t) = \Im[x_1(t)],\ y_2(t) = \Im[x_2(t)] \Rightarrow c_1y_1(t) + c_2y_2(t) = \Im[c_1x_1(t) + c_2x_2(t)]$$

Here c_1 and c_2 can be any complex constant. In other words, if the output of a system for the sum of weighted multiple inputs is the weighted sum of the outputs of each input, such a system is called linear. Based on this definition, a straight line of equation $y = \Im[x] = x + c$ is not linear unless $c = 0$.

As for a system whose output depends on the input values only at the present time and in the past, it is referred to as causal. The following are some examples of causal and noncausal systems:

continuous causal systems: $y(t) = x(t)$, $y(t) = x(t - 1)$
a continuous noncausal system: $y(t) = x(t + 1)$

To define the stability of a system, it is quite difficult to state it in a single term since there are many different types of stability. The details about each kind are beyond the scope of this chapter, but a few of them will be listed without considering them in any further detail. The following are the acronyms of the stability types that will be considered:

SISL: Stability In the Sense of Lyapunov
AS: Asymptotic Stability
ASIL: Asymptotic Stability In Large
BIBS: Bounded Input Bounded State
BIBO: Bounded Input Bounded Output

To briefly describe each category, BIBS stability holds if the state of the system is bounded given that the input to the system is bounded. BIBO stability holds if the output is bounded given that the input to the system is bounded. SISL, AS, and ASIL all define the stability of a system in terms of the Lyapunov function, which is positive definite, continuously differentiable, and nonnegative along the trajectory of the system state. Eigenvalue analysis of the system can be used if the system is represented in a state space representation. The types of stability that are applicable to nonlinear systems are BIBS and BIBO for those with any external input, and SISL, AS, and ASIL for those without any external input. Given a nonlinear system, you can linearize it about an equilibrium point(s), and the types of stability that are applicable to such linearized nonlinear systems are AS and ASIL. (AS and ASIL become identical for linearized systems.) If a system is linear, SISL, AS, ASIL, BIBS, and BIBO can all be applied.

14.1.2 Linear Time Invariant (LTI) System

If a system is nonlinear, which does not satisfy all of the definitions of linearity mentioned previously, you can linearize it about an equilibrium point(s) and treat it as a linear system. Then all of the tools available for linear systems can be applied to the system about the equilibrium point(s). Furthermore, if a system is independent of a time shift in the input, or if the time shift in the input shows up as the same time shift in the output, such a linear system is called linear time invariant (LTI). The discussion in the rest of this chapter will remain valid within the confines of such LTI systems. Also, a continuous or a discrete time system will be used alternately. The variable t is used to indicate continuous time whereas n indicates the indices of discrete time incidents. For example, $x(t)$ and $x[n]$ denote a continuous and a discrete signal.

For the sake of convenience, $x(t)$ or $x[n]$ will be used for the input, $y(t)$ or $y[n]$ for the output, and $h(t)$ or $h[n]$ for the system (response) again.

For LTI systems, their output can be expressed by the convolution sum of the input and the system response. The system response is also called the transfer function.

$$y(t) = x(t) * h(t) = \int_{-\infty}^{+\infty} x(\tau)h(t-\tau)\,d\tau \tag{14.1}$$

$$y[n] = x[n] * h[n] = \sum_{k=-\infty}^{+\infty} x[k]h[n-k] \tag{14.2}$$

The convolution relationship of continuous and discrete time systems is shown in (14.1) and (14.2). The symbol * indicates the convolution in the equations. Note that if the input is the impulse function $\delta(t)$ or $\delta[n]$, the system output shows the characteristics of the system. In other words, the impulse function can be fed into an LTI system if the system response or the transfer function is wanted. The impulse function is defined as follows:

$$\delta(t) = \infty \text{ if } t = 0, \text{ or } 0 \text{ if } t \neq 0$$
$$\delta[n] = 1 \text{ if } n = 0, \text{ or } 0 \text{ if } n \neq 0$$

Note that the convolution does not apply only to LTI systems but to any linear system as well, including linear time variant (LTV) systems. Time variant systems do depend on the time shift in the input, meaning that the time shift in the input does not show up as the same time shift in the output, and it generally alters the nature of the output instead of producing a time shifted version of the original output (the output for the input with no time shift). This chapter will limit discussion to LTI systems, as mentioned earlier. Therefore, if you are interested in the system characteristics, such as the transfer function, and if the system is assumed to be LTI, you can feed the impulse function to the system and the output will be the transfer function of the system. Remember that the discussion of system theory in this chapter is brief and abstract since the intention is to understand LTI systems in general so that you can apply correct data analysis techniques. Those who are interested in the details may refer to many other textbooks available on the system theory. (See Chapter 15 for suggestions for further reading.) Since the goal is the general concept of LTI systems and correct data analysis techniques using LabVIEW, you should make an effort to keep a global view of the topics presented in this chapter instead of focusing on a specific one.

14.2 Stochastic and Deterministic Data Analysis

14.2.1 Deterministic Signals and Systems

Suppose that you have a sine wave of the following equation:

$$x(t) = 3\sin(2t + 0.785) \tag{14.3}$$

Having the expression shown in (14.3), can you tell the value of the function at $t = 100$? The answer is clearly yes since you can substitute 100 for t in (14.3), and $x(t = 100)$ will yield the answer. In other words, you can predict the behavior of the function such as its value at any time incident, and the function is completely specified by the parameters given in (14.3). These types of functions are said to be deterministic, and their nature is not random. (Deterministic signals can have unknown parameters. In such a case, you will still need to estimate them.)

A few examples of analysis techniques for deterministic signals and systems are frequency response analysis, step response analysis, impulse response analysis, Fourier analysis, filtering process, and so forth. (Those techniques can also be applied to stochastic systems with a few modifications. Frequency and Fourier analysis of the stochastic systems are discussed in a later section. The basic concept, however, remains the same for both deterministic and stochastic analyses.) Frequency response of a deterministic system can be obtained by applying an input of known frequency to the system. Observing the frequency spectrum of the output signal can explain to us many aspects about the system, such as phase margin, stability, and robustness. In some cases, one may be interested in the response of a (deterministic) system using different types of input signals, such as the impulse function or the step function. (The step function has two values: zero and one. It remains at zero until a specified time incident, at which the signal level jumps from zero to one and remains at one thereafter.) The responses of the system to those types of input signals are called impulse response or step response. The impulse response shows the profile of the system, as was already explained during the discussion of convolution, and the step response shows the rising time, the effect of damping ratio, the order of the system, and many other important benchmarks of the system.

These techniques are also closely related to Fourier analysis, which represents signals in terms of Fourier series coefficients or frequency components. The Fourier transform analysis is discussed in detail in a later section along with common misconceptions about its application.

Filters (either analog or digital) are commonly used for data analysis by engineers and researchers. A filter is a system that manipulates its input signal to generate an output signal in order to meet certain criteria. A lowpass filter (LPF) can be used to block some frequency components beyond a certain threshold. A highpass filter (HPF) is a complement to the LPF. A bandpass filter (BPF) or a bandstop filter (BSF) can be used to pass or block a range of frequency components. For example, if you want to block the DC noise embedded in your acquired data, an HPF can be used to block the DC component. An in-depth discussion of filters will take place in a later section as well.

14.2.2 Stochastic Processes

Suppose that you have a cassette tape to record your favorite music. If you play a blank tape on a tape player, you will only hear background noise since there is nothing (no music) recorded on the tape. This type of noise is called white noise and is usually assumed to be Gaussian, having zero mean and unit variance. (In case of nonzero mean or nonunity variance, it can easily be shifted or normalized to have zero mean and unit variance.)

Signals like Gaussian white noise are called random or stochastic since their behavior can only be described in a statistical sense. Each signal sample of Gaussian white noise is called a random variable (RV). The formal definition of RVs is a function of elements from a sample space Ω. In other words, RV is a function that maps all of the elements in the sample space Ω into values on the real line if the RV is real. You can also define complex RVs, and you may refer to many other textbooks on the probability and random variables for further discussion of them. (Some suggestions on the textbooks are made at the end of Chapter 15.) If the RV is a function of not only Ω but time t as well, it is called stochastic process. Therefore, $\mathbf{X}$ is an RV if you can write $\mathbf{X}(\omega)$, where $\omega \in \Omega$, and a random process if you can write $\mathbf{X}(\omega, t)$. Note that RVs are a special case of random processes when the time t is fixed.

If a signal is stochastic, how do you represent it while its behavior is random? To answer such a question, first define an expectation operator $\mathcal{E}\{\bullet\}$.

The expectation of an RV $\mathbf{X}$ is defined as follows:

$$\mathscr{E}\{\mathbf{X}\} = \int_{-\infty}^{+\infty} x f_X(x)dx = \sum_{k=1}^{\infty} x_k \mathrm{P}(\mathbf{X} = x_k) \tag{14.4}$$

where $f_X(x)$ is the probability density function (pdf) of a continuous RV $\mathbf{X}$, $P(\mathbf{X} = x_k)$ is the discrete pdf of a discrete RV $\mathbf{X}$, and x and x_k are a sample random value of a continuous and a discrete RV $\mathbf{X}$. More generally, the *generalized moments* of an RV $\mathbf{X}$ is defined by

$$\mathscr{E}\{|\mathbf{X} - \alpha|^n\} \tag{14.5}$$

As you may have recognized, (14.4) implies that the expectation is the mean of an RV $\mathbf{X}$. If α is $\mathscr{E}\{\mathbf{X}\}$ and n is 2, equation (14.5) becomes the variance of $\mathbf{X}$. These two measures (the mean and the variance) are most commonly used to characterize signals of random nature.

If you have stochastic processes, more information than (14.4) and (14.5) is required to describe them. In order to explain stochastic processes properly, we present the following important definitions. Note that this section will consider the continuous time stochastic processes only. The discrete version can easily be derived based on the same concept.

A stochastic process is said to be stationary if its statistical characteristics, such as autocorrelation function, do not depend on time. There are two types of stationary processes: wide-sense stationary (WSS) and strict-sense stationary (SSS) processes. Their definition is as follows: A stochastic process $\mathbf{X}(\omega, t)$ or simply $\mathbf{X}(t)$ is called WSS if its mean is constant and its autocorrelation is a function of time difference; that is,

$$\mathscr{E}\{\mathbf{X}(t)\} = \mu,\ \mathscr{E}\{\mathbf{X}^*(t)\mathbf{X}(t+\tau)\} = R_{XX}(\tau) \tag{14.6}$$

where $\mathbf{X}^*(t)$ is the complex conjugate of $\mathbf{X}(t)$ and $R_{XX}(\tau)$ is the autocorrelation of $\mathbf{X}(t)$. On the other hand, $\mathbf{X}(t)$ is called SSS if (14.6) is true, and its nth-order probability density function satisfies the following:

$$f_X(x_1, \ldots, x_n; t_1, \ldots, t_n) = f_X(x_1, \ldots, x_n; t_1 + \alpha, \ldots, t_n + \alpha) \tag{14.7}$$

for any α. In other words, the nth-order density function is invariant to any time shift α.

As it is shown in (14.6), one of the common measures of stochastic processes is the correlation function, and $R_{XX}(\tau)$ is the autocorrelation function of WSS $\mathbf{X}(t)$. Similarly, the cross-correlation function $R_{XY}(\tau)$ is defined by

$$R_{XY}(\tau) = \mathscr{E}\{\mathbf{X}^*(t)\mathbf{Y}(t+\tau)\} \tag{14.8}$$

If you have a signal or a value of random nature (either a stochastic process or an RV), it can be described in terms of measures given by (14.6) and (14.8). Since the nature of those processes or variables is random, you need to *estimate* their nature. Two simplest probabilistic measures you use daily are the mean and the variance. The mean and the variance are too commonly used to be realized that they are derived from equations (14.4) and (14.5). For example, suppose that an instructor wanted to compute the average of 10 midterm exam scores in his or her class. Then he or she would simply add the 10 scores and divide the sum by 10. In this case, the instructor was using (14.4) without recognizing the random nature of the ten scores. Even now you may ask why the grades are random! Actually, those grades *are* RVs that map a sample space Ω to a number between 0 and 100, assuming that the highest score is 100 points. The sample space Ω would contain the events that indicate the level of each student in the class. Given those 10 RVs, the instructor could *estimate* the mean and the variance of them, assuming the equal likelihood for each event. In this example, the probability of a student receiving a certain score is assumed to be 0.1. In this example, the estimates are *also* RVs since they are the functions of 10 random scores.

14.2.3 Estimation of Random Parameters

There are two types of estimates of random parameters: biased and unbiased. If the estimate $\hat{\theta}$ of θ satisfies the following, it is said to be *unbiased*; otherwise, it is said to be *biased*.

$$\text{Unbiased estimate } \hat{\theta} \text{ of } \theta\text{: } \mathscr{E}\{\hat{\theta}\} = \theta \tag{14.9}$$

Therefore, if an estimate is unbiased, such an estimate converges to the true value in a probabilistic sense. Given such a definition, consider the following interesting facts: If M number of RVs $\mathbf{X}_i$ are assumed to be independent and Gaussian distributed, the maximum likelihood estimate (MLE) of their mean and variance satisfy the following:

1. If the mean μ is known, but not the variance σ^2, the MLE of the variance, $\hat{\sigma}^2$, is

$$\hat{\sigma}^2 = \frac{1}{M}\sum_{i=1}^{M}(\mathbf{X}_i - \mu)^2 \tag{14.10}$$

   This estimate is unbiased.

2. If the variance σ^2 is known, but its mean μ is not, the MLE of the mean, $\hat{\mu}$, is

$$\hat{\mu} = \frac{1}{M}\sum_{i=1}^{M} \mathbf{X}_i \tag{14.11}$$

This estimate is unbiased.

3. If both the mean and the variance are unknown, their MLEs $\hat{\mu}$ and $\hat{\sigma}^2$ are

$$\hat{\mu} = \frac{1}{M}\sum_{i=1}^{M} \mathbf{X}_i,\ \hat{\sigma}^2 = \frac{1}{M}\sum_{i=1}^{M} (\mathbf{X}_i - \hat{\mu})^2 \tag{14.12}$$

Note that $\hat{\mu}$ is an unbiased estimate of the mean, but $\hat{\sigma}^2$ is *biased*. The unbiased version that is not MLE is

$$\hat{\sigma}^2 = \frac{1}{M-1}\sum_{i=1}^{M} (\mathbf{X}_i - \hat{\mu})^2 \tag{14.13}$$

The reason why those points are mentioned is that RVs are often assumed to be normally (Gaussian) distributed, and statistically independent, and you should be aware of what kind of estimates you implement for the mean and the variance. RVs that are independent and identically distributed are said to be i.i.d., and MLE is an estimate which is most likely accurate in a statistical sense. For further details, you may refer to the textbooks suggested at the end of Chapter 15. The details about MLE and i.i.d. will not be discussed beyond this scope. However, you should be aware of what type of estimators to be used and why, based on their biases.

14.2.4 Hypothesis Testing

One interesting field of stochastic data processing is hypothesis testing. Consider a case where a source transmits a binary signal $s(t)$ through a noisy channel where additive Gaussian noise $\mathbf{N}(t)$ is present (Figure 14.2). This means that $s(t)$ has two TTL states, and such a signal is commonly used in digital communication systems to send binary bit patterns.

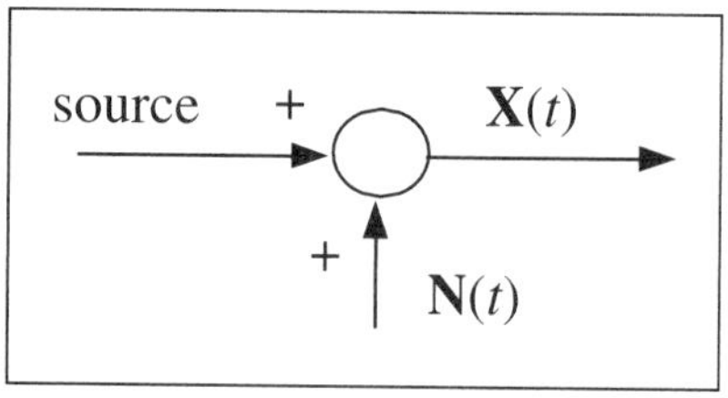

Figure 14.2 Simple communication channel model.

Since the noise $\mathbf{N}(t)$ is Gaussian distributed, and $s(t)$ is a binary signal, the received signal $\mathbf{X}(t)$ will be distributed as $N[m_0, \sigma_0^2]$ when $s(t)$ is absent and $N[m_1,\sigma_1^2]$ when $s(t)$ is present. The notation $N[m_0, \sigma_0^2]$ indicates a Gaussian distribution with the mean m_0, and the variance σ_0^2. Call the hypothesis when the signal is present H_1, and H_0 when it is absent. Then

$$\mathbf{X}(t) = \begin{Bmatrix} s(t) + \mathbf{N}(t) : \mathbf{H}_1 \\ \mathbf{N}(t) : \mathbf{H}_0 \end{Bmatrix} \tag{14.14}$$

where

$$H_0 \sim \mathbf{X}(t): N[m_0, \sigma_0^2]$$
$$H_1 \sim \mathbf{X}(t): N[m_1, \sigma_1^2]$$

This example is called binary hypothesis testing, and your object is to decide if the signal is present or not based on your observation x in (14.15) using its distribution function. It is shown mathematically in (14.15) that you can decide if the case is either H_0 or H_1 using the threshold k, based on the observation x. (x is a sample of the stochastic process $\mathbf{X}(t)$.) Therefore, you need to compute k based on the *a priori* information, such as Probability of False Alarm (P_{FA}). P_{FA} is the probability of a case where H_1 is chosen while the truth is H_0.

$$l(x) = \frac{f_{H_1}(x)}{f_{H_0}(x)} = \frac{\sigma_0 \exp\left\{-\frac{1}{2\sigma_1^2}(x-m_1)^2\right\}}{\sigma_1 \exp\left\{-\frac{1}{2\sigma_0^2}(x-m_0)^2\right\}} \underset{H_0}{\overset{H_1}{\gtrless}} k \tag{14.15}$$

In order to compute the threshold, proceed to rewrite (14.15) after some simplification as

$$L(x) \underset{H_0}{\overset{H_1}{\gtrless}} \eta \tag{14.16}$$

where

$$L(x) = \frac{1}{2}\left\{\frac{1}{\sigma_0^2}(x-m_0)^2 - \frac{1}{\sigma_1^2}(x-m_1)^2\right\} \tag{14.17}$$

$$\eta = \ln(k) - \text{In}\left(\frac{\sigma_0}{\sigma_1}\right) \tag{14.18}$$

Assuming that the variance is $\sigma_0^2 = \sigma_1^2 = \sigma^2$, further simplify (14.17) as

$$L(x) = \sigma^{-2} (m_1 - m_0)\{x - 0.5 (m_1 + m_0)\} \tag{14.19}$$

Since $L(x)$ is a function of x that is Gaussian distributed, $L(x)$ is also Gaussian, and its mean and variance for H_0 and H_1 can be obtained simply as follows:

$$H_0 \sim L(x):\ N\left[-\frac{d^2}{2},\ d^2\right]$$

$$H_1 \sim L(x):\ N\left[\frac{d^2}{2},\ d^2\right]$$

where

$$d = \frac{m_1 - m_0}{\sigma} \tag{14.20}$$

If $\Phi(z)$ is defined by

$$\Phi(z) = \frac{1}{\sqrt{2\pi}} \int_{-\infty}^{z} \exp\left(-\frac{\zeta^2}{2}\right) d\zeta \tag{14.21}$$

then it can be shown that the Probability of False Alarm (P_{FA}) and Probability of Detection (P_D) are related to $\Phi(z)$ as follows:

$$P_{FA} = 1 - \Phi(z), \text{ and } P_D = 1 - \Phi(z - d) \tag{14.22}$$

where

$$z = \frac{\eta + \frac{d^2}{2}}{d}$$

P_D is the probability of a case where H_1 is chosen while the truth is also H_1. To summarize, the binary hypothesis testing for the model in (14.14), where $\mathbf{X}(t)$ is assumed to be distributed as $\mathbf{N}[m_0, \sigma^2]$ or $\mathbf{N}[m_1, \sigma^2]$ for H_0 and H_1, respectively, can be achieved by the following steps:

Step 1: Observe a sample x from the stochastic process $\mathbf{X}(t)$.

Step 2: Compute $L(x)$ from (14.19).

Step 3: Compute η from (14.22) for given P_{FA}, using d and $\Phi(z)$ from (14.20) and (14.21), respectively.

Step 4: Decide either H_0 or H_1 by (14.16).

For Step 3, the error function erf(x) can be useful.

$$\mathrm{erf}(x) = \frac{2}{\sqrt{\pi}} \int_0^x \exp\left(-u^2\right) du \tag{14.23}$$

Then $\Phi(z)$ is related to erf(x) as follows:

$$\Phi(z) = \begin{cases} 0.5 + \frac{1}{2}\mathrm{erf}\left(\frac{z}{\sqrt{2}}\right), & z \geq 0 \\ 0.5 - \frac{1}{2}\mathrm{erf}\left(\frac{-z}{\sqrt{2}}\right), & z \leq 0 \end{cases} \tag{14.24}$$

An example VI of the binary hypothesis testing is shown in Figure 14.3. The diagram window of the sub VI **Phi(z).vi** is also shown in Figure 14.4. The input and the output of **Phi(z).vi** are **z** and **Phi(z)**, respectively.

14.2.5 Data Analysis with Filters

Probably, the filtering operation is the most commonly used data analysis technique. The ideas behind filters are so diverse that their design and implementation can be very easy and intuitive or very complicated. Filters can roughly be categorized into two types: those for deterministic signals and those for stochastic signals. Filters for deterministic signals are designed and implemented assuming that the behavior of the input data is completely known; for example, you want to pass or block a sinusoidal noise of a certain frequency present in the signal. If the signals of interest are stochastic, however, you may no longer apply those filters that are designed for deterministic signals due to the various nature of random signals, such as the white spectrum of white noise.

LabVIEW provides numerous kinds of filters, such as Butterworth, Chebyshev, Elliptic, Equi-Ripple, and many others. They are all available in **Functions >> Analysis >> Filters**, and most of them are applicable to deterministic models, where the input signals are deterministic. As for stochastic models, all of the properties of random nature will need to be considered, using the measures introduced previously, such as expectation, moments, autocorrelation function, and cross-correlation function. Since the nature of

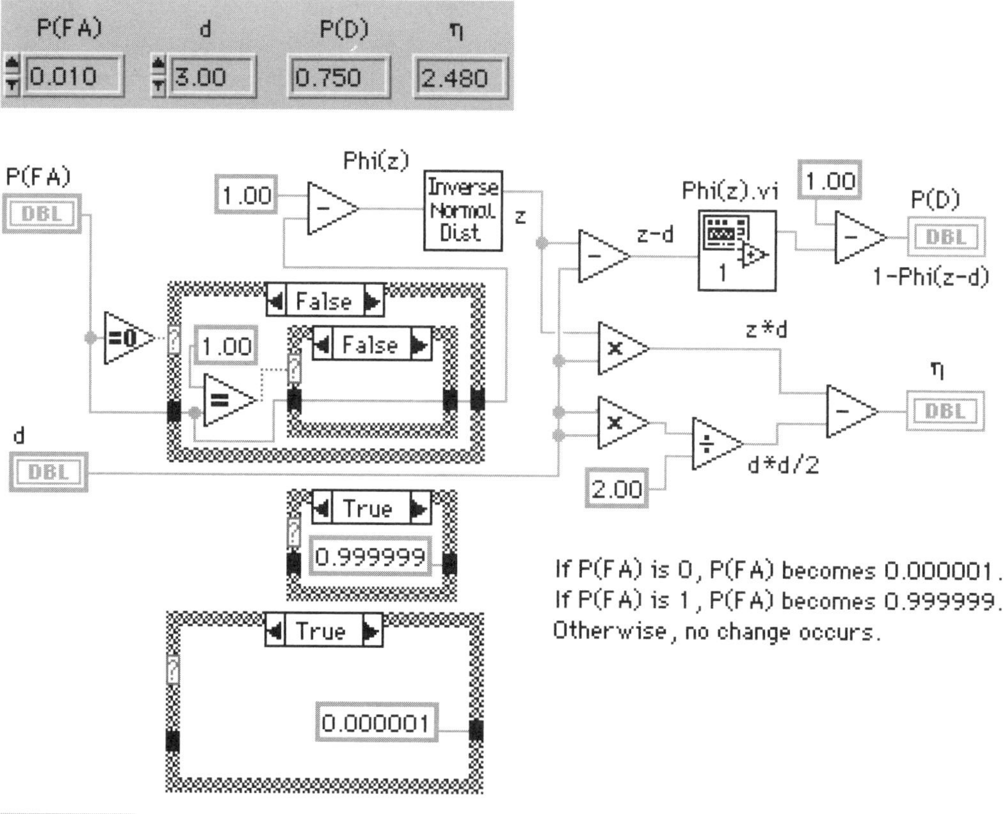

Figure 14.3
The front panel and the diagram window of a VI that computes the threshold for the binary hypothesis testing. **Phi(z).vi** is a sub VI that computes $\Phi(z)$, which performs the reverse process of **Inv Normal Distribution.vi**.

stochastic models is random and mostly unpredictable, filters are usually designed using adaptive algorithms so that they can *adapt* their processing method to randomly varying stochastic input signals. A few standard examples of such filters for stochastic processes are Kalman filter, Wiener filter, eigenfilter, and linear prediction filter. Typically, the goal in the filter design for a stochastic process is to minimize the mean square error in the estimate. Using the well-known Wiener-Hopf equation and the principle of orthogonality between the input, the output, and the error term, the weights or coefficients of those filters can be obtained.

Filters such as Butterworth and Chevyshev can eliminate noise components that are distinctly separated from the signals of interest, but what if the

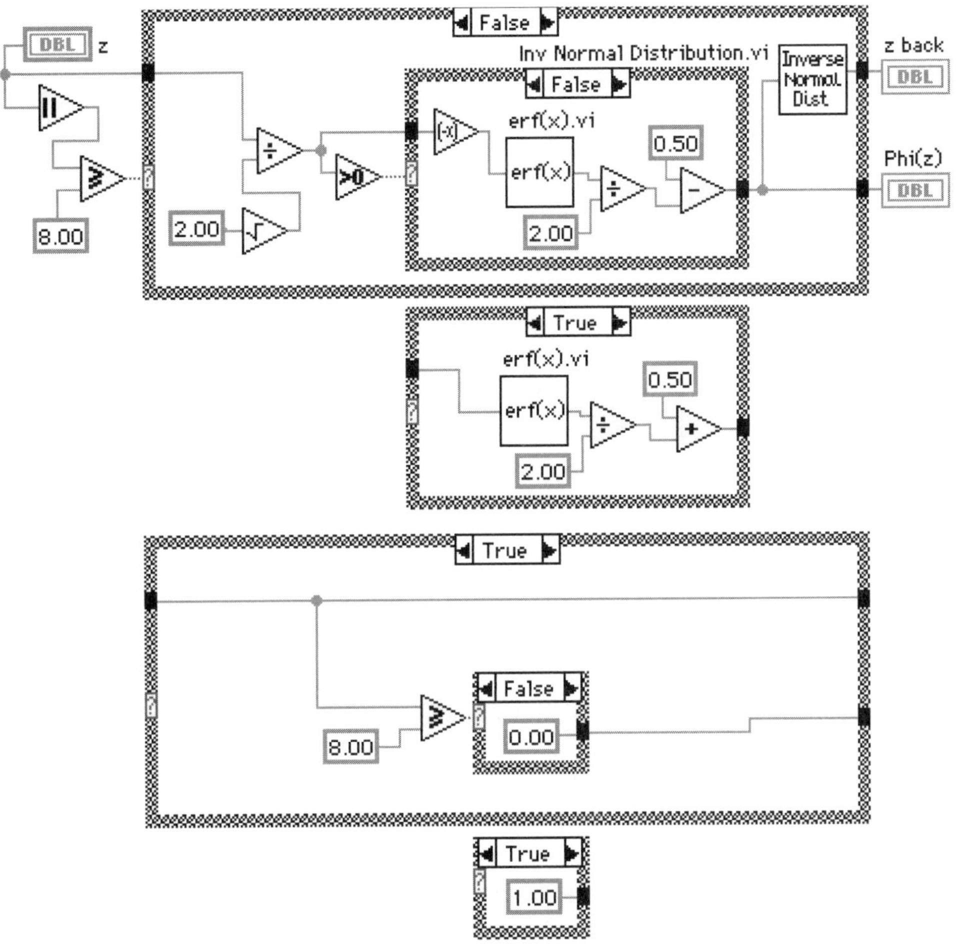

Figure 14.4
Diagram window of **Phi(z).vi**.

noise and the signals share the same frequency range? Obviously, those filters will fail to separate them, and, therefore, you need to implement stochastic models to *optimize* the filtering process. Even though the Wiener filter does provide useful solutions to separate noise that is spread over the signal frequency range, it is difficult to expand it to discrete time models, or time-varying systems, or multiple-input and multiple-output systems. The Kalman filter was then introduced in 1960 and was quickly adopted by engineers, who saw the practical aspect of it using the state space representation. The Kalman filter fits both continuous and discrete time models well, and its inherent adaptive structure can easily be implemented for stochastic models. However, since the Kalman filter is model dependent, the modeling process must be performed with caution. Both the Wiener filter and the Kalman filter provide a filtering solution to separate signals from noise in the minimum mean square error sense. The eigenfilter uses the eigenvectors of the data correlation matrix for its frequency response coefficients, and its design includes the following:

1. eigendecomposition of the data correlation matrix $\mathbf{R} = \mathcal{E}\{\mathbf{x}[n]\mathbf{x}^H[n]\}$ where $\mathbf{x}^T[n] = [x[n], x[n-1], \ldots, x[n-M+1]]$. The superscript ^{T} indicates transpose; $\mathbf{x}[n]$ is a vector whose elements are discrete random variables $x[n], x[n-1], \ldots, x[n-M+1]$; and M is the length of the vector. (Vector representation of data is presented in a later section in this chapter.) In order to avoid any confusion, boldface letters are not used in this case for RVs.
2. Choose the eigenvector that corresponds to the largest eigenvalue as the impulse response of the eigenfilter. The maximum output signal-to-noise ratio (SNR) becomes the ratio of maximum eigenvalue to the noise variance.

The linear prediction filter uses $M-1$ samples to estimate the latest or the oldest sample and minimizes the mean square error between the estimate and the actual sample. For example, suppose that you have M data samples that are sampled from a stochastic process and denote them as $x[n], x[n-1], \ldots, x[n-M+1]$. (Again, boldface letters are not used for RVs in the discussion of the linear prediction filter.) Forward linear prediction (FLP) filters estimate $x[n]$ as a linear combination of the rest of the $M-1$ samples and minimize the mean square error in the estimate. Backward linear prediction (BLP) filters operate the opposite way: They estimate the last (oldest) value $x[n-M+1]$ as a linear combination of the previous $M-1$ samples and minimize the mean square error in the estimate.

In many applications, stochastic models play more important roles than deterministic ones do, and those stochastic model filtering tools can be built upon the VIs in LabVIEW. Note that it is usually difficult to standardize an algorithm for stochastic models since it depends heavily on the models you select, and precise understanding about the models is required to build those filters for stochastic models. Also, you should use your discretion in deciding when and where to use a stochastic model or a deterministic model after first scrutinizing the type of input signals or input data samples. Determining the correct model is often the most important step in stochastic model filtering.

14.3 Time and Frequency Domain Data Analysis

14.3.1 Time Domain Analysis

Data analysis in the time domain is very common, probably because it seems more intuitive and easier to understand the analysis results as a function of time, and you are more used to thinking in terms of time rather than frequency. An example of such an analysis is data fitting, where the objective is to obtain a mathematical model that fits best to the given data samples. Here *best* can have different implications, and minimal mean square error is the category that is most often used.

Figure 14.5 shows a data fitting example using **Linear Fit.vi**, which is available from **Functions >> Analysis >> Curve Fitting**. This example uses an **XY Graph** to display two sets of data: x-axis and y-axis values. **Linear Fit.vi** also returns the coefficients of the first order polynomial so that unknown values at any x can be estimated.

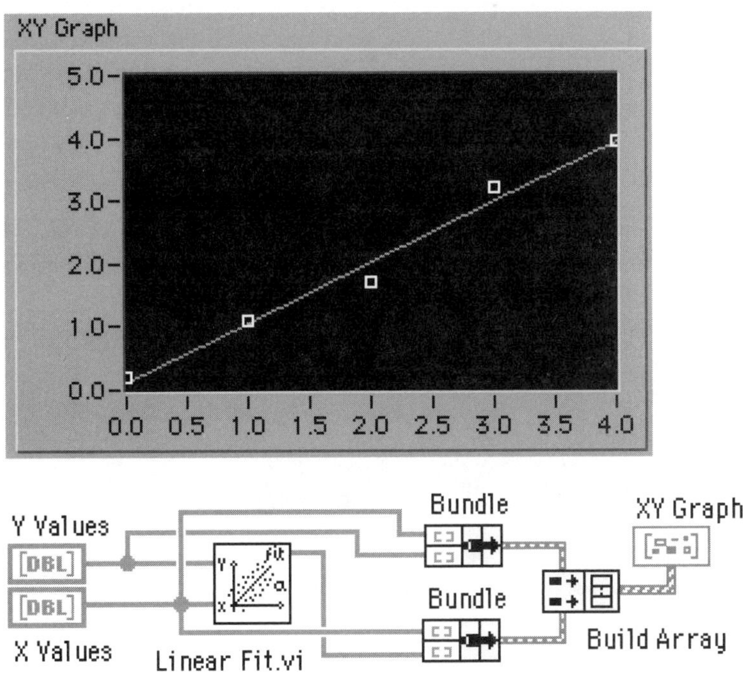

Figure 14.5
Example of data fitting using **Linear Fit.vi**.

If you are interested in quantities such as the area under given data samples or the roots of a polynomial, **Numeric Integration.vi** or **Complex Polynomial Roots.vi** can be used to compute the answer. Those VIs are available in **Functions >> Analysis >> Additional Numerical Methods.** (The VIs in the subpalette **Analysis** are stored in the new two subpalettes, **Signal Processing** and **Mathematics**, in LabVIEW 5.1 or higher. See your LabVIEW package for more information about the change.) The example in Figure 14.6 calculates the area under the given waveform samples using **Numeric Integration.vi**. There is another VI that performs similar numeric integration, and it is **Functions >> Analysis >> Digital Signal Processing >> Integral x(t).vi**. Even though both VIs perform numeric integration, their results are far different. Consider the two equations (14.25) and (14.26):

$$I = \int_a^b f(x)\,dx \tag{14.25}$$

$$x(t) = \int_a^t f(x)\,dx \tag{14.26}$$

Equation (14.25) returns a scalar as the result of integrating the function $f(x)$ over the interval $[a, b]$, and a and b are constants. However, equation (14.26) is a function of t and will return multiple values for multiple values of t. **Numeric Integration.vi** performs the numerical version of (14.25), and **Integral x(t).vi** performs that of (14.26). Therefore, **Numeric Integration.vi** returns a scalar whereas **Integral x(t).vi** returns an array, and both VIs take an array as their input. In other words, **Integral x(t).vi** is a running integral using the samples up to the value t. To be exact, **Integral x(t).vi** computes

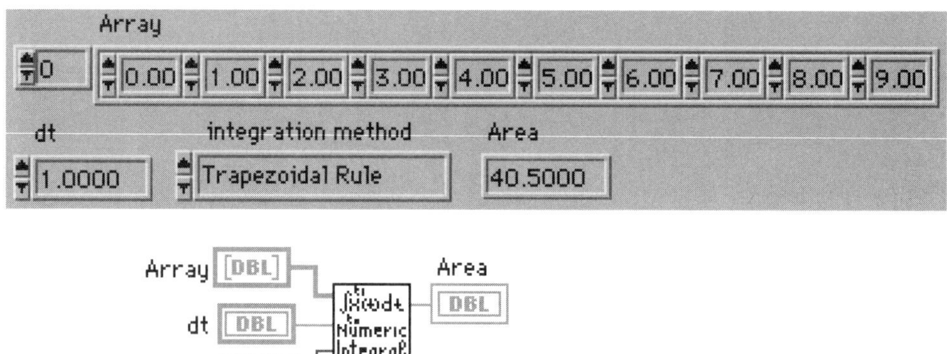

Figure 14.6
Example of numerical integration using **Numeric Integration.vi**.

the area $x[i]$ based on the previous running integral $x[i-1]$ and three data samples $f[i-1]$, $f[i]$, and $f[i+1]$. (The continuous variable t has been replaced with the discrete variable i in (14.26) since **Integral x(t).vi** performs numerical integration on the discrete data samples $f[i]$.)

The comparison between those two VIs is shown in Figure 14.7. The plot of the resulting array from **Integral x(t).vi** shows a parabolic function. The input samples are obtained from a function $y = x$. On the other hand, **Numeric Integration.vi** returns a single value, which is the area under the samples of $y = x$.

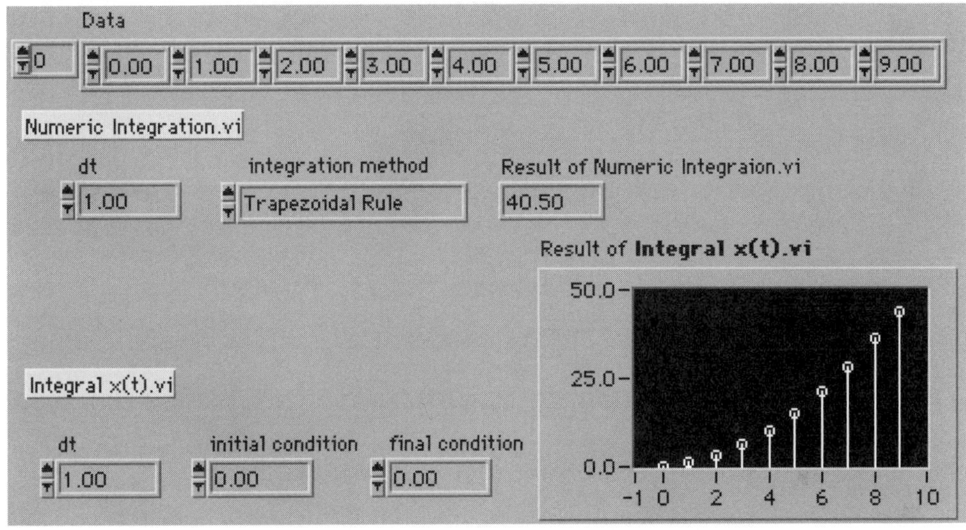

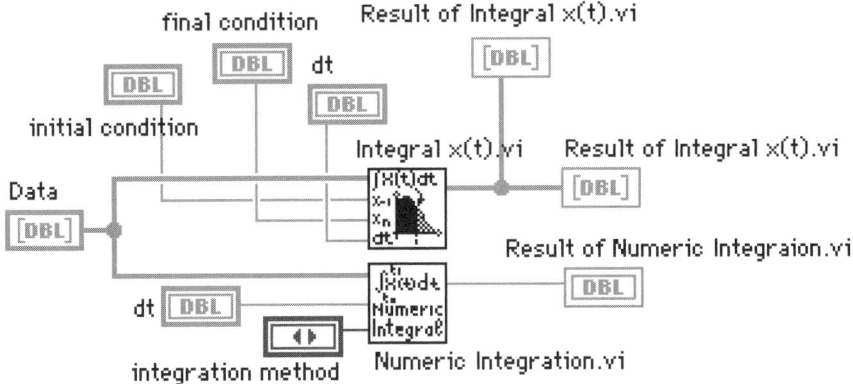

Figure 14.7
Comparison of **Numeric Integration.vi** with **Integral x(t).vi**. **Numeric Integration.vi** returns a single value area under the input data samples. **Integral x(t).vi**, however, returns an array while performing a running integral on the input data array.

Another common analysis technique in time is averaging, which can be applied to alleviate the effect of noise, especially in slowly varying signals such as temperature, strain gauge, or RTD measurements. This section will introduce two types of averaging techniques: ensemble averaging and moving averaging. (One more type *running averaging* is considered in Problem 14.9.) The ensemble averaging is performed across snapshots of data samples (usually in a vector format) whereas the moving average is performed across time with overlaps.

Ensemble averaging and moving averaging are illustrated in Figure 14.8 with a window size of 3 for both averaging techniques. The ensemble average vector $\mathbf{e}_p$ is the average of data vectors $\mathbf{x}[3p-3]$, $\mathbf{x}[3p-2]$, and $\mathbf{x}[3p-1]$, and the scalar moving average $m_k[i]$ for the kth sensor at time index i is the average of $x_k[i-2]$, $x_k[i-1]$, and $x_k[i]$. In ensemble averaging, there is no time overlap, but there is in moving averaging. Also, if the number of sensors is 1, the ensemble average vector becomes a scalar.

The ensemble averaging technique is often applied to image processing or array signal processing, where an array of sensors returns a stream of data vectors $\mathbf{x}[i]$. Then ensemble averaging is applied as shown in Figure 14.8,

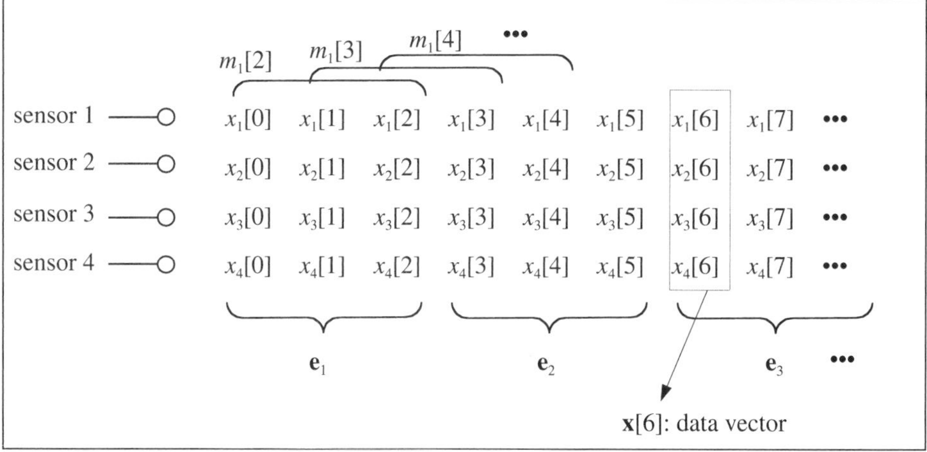

Figure 14.8

Ensemble average $\mathbf{e}_i$, and moving average $m_k[i]$ for the kth sensor at time index i. $\mathbf{e}_i$ is a vector and $m_k[i]$ is a scalar. Window size for both averaging is 3. The moving average for the kth sensor at time index 0 is $m_k[0]$, which is the average of $x_k[0]$ and two zeros, while $m_k[1]$ at time index 1 is the average of $x_k[0]$, $x_k[1]$ and a zero. The ensemble average $\mathbf{e}_1$ is the average of three data vectors $\mathbf{x}[0]$, $\mathbf{x}[1]$, and $\mathbf{x}[2]$, and $\mathbf{e}_2$ is the average of $\mathbf{x}[3]$, $\mathbf{x}[4]$, and $\mathbf{x}[5]$, and so forth. Note that there is no time overlap in ensemble averaging, unlike moving averaging.

and the result is usually used to compute the covariance matrix of the data samples. Moving averaging is a common method for filtering out the noise in thermocouple or strain gauge readings. Generally speaking, if the sampling frequency is relatively low, ensemble averaging can be applied. Note that the ensemble average can be a scalar, too. Therefore, you can acquire 100 thermocouple readings from a single channel at each sampling incident and average them (ensemble averaging) to generate a scalar average value, and such steps can repeat continuously. If the sampling frequency is relatively high, moving averaging can be more appropriate since it is a continuous operation with less overhead than ensemble averaging can impose. A sample realization of moving averaging in LabVIEW is shown in Figure 14.9. In this case, you can change the window size in real time without losing any data sample for averaging. The sub VI **New X.vi** is shown in Figure 14.10. As for ensemble averaging, it is left as an exercise at the end of this chapter.

14.3.2 Frequency Domain Analysis

Analysis of data in the frequency domain provides a great deal of detailed information about signals such as frequency component distribution and types of embedded noise. In order to understand different techniques of frequency analysis, this section will review a few fundamental transformation methods: Fourier series (FS) expansion, Fourier transform (FT), discrete Fourier transform (DFT), and discrete time Fourier transform (DTFT).

Fourier series (FS) expansion is defined by (14.27) for a deterministic and periodic continuous time signal $x(t)$, which satisfies the following three conditions: (1) $\int_{T_0} |x(t)|\,dt < \infty$, (2) $x(t)$ has a finite number of maxima and minima in any finite time interval, and (3) $x(t)$ has a finite number of discontinuities in any finite time interval where each discontinuity, too, is finite. Note that T_0 is the period of the signal $x(t)$, ω_0 is the fundamental frequency $2\pi/T_0$ in radian/sec, and $j = \sqrt{-1}$.

$$x(t) = \sum_{k=-\infty}^{\infty} a_k e^{jk\omega_0 t}, \quad a_k = \frac{1}{T_0}\int_{T_0} x(t)e^{-jk\omega_0 t}\,dt \tag{14.27}$$

Fourier transform (FT) is defined by (14.28) for a deterministic and aperiodic continuous time signal $x(t)$, which satisfies the same three conditions described for FS, except for the first one with different integral limit such as $\int_{-\infty}^{\infty} |x(t)|\,dt < \infty$.

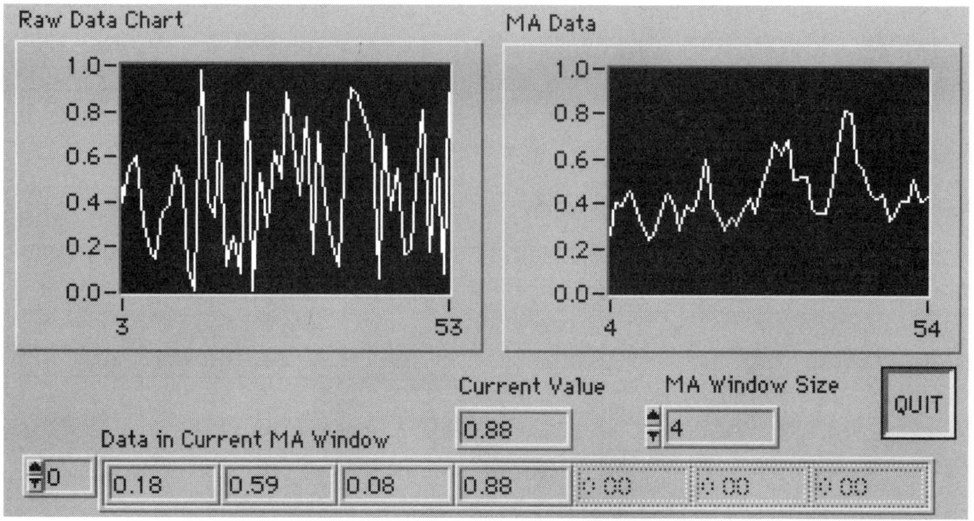

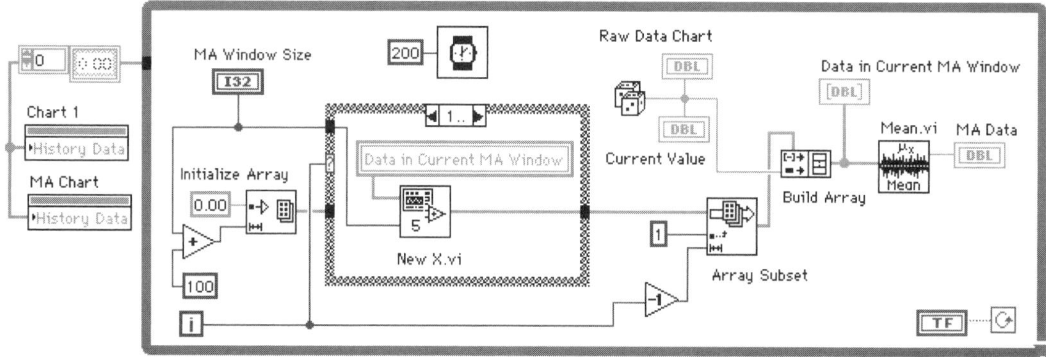

Figure 14.9(a)

The front panel and the diagram window of an example of moving average. You can change the window size in real time without losing any data samples in this example.

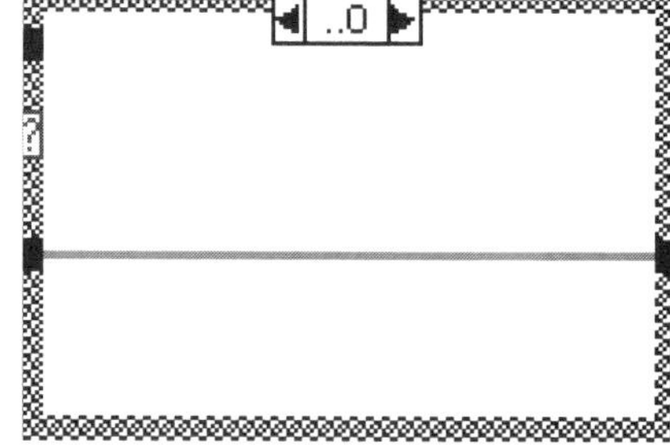

Figure 14.9(b) The other frame of the **Case** structure in (a).

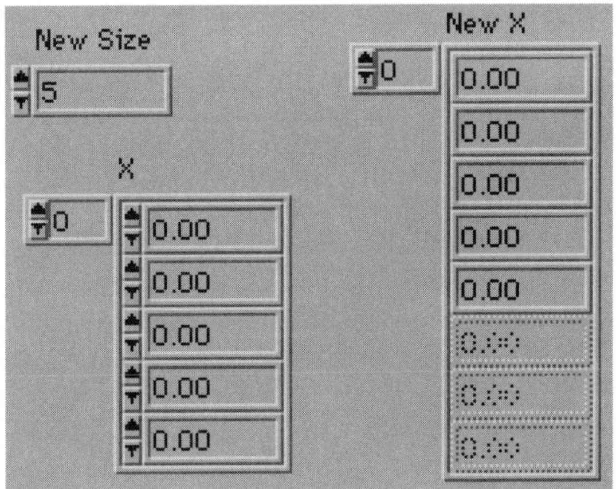

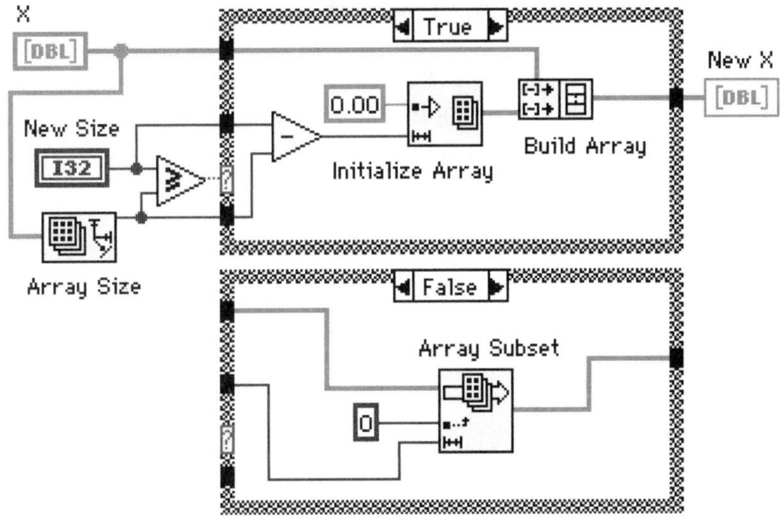

Figure 14.10
Sub VI **New X.vi**, which is used in Figure 14.9.

$$x(t) = \frac{1}{2\pi}\int_{-\infty}^{\infty} X(\omega)e^{j\omega t}d\omega, X(\omega) = \int_{-\infty}^{\infty} x(t)e^{-j\omega t}dt \tag{14.28}$$

Discrete Fourier transform (DFT) or discrete time Fourier series (DTFS) is defined by (14.29) for a deterministic and periodic discrete time signal $x[n]$, which satisfies $\sum_{n=0}^{N-1}|x[n]| < \infty$, where N is the period of $x[n]$.

$$x[n] = \frac{1}{N}\sum_{m=0}^{N-1} X[m]e^{-j\frac{2\pi}{N}mn}, X[m] = \sum_{n=0}^{N-1} x[n]e^{-j\frac{2\pi}{N}mn} \tag{14.29}$$

Discrete time Fourier transform (DTFT) is defined by (14.30) for a deterministic and aperiodic discrete time signal $x[n]$, which satisfies $\sum_{n=-\infty}^{\infty} |x[n]| < \infty$.

$$x[n] = \frac{T}{2\pi}\int_{\frac{2\pi}{T}} X'(\omega)e^{j\omega nT}d\omega,\ X'(\omega) = \sum_{n=-\infty}^{\infty} x[n]^{-j\omega nT} \tag{14.30}$$

The characteristics of signals for each transformation method are summarized in Table 14.1. Note that DTFT of $x[n]$, $X'(\omega)$, can also be obtained by substituting $e^{j\omega nT}$ for z in the z-transform of $x[n]$, $X(z)$.

It was assumed in the definition of all four transformation methods that the signal was deterministic, not stochastic. If the signal is stochastic, the same transformation can still be applied, but the correlation function of the signal should be used instead of the original signal. If the signal is a WSS stochastic process $\mathbf{X}(t)$, and the system $h(t)$ is linear, the output $\mathbf{Y}(t)$ is also a WSS stochastic process, and they are related as

$$S_{YY}(\omega) = S_{XX}(\omega)\,|\,H(\omega)\,|^2 \tag{14.31}$$

where $S_{XX}(\omega)$ and $S_{YY}(\omega)$ are the *power spectrum* or *spectral density* of WSS stochastic processes $\mathbf{X}(t)$ and $\mathbf{Y}(t)$, which is the FT of their autocorrelation functions $R_{XX}(\tau)$ and $R_{YY}(\tau)$, respectively. The autocorrelation function was defined by (14.6).

If the signal $x(t)$ is continuous, aperiodic, deterministic, and Fourier transformable, the FTs of $x(t)$ and the output $y(t)$ are related as

$$Y(\omega) = X(\omega)\,H(\omega) \tag{14.32}$$

Table 14.1. *Comparison of each method in the time and the frequency domain.*

	Time Domain	**Frequency Domain**
FS	continuous, periodic	discrete, aperiodic
FT	continuous, aperiodic	continuous, aperiodic
DFT or DTFS	discrete, periodic	discrete, periodic
DTFT	discrete, aperiodic	continuous, periodic

where $H(\omega)$ is the FT of the system impulse response $h(t)$. (Stochastic processes are denoted in boldface, and deterministic signals are italic. Therefore, $\mathbf{X}(t)$ is a stochastic process whereas $x(t)$ is a deterministic signal.) In other words, the FT of the output is the multiplication of the FT of the input and the system response. Recall that the output of a linear system was the convolution of the input and the system response (or impulse response) where the input is a deterministic signal. This implies that convolution in the time domain corresponds to the multiplication in the frequency domain as given by (14.32).

The FT of an impulse signal is white in the frequency domain, implying that its frequency components are uniformly distributed throughout the entire frequency range. Therefore, its FT will be flat in the frequency domain. If you multiply the frequency contents in $H(\omega)$ by uniformly distributed frequency contents in the impulse input signal, $Y(\omega)$ will be merely a scaled version of $H(\omega)$. If the scaling factor is assumed to be 1, then $Y(\omega)$ becomes $H(\omega)$. This means that if you want to know the characteristics of a linear system in the frequency domain, you need to apply an impulse signal to the system and observe the FT of the output signal $Y(\omega)$. This was also discussed previously using convolution instead of FT. As for stochastic input signals, you apply a white noise whose power spectrum (FT of its autocorrelation function) is also flat in frequency. Therefore, if you want to know the characteristics of a linear system in the frequency domain using a stochastic process, you can apply a white noise to the system and observe the FT of the autocorrelation of the output signal ($S_{YY}(\omega)$). Then $H(\omega)$, FT of the system response $h(t)$, will satisfy $|H(\omega)|^2 = S_{YY}(\omega)$.

Now, apply the system theory that has been discussed so far to an LTI system Butterworth LPF. Apply the impulse signal to the system (Butterworth LPF), and study the FT of the output to observe the frequency response of the filter. Since your system and signals are all discrete in this example, apply DFT instead of FT. This example can be useful if you do not know which filter to choose from many filter VIs in LabVIEW. In such a case, you can easily check the frequency response of each filter VI by using this example.

In Figure 14.11, the front panel and the diagram window of such an example are shown. This example uses **Real FFT.vi** to perform DFT on the output of the LTI system Butterworth filter using the impulse signal generated by **Impulse Pattern.vi** as the input. The graph in the front panel displays the frequency response of the Butterworth LPF. Note that the sampling frequency is 100 Hz, the cutoff frequency of the LPF is 25 Hz, and the order of filter is 6. This section will study this example in detail since it contains interesting and important information about FT.

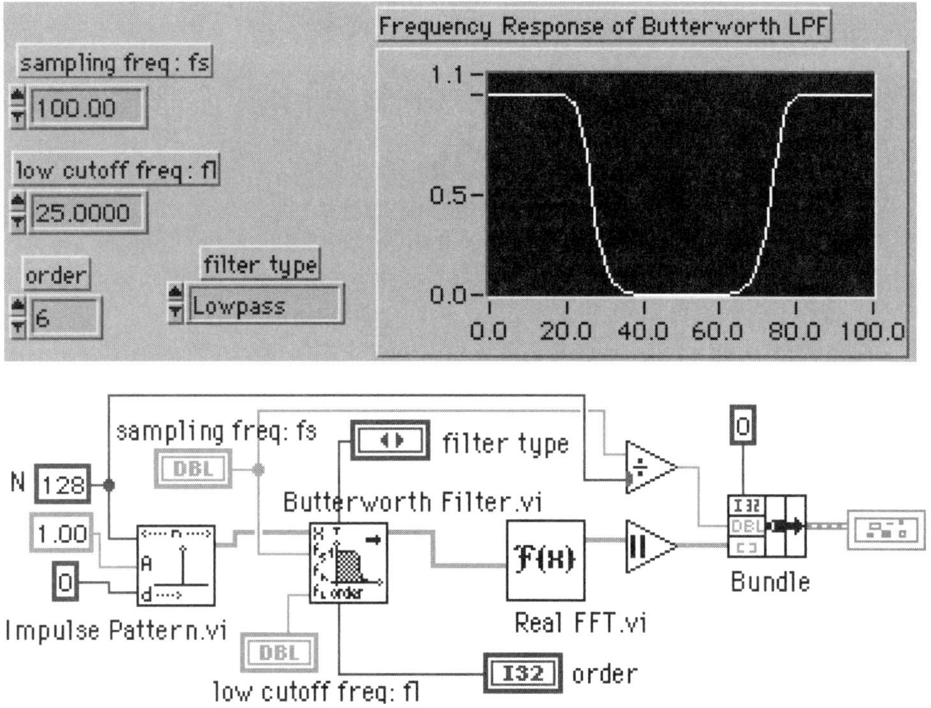

Figure 14.11
Frequency response of the Butterworth LPF using the impulse function.

First, you are using **Bundle** to specify the initial value and the interval of x-axis, which represents the frequency in Hz. Note that the interval is f_s/N, where f_s is the sampling frequency in Hz, and N is the total number of data samples. (128 samples are used in this example.) You are applying DFT assuming that the discrete output signal (the impulse response of the system) is periodic. In other words, you are assuming that the 128 samples of both the discrete input signal and the discrete output signal are one period of the periodic version of both input and output signals. Then the use of DFT is justified. In fact, you are using a fast Fourier transform (FFT) algorithm, which is an improved version of DFT to optimize the computation.

Second, the frequency response is two sided (i.e., the frequency content beyond 50 Hz (half the sampling frequency) is a mirror image of that in 0–50 Hz). Therefore, you need to display the result of FFT only up to the half the sampling frequency.

The interval of the result of DFT or DTFS is f_s/N, where f_s is the sampling frequency and N is the total number of data samples. Also, the upper half of the frequency response is simply a mirror image of the lower half; therefore, only the first half needs to be displayed.

Further discussion about the VIs of DFT is presented in a later section of this chapter. Before moving on to the next section, here is a summary of the frequency analysis techniques that have been covered so far:

1. The relationship between the input and the output in the frequency domain is important. If the input is deterministic, (14.32) describes the relationship. If the input is stochastic, (14.31) explains it.
2. The frequency response of an LTI system can be obtained by either a deterministic input or a stochastic process. The impulse signal and the white noise are the inputs for each case.
3. The interval of the frequency in the display of the result of FFT (or DFT or DTFS) is f_s/N, where f_s is the sampling frequency and N is the total number of samples.
4. The upper half of the result of FFT is merely a mirror image of the lower half. Therefore, only the lower half needs to be displayed.

14.3.3 Digital Filter Types

A filter is a system that modifies its input in order to generate a desired output. For example, you may want to average the input samples using a moving average filter. An LPF can be used to block frequency components above a threshold. A filter can even be spatial instead of temporal. For example, a filter can block some signal components arriving from a certain direction.

This section will review different types of filters and their design methods. Since the goal in this textbook is to learn G-programming in LabVIEW, the discussion will be limited to digital filters rather than analog filters.

One way to categorize filters is based on their impulse response. If a filter has a finite duration of impulse response, it is said to be finite impulse response (FIR) type. If a filter has an infinite duration of impulse response, it is said to be infinite impulse response (IIR) type. FIR filters are inherently stable since they only have zeros and no poles. Another major advantage of FIR type over IIR type is linear phase. If the impulse response is even or odd symmetric, the phase information becomes linear with respect to the amount of shift in the impulse response of an FIR filter. This is an important feature if you need to analyze noncausal systems. Since the time shift in the impulse response would show up as a linear phase, noncausal systems can easily be analyzed using the causal version of the noncausal systems, and the proper phase shift information. IIR filters, however, do not provide such a linear phase feature. Also, FIR filters can easily be implemented adaptively whereas IIR filters are relatively difficult to be realized using an adaptive algorithm.

The IIR filters have an infinite duration of impulse response, but due to their recursive nature, the implementation of IIR filters is much simpler than that of FIR filters, which will increase the filtering speed and decrease the computational load. Also, the order of IIR filters is relatively lower than that of FIR filters for the same performance requirement. The IIR filters have both poles and zeros; therefore, they can be unstable if any pole is not within the unit circle in the z-domain, or even oscillate.

The common FIR filter design methods are (1) the window and truncation method, (2) the frequency sampling design method, and (3) the minimax error design method. The window and truncation method is the easiest one and provides a closed-form equation for the filter coefficient calculation. This method computes the filter coefficients as follows: A desired continuous frequency response is given. Consider such a response as the result of DTFT. Note that DTFT provides a continuous frequency spectrum, as shown in Table 14.1. Then, using the inverse DTFT, you can calculate the infinite and discrete impulse response of the filter. Since you are interested in designing an FIR filter, the impulse response of infinite length needs to be truncated. This process is called windowing, and there are numerous types of windows that can be applied. Some examples are Rectangular, Hanning, Hamming, Blackman, Fejer-Cesaro, Dolph-Chevyshev, Papoulis, Kaiser, Turkey, and Parzen windows. Each has advantages and disadvantages; therefore, the task is to choose the best window for your application. After a window is applied to the infinite impulse response obtained from the inverse DTFT, you will have an FIR filter.

On the other hand, the frequency sampling design method uses DFT. Given a desired continuous frequency response, sample N points in frequency. Treat those N frequency samples as the result of DFT since DFT produces discrete samples in the frequency domain, as shown in Table 14.1. Then you can compute the impulse response of length N by the inverse DFT.

Note that this will introduce errors since you are *sampling* from the desired continuous frequency response. Therefore, the samples should have a reasonable resolution to represent the response fairly accurately; otherwise, the resulting filter can be dramatically off the specs desired.

The last category, the minimax method, is to design an FIR filter while minimizing the error (mean square error) and maximizing the output signal to noise ratio. This type of filter is often employed in adaptive filter applications, and it will continuously change (adapt) its impulse response to maintain the design criteria met.

A few common design methods of IIR filters are (1) the bilinear transformation method, (2) the impulse invariant transformation method, and (3) the frequency transformation method. The first two methods convert an analog filter to a digital one. The bilinear transformation method uses an analog filter that has the desired frequency response in the s-domain (Laplace transform). Given the frequency response in the s-domain, convert it to the z-domain using bilinear transformation. This conversion will yield a frequency response in the z-domain, and its discrete impulse response can be obtained by the inverse DTFT, which will provide you with an impulse response of an infinite duration (IIR filter). The second method is also based on an analog filter, but the impulse invariant transformation method uniformly *samples* the analog impulse response to obtain discrete impulse response samples. The analog impulse response $h(t)$ can be obtained by the inverse Laplace transform of the frequency response in the s-domain ($H(s)$). (Recall that the desired frequency response $H(s)$ is assumed to be given in the s-domain (Laplace transform).) Since you are *sampling* the resulting continuous impulse response $h(t)$ to obtain $h[n]$, the frequency response (DTFT of $h[n]$) of the discrete impulse response $h[n]$ would be a repeated version of the original frequency response $H(s)$. (Recall that the sampling process creates repeated replicates of the original frequency response.)

As shown in Figure 14.12, you will have aliasing inevitably at both ends of each replicated frequency responses. This implies that impulse invariant transformation is mostly suitable for lowpass filtering since the aliasing will degrade the accuracy in the frequency response on its tails.

Lastly, the frequency transformation method directly or indirectly converts the continuous frequency response from the s-domain to the z-domain using a conversion table. The types of resulting IIR filters from those various methods are (1) Butterworth type, which provides a maximally flat passband frequency response; (2) Bessel type, which provides a maximally flat group delay; (3) Chebychev type, which produces an equi-ripple either in the passband or in the stopband; and (4) elliptic type, which causes an equi-ripple

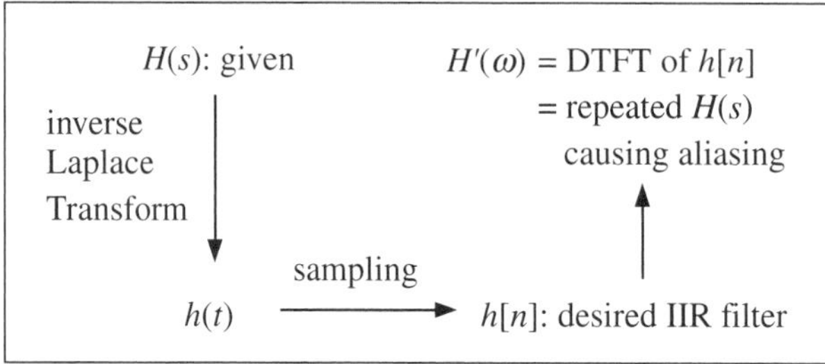

Figure 14.12
Impulse invariant transformation method to design an IIR filter.

in both the passband and the stopband. In order to choose the right type of filters in LabVIEW, you should first understand the advantages and disadvantages of FIR and IIR types.

14.3.4 Decimation and Interpolation

The Nyquist sampling theorem states that if you do not sample a bandlimited signal as fast as twice the frequency bandwidth of the signal, you will experience aliasing. Then what would be the effect of interpreting the data samples using a different sampling frequency from what was originally used? In audio processing applications, the sampling rate often changes at different stages of processing. For example, suppose that your music CD contains music data sampled at 44.1 kHz, but your CD player can only play the CD using 22 kHz sampling frequency. Then sampling rate conversion must be performed before playing the music recorded at 44.1 kHz.

There are two types of sampling rate conversion: decimation and interpolation. The effects of decimation and interpolation in the time domain are shown in Figure 14.13. Note that the interpolation process does not imply that it estimates the missing values between known values. (This is usually how *interpolation* is understood.) In digital signal processing, the interpolation process is a zero stuffing operation between samples. The factor of decimation and interpolation in Figure 14.13 is 3. Then the following summaries are given without a proof. Note that the frequency contents of *sampled* data have the original spectrum (the spectrum of the continuous signal being sampled) and replicates of it. The center frequencies of those replicates are the multiples of sampling frequency.

1. The decimation process stretches the frequency components by the decimation factor D. This will cause an overlapping at both ends of each

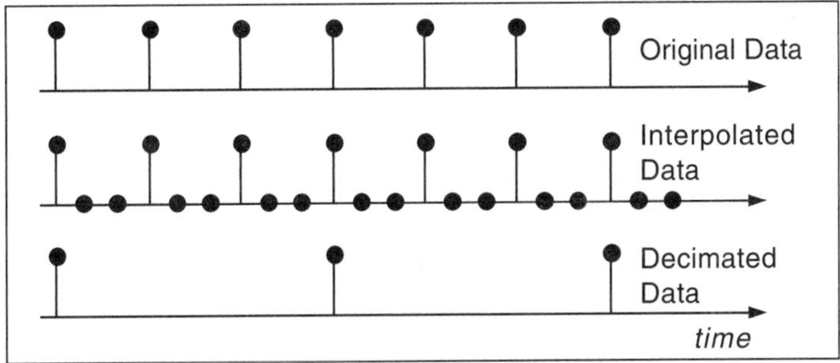

Figure 14.13
Interpolation and decimation process.

frequency content. Therefore, an LPF whose cutoff frequency is π/D should be applied *before* the decimation stage in order to avoid such aliasing. Note that the range of the original frequency content is assumed to be from $-\pi$ to π.

2. The interpolation process squeezes the frequency replicates about the zero frequency by the interpolation factor I. This will give you more than the original spectrum in its frequency range; therefore, an LPF whose cutoff frequency is π/I should be applied *after* the interpolation process to block any unnecessary duplicates of the original spectrum. Again, the range of the original frequency content is assumed to be from $-\pi$ to π.

Therefore, in order to convert the sampling frequency from one to another, a different combination of decimation rates and interpolation rates may be implemented while an LPF with a proper cutoff frequency accompanies before or after each stage. The decimation and interpolation processes are common in audio signal processing and third octave analysis.

14.4 Matrix and Vector Representation of Data

If more than a single data sample is acquired during the data acquisition process, carrying and processing multiple data samples as a set would be much more efficient than treating each sample separately. A matrix is a data structure that has columns and rows of multiple data samples, and a vector is a special case of the matrix where there is only a single column or a single row. In this section, the convention of a vector will be a column vector.

$$\mathbf{X} = \begin{bmatrix} 1 & -3 & 2 \\ 0 & 7 & -5 \end{bmatrix} = [\mathbf{x}_1 \; \mathbf{x}_2 \; \mathbf{x}_3]$$
$$\mathbf{x}_1 = \begin{bmatrix} 1 \\ 0 \end{bmatrix}, \; \mathbf{x}_1^T = [1 \; 0], \; \mathbf{x}_2 = \begin{bmatrix} -3 \\ 7 \end{bmatrix}, \; \mathbf{x}_3 = \begin{bmatrix} 2 \\ -5 \end{bmatrix} \tag{14.33}$$

Equation (14.33) shows an example of a matrix and column vectors. Matrices will use bold uppercase letters, and vectors will use bold lowercase letters. The superscript ^{T} indicates the transpose operation.

If you collect M real RV data sample vectors $\mathbf{x}_i$ of length N and construct an RV data sample matrix $\mathbf{X}$ of size N by M, such as $\mathbf{X} = [\mathbf{x}_1 \; \mathbf{x}_2 \; . \; . \; . \; \mathbf{x}_M]$, the correlation matrix $\mathbf{R}$ can be obtained by

$$\mathbf{R} = \mathscr{E}\{\mathbf{x}_i\mathbf{x}_i^T\} \approx \mathbf{X}\mathbf{X}^T = \sum_{i=1}^{M} \mathbf{x}_i\mathbf{x}_i^T \tag{14.34}$$

This matrix is often encountered when your application involves the stochastic processes. Then $\mathbf{R}$ is said to be positive semidefinite if $\mathbf{w}^T\mathbf{R}\mathbf{w} \geq 0$ for all nonzero vector $\mathbf{w}$. Also, $\mathbf{R}$ is said to be symmetric if it satisfies $\mathbf{R}^T = \mathbf{R}$, which is the case.

Symmetric matrices can be decomposed into $\mathbf{LDL}^T$, where $\mathbf{L}$ is a lower triangular matrix and $\mathbf{D}$ is a diagonal matrix, and the lower triangular matrix $\mathbf{L}$ is not necessarily unique. If $\mathbf{R}$ is positive definite (i.e., $\mathbf{w}^T\mathbf{R}\mathbf{w} > 0$ for all nonzero vector $\mathbf{w}$), the Cholesky factorization decomposes it into $\mathbf{R} = \mathbf{\Delta}^T\mathbf{\Delta}$, where $\mathbf{\Delta}$ is an upper triangular matrix, and furthermore it is unique. If $\mathbf{x}_i$ contains complex data and $\mathbf{R}$ satisfies $\mathbf{R}^H = \mathbf{R}$, where the superscript ^{H} indicates complex conjugate transpose, $\mathbf{R}$ is said to be Hermitian. Hermitian matrices can be factored into $\mathbf{R} = \mathbf{E}\mathbf{\Lambda}\mathbf{E}^H$ using eigendecomposition, where $\mathbf{E}$ is the eigenmatrix and $\mathbf{\Lambda}$ has eigenvalues on its diagonal axis and zeros elsewhere. This implies that $\mathbf{R}\mathbf{e}_i = \lambda_i\mathbf{e}_i$ for all i, where $\mathbf{E} = [\mathbf{e}_1 \; \mathbf{e}_2 \; . \; . \; . \; \mathbf{e}_M]$ and $\mathbf{e}_i$ are the eigenvectors for the eigenvalues λ_i. Also, the data sample matrix $\mathbf{X}$ can be decomposed into an orthonormal matrix $\mathbf{Q}$ ($\mathbf{Q}^H\mathbf{Q} = \mathbf{Q}\mathbf{Q}^H = \mathbf{I}$) and an upper triangular matrix $\mathbf{\Delta}$ satistying $\mathbf{X} = \mathbf{Q}\mathbf{\Delta}$. Note that the dimension of $\mathbf{Q}$ is N by N (a square matrix), $\mathbf{\Delta}$ is N by M (a rectangular matrix), and $\mathbf{I}$ is the identity matrix that has 1's on its diagonal axis and 0's elsewhere. The upper triangular matrix $\mathbf{\Delta}$ in this case is sometimes labeled $\mathbf{R}$, and this factorization is called QR factorization. The singular value decomposition (SVD) of $\mathbf{R}$ (the correlation matrix) is $\mathbf{U\Sigma V}^H$, where $\mathbf{U}$ is an N by N orthogonal matrix and $\mathbf{V}$ is an M by M orthogonal matrix. The diagonal elements of $\mathbf{\Sigma}$ contains min(N, M) number of singular values in descending order and max(N, M) – min(N, M) number of zeros.

The rank of a matrix is the number of linearly independent rows or columns. An N by M matrix is said to be full rank or nonsingular if the rank is min(N, M) and the inverse of a matrix only exists for nonsingular square matrices. (An inverse of a square matrix $\mathbf{A}$ is denoted as $\mathbf{A}^{-1}$ and satisfies $\mathbf{A}\mathbf{A}^{-1} = \mathbf{A}^{-1}\mathbf{A} = \mathbf{I}$.) As for nonzero rectangular matrices, their inverse can be computed by the pseudo inverse, which is the least mean square solution. The pseudo inverse of $\mathbf{X}$ is $(\mathbf{X}^H\mathbf{X})^{-1}\mathbf{X}^H$, and it exists for any nonzero matrix. All these important aspects about matrix representation are summarized as follows:

1. If $\mathbf{R}$ is symmetric, $\mathbf{R} = \mathbf{L}\mathbf{D}\mathbf{L}^T$, where $\mathbf{L}$ is a lower triangular matrix and $\mathbf{D}$ is a diagonal matrix.
2. If $\mathbf{R}$ is positive definite, $\mathbf{R} = \mathbf{\Delta}^T\mathbf{\Delta}$, where $\mathbf{\Delta}$ is an unique upper triangular matrix. (Cholesky factorization)
3. If $\mathbf{R}$ is Hermitian, $\mathbf{R} = \mathbf{E}\mathbf{\Lambda}\mathbf{E}^H$, where $\mathbf{E}$ is the eigenmatrix and $\mathbf{\Lambda}$ has eigenvalues on its diagonal axis and zeros elsewhere. (eigendecomposition)
4. For any matrix $\mathbf{X}$, $\mathbf{X} = \mathbf{Q}\mathbf{\Delta}$, where $\mathbf{Q}$ is an orthonormal matrix and $\mathbf{\Delta}$ is an upper triangular matrix. $\mathbf{Q}$ is square and $\mathbf{\Delta}$ is rectangular. (QR factorization)
5. If $\mathbf{X}$ is rectangular, $\mathbf{X} = \mathbf{U}\mathbf{\Sigma}\mathbf{V}^H$, where $\mathbf{U}$ is an N by N orthogonal matrix and $\mathbf{V}$ is an M by M orthogonal matrix. (singular value decomposition)
6. Rank of a matrix is the number of linearly independent column or row vectors. A matrix is singular if it does not have an inverse.
7. The inverse of a matrix only exists for nonsingular square matrices.
8. Pseudo inverse exists for any nonzero matrix $\mathbf{X}$ and is $(\mathbf{X}^H\mathbf{X})^{-1}\mathbf{X}^H$.
9. Eigenvectors of a Hermitian matrix are orthogonal.
10. Eigenvalues of a Hermitian matrix are real.
11. Eigenvalues of the correlation matrix are real and nonnegative.
12. The correlation matrix of a discrete-time WSS process is Hermitian, Toeplitz, and positive semidefinite.
13. The correlation matrix of a discrete-time WSS process is almost always positive definite.

14.5 Analysis Subpalettes

LabVIEW provides various tools for data analysis. Under **Functions** palette >> **Analysis**, there are 10 subpalettes: **Signal Generation**, **Digital Signal Processing**, **Measurement**, **Filters**, **Windows**, **Curve Fitting**, **Probability**

and Statistics, **Linear Algebra**, **Array Operations**, and **Additional Numerical Methods**. (The subpalette **Analysis** has been divided into two new ones, **Signal Processing** and **Mathematics**, in LabVIEW 5.1 or higher. See your LabVIEW package about the changes.) Each subpalette contains VIs that perform tasks of each category, and this section will discuss some of the useful VIs among them. However, depending on the version of LabVIEW, the name and the location of those VIs in your LabVIEW may differ from those listed in this section. (This book uses the names and the locations of VIs based on LabVIEW 5.0.) If that is the case, then you should refer to the upgrade information, which describes the differences among different versions of LabVIEW. Nonetheless, all of the examples and illustrations do not depend on the versions of LabVIEW; therefore, nothing would become obsolete in this book unless LabVIEW makes a dramatic change! The only extra work you need to do is finding the new location of the VIs, and such information is provided in your LabVIEW package. Keeping this in mind, consider the first subpalette **Signal Generation.**

Signal Generation contains VIs that generate different kinds of stochastic and deterministic signal samples, such as sine, square pulse, impulse, ramp, white noise, periodic random noise, and even some arbitrary waveforms. Each VI takes proper input parameters such as amplitude, phase, or number of samples, and returns a waveform in a 1-D array. Note that the impulse function or the white noise can be used to obtain the frequency response of a linear system.

Digital Signal Processing contains VIs for processing discrete signal samples. **Convolution.vi** returns the convolution of two input arrays. If the lengths of the two input arrays are N and M, the output array will be of length $N + M - 1$, and this is the result of linear convolution. The other type is circular convolution, which returns a 1-D array of the same length as the two input arrays. The two input arrays must be of the same length for the circular convolution. If the lengths of input arrays are N and M ($N \neq M$), you can create another two input arrays of length $N + M - 1$, where $M - 1$ and $N - 1$ zeros are added at the *end* of each array of length N and M, respectively. Then the output of the circular convolution will be an array of length $N + M - 1$. (However, LabVIEW does not provide a VI for circular convolution.) The other VIs in this subpalette are self-explanatory. Fast Fourier transform (FFT) is a computationally improved version of DFT or DTFS.

The Hilbert transform is useful if you want to represent (1) a complex analytic signal, or (2) the pre-envelope of a real-valued bandpass signal whose frequency contents are concentrated in a narrowband frequency range in the vicinity of a frequency $\pm f_c$ (usually modulating frequency). The Hilbert transform is also useful for bandpass sampling. You know from the Nyquist sampling theorem that you must sample faster than twice the maximum frequency of the (baseband) signal to be sampled to avoid aliasing. Suppose that the signal to be sampled was a modulated signal with carrier frequency of 850 MHz, with a bandwidth of 20 kHz (an example of the cellular communication). In order to sample this modulated signal, would you have to find a sampling circuitry that can sample faster than twice 850 MHz + 10 kHz? Obviously, sampling at that fast of a frequency is not practical, and the correct answer would be to downconvert the complex analytic signal or the pre-envelope of the real-valued modulated signal to obtain the equivalent lowpass (baseband) representation, and sample it faster than twice the 20-kHz bandwidth.

Derivative x(t).vi and **Integral x(t).vi** perform running derivative and running integral operations. Numeric integration can be achieved by using **Numeric Integration.vi** in the subpalette **Additional Numerical Methods**. (Refer to Figure 14.7 for the comparison between **Integral x(t).vi** and **Numeric Integration.vi**.)

FHT.vi is similar to the FFT routine except that it requires less computation than FFT. However, the number of samples in the input of **FHT.vi** must be a power of 2. As for FFT, there are two types of VIs: **Real FFT.vi** and **Complex FFT.vi**, and each name refers to the input type. In other words, **Real FFT.vi** is for real input data, and **Complex FFT.vi** is for complex input data. However, the output of any FFT routine is always complex, consisting of magnitude and phase information. Also, the interval of frequency axis (resolution) is f_s/N where f_s is the sampling frequency and N is the total number of data samples. When you display the result of FFT, only the lower half would need to be displayed since the other upper half is a mirror image of the lower half. Figure 14.4 shows an example of FFT setup.

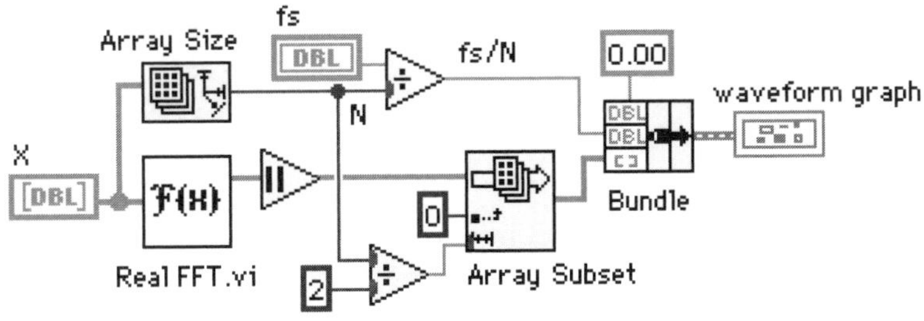

Figure 14.14
Typical setup of FFT process of the input data array **X**. Note the interval of frequency axis and absolute value of the complex output of **Real FFT.vi**. Only the lower half of the result from **Real FFT.vi** is displayed.

One last point about FFT VIs is that they return phase information that repeats (wrapped phase). In order to undo this process, you may use **Unwrap Phase.vi** to obtain a nonrepeating phase.

Measurement contains VIs for higher-level data processing, and those VIs use the VIs in the subpalette **Digital Signal Processing**. You can obtain the frequency response of a linear system, compute the power spectrum of stochastic processes, or estimate the amplitude and the phase using the VIs in this subpalette.

Filters contains numerous types of FIR and IIR filters, such as Butterworth, Chebyshev, elliptic, and Bessel filters. The unit of frequency in the inputs of those VIs is Hz, not a normalized frequency by the sampling frequency. In a normalized frequency unit, 1 is the sampling frequency and 0.5 is the half of the sampling frequency. Since DFT returns a frequency content including a mirror image, you would only need to display the output up to 0.5 in normalized frequency, since the other half between 0.5 and 1 is just the mirror image.

It is important to understand the trade-off between the filter order and the filtering speed. Recall the convolution equation for discrete linear system. For convenience, it is shown in (14.35) again.

$$y[n] = x[n] * h[n] = \sum_{k=-\infty}^{+\infty} x[k]h[n-k] \tag{14.35}$$

The steps of convolution are (1) flip either signal sample about the time origin, and (2) slide the flipped signal (either $x[n]$ or $h[n]$) to the right while multiplying the values of the overlapped samples, and adding the results. The resulting sum becomes the output at that particular time index.

Since the filtering process is a convolution between the input data sequences ($x[n]$) and the impulse response sequence ($h[n]$) of a linear system (filter), the filtering process can be illustrated using Figure 14.15. Suppose $h[n]$ has an impulse response, as shown in Figure 14.15. It has four identical values at n = 0, 1, 2, and 3. This kind of filter is called a moving average (MA) filter with unity coefficients. Then you first choose to flip the impulse response of the MA filter and start sliding it in to the right as you multiply overlapped samples with the input samples, add the results, and assign the total sum to $y[n]$. As you can see, the complete overlap between $h[n-k]$ and $x[k]$ does not occur until n is 3, which is $M-1$ where $M-1$ is the *order* of the MA filter ($M = 4$). What if M was very large such as 20? Then the complete filtering process will not occur until 19 shifts are made. (We are assuming that the length of $x[n]$ is sufficiently larger than M.) Filters of higher order will have sharper and more accurate frequency responses, which will lead to better performance, but the transition time until the complete filtering process occurs will be much longer than that of lower-order filters. There-

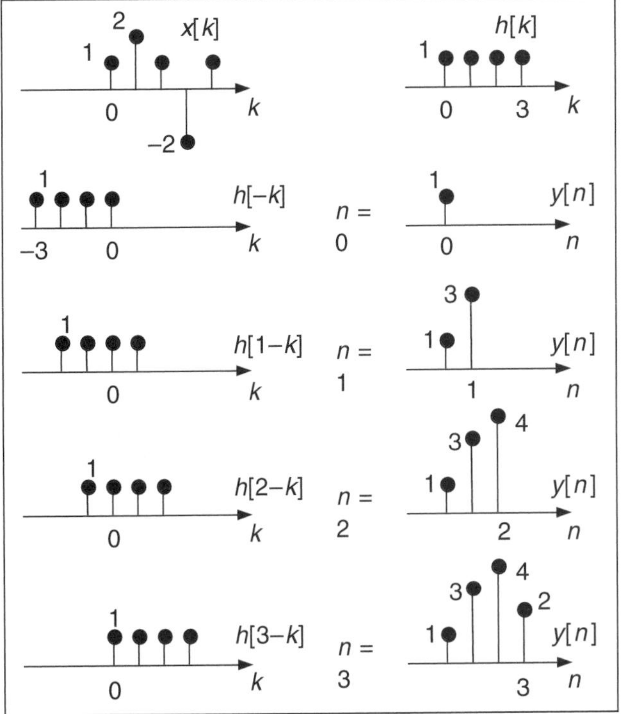

Figure 14.15
Convolution operation.

fore, the order of filters must be carefully chosen. Too low an order may return unsatisfactory filtering results, and too high an order may take too long a transition time until the complete filtering process begins.

The input **init/cont (init:F)** is also closely related to such trade-offs. If you initialize the filter every time it repeats for a continuous filtering process where data stream in continuously, the time needed for filters to settle down will be consumed at each iteration. That will obviously affect the speed of filtering as well as the overall performance. The input **init/cont (init:F)** is similar to the input **iteration** of data acquisition VIs, which controls the hardware configuration. Therefore, you need to initialize a filter only once at the first iteration when continuous filtering is performed. Figure 14.16 shows how to acheive such a task.

Windows provides coefficients of different types of window functions, such as Hamming, Hanning, Blackman, and many others. As mentioned earlier, window functions are useful in the FIR filter design, but they can also enhance the results of DFT. The simplest example of the use of

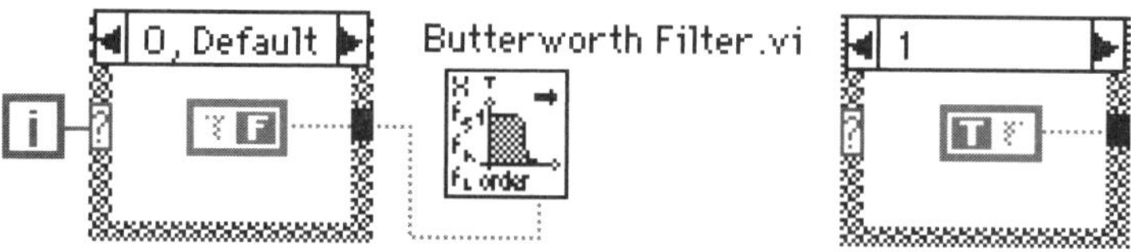

Figure 14.16
Controlling the initialization of a filter for a continuous filtering application. The iteration terminal of loops such as **For Loop** or **While Loop** is wired to a **Case** structure with two frames.

window functions can be seen by asking the following question: Can you have a signal of infinite time duration? The answer is no, since all of the data you collect and store has a finite length. (You must stop acquiring data at some point.) This implies that you are discarding or ignoring some of the beginning and some of the end portion of the original infinitely long signal by sampling it. For example, if you have the samples of sine wave of a certain frequency, the acquired data is a finite portion of the true sine wave ranging from $-\infty$ to ∞. If you use $x[n]$ to denote the finite length data, and $s[n]$ for the true sine wave, $x[n]$ can be expressed as follows:

$$x[n] = s[n]w[n] \tag{14.36}$$

where $w[n]$ has 1's for the same finite duration as $x[n]$, and 0's outside the duration. Such a function $w[n]$ is called a window and is a rectangular window in this example. Since a square pulse is a sinc function in the frequency domain, and multiplication in time corresponds to convolution in frequency, the result of a windowing operation using a rectangular window will be the convolution of the frequency content of the true signal $s[n]$ and the sinc function. This will obviously blur the original frequency content, which is the expected result of rectangular windowing.

This example also shows that you have been processing data samples of finite length without knowing about the existence of rectangular window, which prohibits the accurate display of the frequency content of the original signal. Such a blurring effect on the spectrum display can be enhanced or improved by using different types of window functions.

Curve Fitting provides VIs for data fitting as well as data interpolation. Those VIs accept an array of data to be fit and another array whose elements are x coordinates of each value, and return either the coefficients of a polynomial for a specified order or the parameters for different optional equations.

Probability and Statistics contains VIs that provide statistical analysis tools; for example, the probability density functions (pdf's)

for different types of distributions, such as chi square, F, and T distributions. Also, this palette contains VIs for the inverse of those pdf's. In other words, given the probability, the inverse VIs return the original observation value. For example, the cumulative density function (cdf) returns the cumulative probability for a given observation x such that $\mathcal{P}\{\omega \in \Omega \mid \mathbf{X}(\omega) < x\}$, where $\mathbf{X}(\omega)$ is an RV mapping the outcome ω in the space Ω to a real number $\mathbf{X}(\omega)$. Then the inverse VIs will return x, given the cumulative probability $\mathcal{P}\{\omega \in \Omega \mid \mathbf{X}(\omega) < x\}$. Knowing that $\lim_{x\to\infty} \mathcal{P}\{\omega \in \Omega \mid \mathbf{X}(\omega) < x\} = 1$, the probability of the complement set $\mathcal{P}\{\omega \in \Omega \mid \mathbf{X}(\omega) \geq x\}$ can be calculated easily.

The mean square error (MSE) is a common optimization category and is defined by

$$\mathrm{MSE} = \mathcal{E}\{\,|\mathbf{X} - \hat{x}|^2\} \tag{14.37}$$

where $\mathbf{X}$ is an RV and $\hat{x}$ is its estimate. Note that $\mathbf{X} - \hat{x}$ is the error in the estimate. Another common measure in parameter estimation is the root mean square (RMS), and it is is defined by

$$\mathrm{RMS} = \sqrt{\mathcal{E}\{|\mathbf{X}|^2\}} \tag{14.38}$$

If you assume that the discrete probability density function of a discrete RV $\mathbf{X}_i$ is equally likely, then given N observations $\mathbf{X}_1, \ldots, \mathbf{X}_N$, expectation operator $\mathcal{E}$ can be replaced by a summation of each observation weighted by $1/N$. Therefore, for example, (14.38) can be written as

$$\mathrm{RMS} = \sqrt{\frac{1}{N}\sum_{i=1}^{N}|\mathbf{X}_i^{\,2}|} \tag{14.39}$$

and (14.37) can similarly be rewritten in terms of summation and N.

Linear Algebra provides VIs for matrix and vector manipulation and decomposition such as SVD, LU, QR, eigendecomposition, and many others. **Complex Linear Algebra** subpalette contains the VIs for complex matrices and vectors.

Array Operations contains VIs for tasks with 1-D arrays (vectors) and 2-D arrays (matrices), such as normalizing an array, scaling an array, evaluating a straight line equation for a given 1-D or 2-D array, or coordinate conversion between polar and rectangular. One useful usage of **Quick Scale 1D.vi** is shown in Figure 14.17.

Additional Numerical Methods contains four VIs that are ready to be used for some canned applications, such as peak detection, root

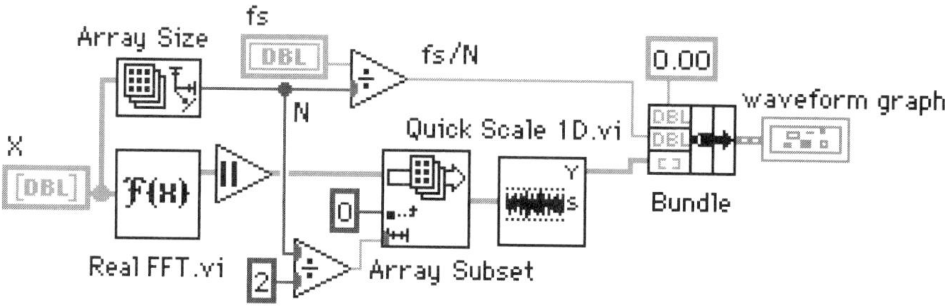

Figure 14.17
Usage of **Quick Scale 1D.vi** to normalize the absolute magnitude of FFT without finding the maximum absolute value in the FFT output manually.

calculation of complex polynomials, and numerical integration. Detailed discussion of numerical integration is presented in Section 14.3.1.

PROBLEMS

14.1 Identify the causality of the following systems.

(a) $y[n] = x[n] - x[n-1]$
(b) $y[n] = ax[n], a \neq 0$
(c) $y[n] = x[n] + 3x[n+4]$
(d) $y[n] = x[n^2]$
(e) $y[n] = x[2n]$
(f) $y[n] = x[-n]$
(g) $y[n] = \sum_{k=n-1}^{n+3} x[k]$

14.2 Identify the linearity of the following systems.

(a) $y[n] = nx[n]$
(b) $y[n] = x[n^2]$
(c) $y[n] = x^2[n]$
(d) $y[n] = ax[n] + b, a \neq 0, b \neq 0$
(e) $y[n] = e^{x[n]}$
(f) $y[n] = x[-n]$

14.3 Compute the convolution output of the following $x[n]$ and $h[n]$. Note that $h[n]$ is an LTI system. The symbol • indicates $n = 0$.

(a) $x[n] = \{\ldots, 0, 0, \overset{\bullet}{1}, 2, 3, 2, 1, 0, 0, \ldots\}$
$h[n] = \{\ldots, 0, 0, \overset{\bullet}{1}, 1, 1, 1, 1, 0, 0, \ldots\}$

(b) $x[n] = \{\ldots, 0, 0, \overset{\bullet}{1}, 0, 0, \ldots\}$
$h[n] = \{\ldots, 0, 0, \overset{\bullet}{1}, -1, 1, -1, 1, 0, 0, \ldots\}$

(c) $x[n] = \{\ldots, 0, 0, \overset{\bullet}{1}, 2, 3, 4, 5, 6, 0, 0, \ldots\}$
$h[n] = \{\ldots, 0, 0, \overset{\bullet}{1}, 2, 3, -3, -2, -1, 0, 0, \ldots\}$

14.4 Verify your answers in Problem 14.3 using **Convolution.vi**, and save the VI as **P14_04.vi**.

14.5 Design a VI that runs continuously until a button **QUIT** is pressed while executing the following tasks:

(a) The VI displays the frequency response of all three impulse responses in Problem 14.3.

(b) The VI displays the frequency response of one additional $h[n]$, which is as follows:

$$h[n] = \{\ldots, 0, 0, \overset{\bullet}{1}, 2, 3, 3, 2, 1, 0, 0, \ldots\}$$

(c) Use a menu ring to select an impulse response from the four choices. In order to add more items to the menu ring, the key combination **Shift-Enter** may be used.

(d) Use 1 Hz for the sampling frequency; therefore, you will be using the normalized frequency.

(e) Use *zero padding* to increase the resolution of the frequency response display.

The front panel of a sample solution is shown in Figure P14.5. Based on the frequency response, which system is an LPF? Also, which one is an HPF? Can you tell the type of the filter by observing the impulse response?

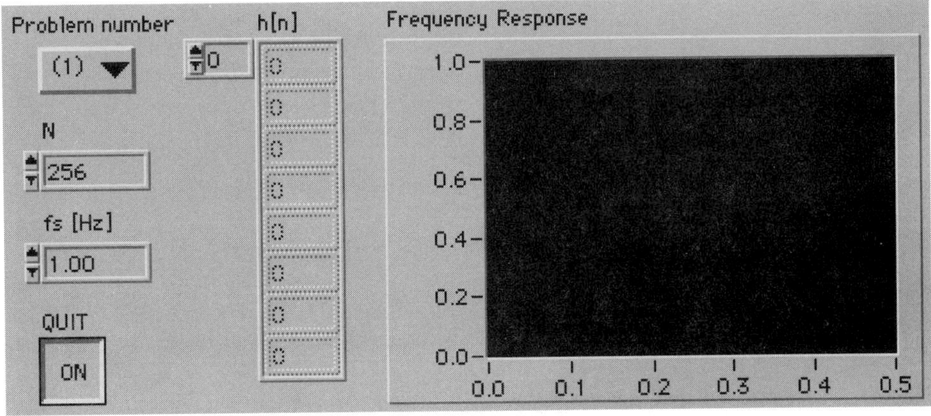

Figure P14.5

Zero padding: Recall that when you display the result of FFT, the interval of the frequency axis is f_s/N, where f_s is the sampling frequency and N is the total number of samples. This implies that N controls the resolution of the frequency response but has no effect on the shape of the response itself. Therefore, if you have a small number of data samples and wish to view the spectrum distribution, you can add additional zeros at the end of the original data samples to increase the resolution of the spectrum display. For example, the impulse response $h[n]$ in the first problem in Problem 14.3 has only five nonzero values. Then, by adding additional 123 zeros at the end, you can use $N = 128$ to compute the spectrum using FFT algorithm.

14.6 In this problem, you will prove the following relationship:

$$S_{YY}(\omega) = S_{XX}(\omega)\,|H(\omega)|^2$$

Follow these steps to design **P14_06.vi**, which proves the foregoing relationship.

(a) Generate N number of white Gaussian noise samples, $x[n]$, using **Gaussian White Noise.vi**. Compute its discrete autocorrelation function $R_{XX}[n]$ using **AutoCorrelation.vi**. What is the length of the output $R_{XX}[n]$?

(b) Generate $h[n]$ of length $2N - 1$, where the first N samples are 1's, and the other $N - 1$ samples are 0's. This is an MA filter of order $N - 1$ with unity coefficients.

(c) Convolve $x[n]$ with $h[n]$ to generate the output $y[n]$ using **Convolution.vi**. What is the length of $y[n]$, which is the result of the linear convolution of $x[n]$ with $h[n]$? Note that $x[n]$ is of length N, and $h[n]$ is of length $2N - 1$.

(d) Compute the discrete autocorrelation function $R_{YY}[n]$ using **AutoCorrelation.vi**. What is the length of the output?

(e) Compute $S_{YY}(\omega)$ using **Real FFT.vi** and **Zero Padder.vi**. The VI **Zero Padder.vi** will insert 0's at the end of its input array to make the size of the resulting array the next higher power of 2.

(f) Zero pad both $R_{XX}[n]$ and $h[n]$ to match their length to that of zero padded $R_{YY}[n]$, which is the output of **Zero Padder.vi** in step (e). Having the same resolution in both $S_{YY}(\omega)$ and $S_{XX}(\omega)\,|H(\omega)|^2$ is important since you will be calculating the error by subtracting each frequency sample.

(g) Compute $S_{XX}(\omega)$ and $H(\omega)$ using **Real FFT.vi**. Note that both results must be of the same length as $S_{YY}(\omega)$.

(h) Multiply $S_{XX}(\omega)$ and $|H(\omega)|^2$ array element by array element.

(i) Plot $S_{YY}(\omega)$ and $S_{XX}(\omega)\,|H(\omega)|^2$, which are the results from steps (e) and (h), using two **Waveform Graphs**. Use the normalized frequency for x-axis.

(j) Compute the maximum of the two arrays $S_{YY}(\omega)$ and $S_{XX}(\omega)\,|H(\omega)|^2$. Are they identical?

(k) Compute the mean square error (MSE) of the two arrays $S_{YY}(\omega)$ and $S_{XX}(\omega)\,|H(\omega)|^2$ using **MSE.vi**. What is the MSE?

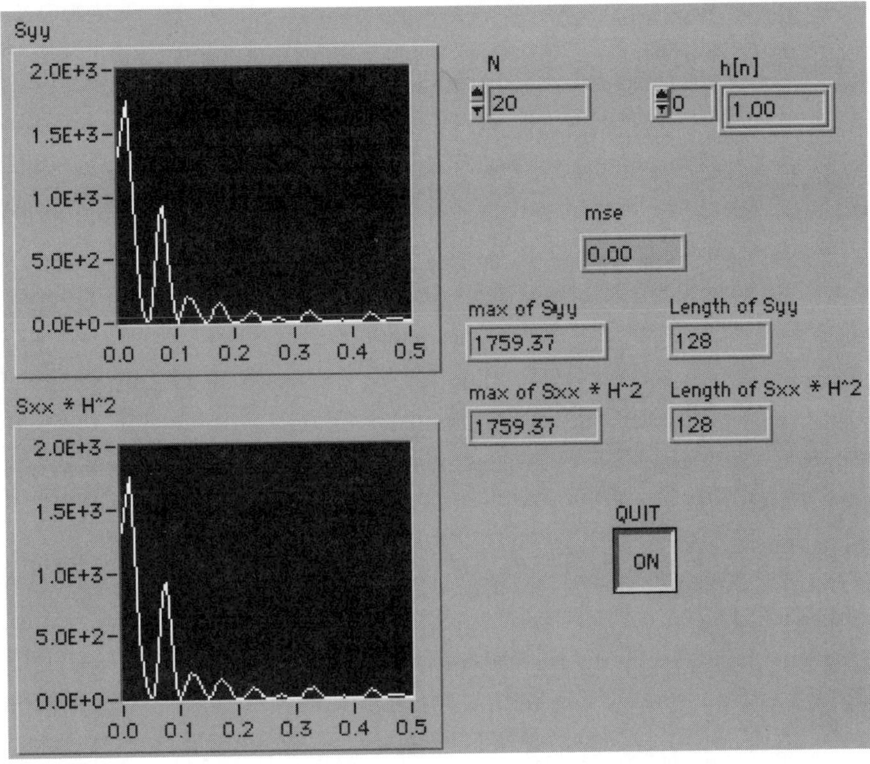

Figure P14.6

(l) Place the entire content from the previous steps in a **While Loop**, and make it iterate every 1 second until the Boolean button **QUIT** is pressed. The front panel of a sample solution is given in Figure P14.6.

14.7 (a) Design a VI that computes an ensemble average of data vectors of size 4, using the window size 3. In other words, you have a stream of incoming data vectors **x**[k] of size 4 and average every 3 vectors continuously. Therefore, the first ensemble average will be the average of **x**[0], **x**[1], and **x**[2]; the second ensemble average will be the average of **x**[3], **x**[4], and **x**[5], and so forth. Save the VI as **P14_07a.vi**.

(b) Modify **P14_07a.vi** so that you can change the size of the window as well as the size of the data vector in real time without losing any data samples for averaging. Save the VI as **P14_07b.vi**.

14.8 Generate N number of random numbers between 0 and 1 every 250 msec, and filter them continuously using **Butterworth Filter.vi**. Make sure not to initialize the filter repeatedly. While filtering the samples, display separately the results of FFT both before and after filtering in dB scale. Use the normalized frequency for the x-axis. Design the VI so that you can change the following parameters while running the VI:

Filter type (lowpass or highpass)

Order of the filter

Number of data samples (N)

Save the VI as **P14_08.vi**. Run the VI while changing the parameters to see the effect of them. The front panel of a sample solution is shown in Figure P14.8.

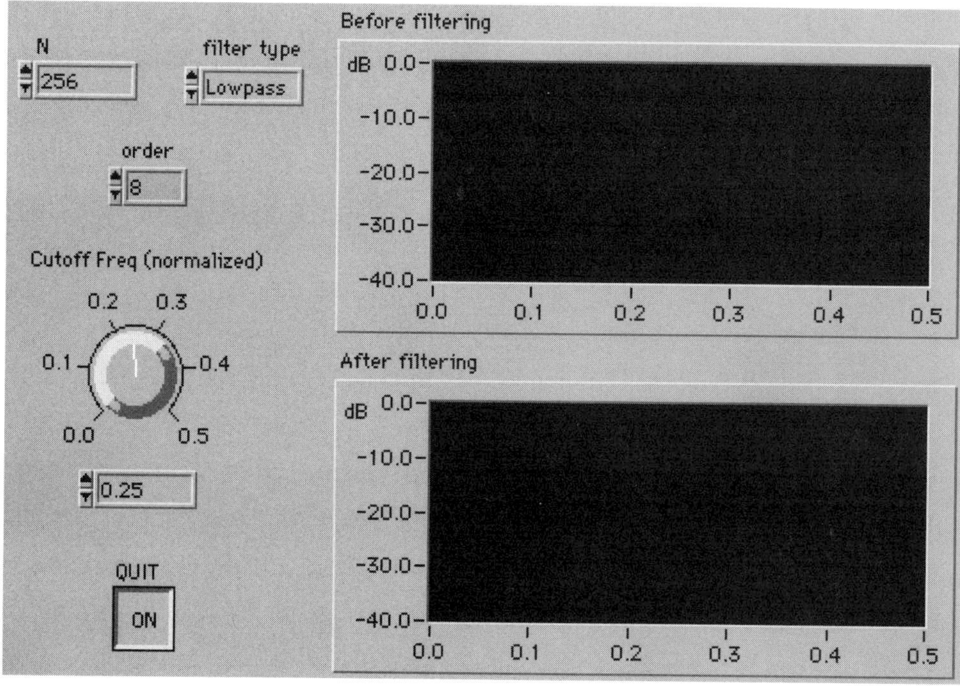

Figure P14.8

14.9 Suppose that you are acquiring a data sample at each iteration in a **While Loop** and want to have the latest average at all time. Such a task can be achieved by the following equation, which is often called *running average*:

$$s_k = \frac{x_k + ks_{k-1}}{k+1} \tag{14.40}$$

where s_k is the average of $x_0, x_1, \ldots, x_k$, and x_k is the data sample acquired at the $(k + 1)$th iteration, and $k = 0, 1, \ldots$.

(a) Prove that (14.40) performs such a task.

(b) What would be the advantage of this equation over the conventional method, which is described by the following equation (14.41)?

$$s_k = \frac{1}{k+1}\sum_{i=0}^{k} x_i \tag{14.41}$$

(c) Design a VI that performs (14.40). Use **Random Number (0-1)** to generate the data sample.

Building an Application

15

This chapter will cover the topics that pertain to creating your own application using all of the techniques that have been covered in the previous chapters. Some of those topics may play a critical role in writing an application, such as memory management, and making the executable efficient in terms of speed and overheads. Some of the topics apply to stand-alone applications only, and they will be denoted as such at the time of discussion. For example, calling sub VIs dynamically increases the efficiency of a VI run very much only in the LabVIEW environment. Also, practical examples are provided and studied to explain each topic. This chapter is indeed a very important one since your ultimate goal is to write an efficient application, not something that merely works. At the end of this chapter, suggestions for further reading are made for those who would like to know more details about the topics presented in this book. The chapter begins with the customization of the **Controls** and the **Functions** palettes.

15.1 Palette Customization

You have been studying the VIs and functions in the **Controls** palette and the **Functions** palette that can be brought up by right clicking in the front panel and the diagram window, respectively. LabVIEW allows you to have your own controls and VIs appear in each palette just like the other LabVIEW ones. This can be very useful if you frequently use your customized objects (controls and indicators) and VIs (sub VIs), and having them in the two LabVIEW palettes will make it easy to access them. This section will illustrate how to do this task with the following example.

Suppose that you want to create a Boolean button with an image as shown in Figure 15.1.

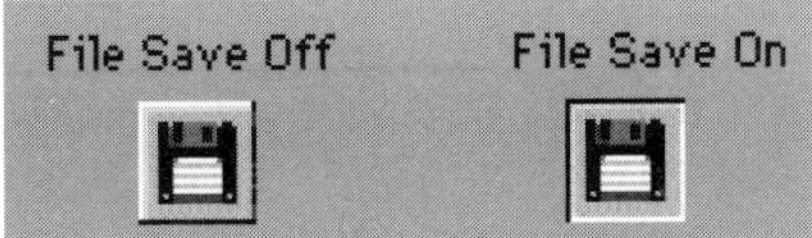

Figure 15.1 Customized Boolean button to control file save operation.

To create such a button, follow these steps:

Step 1: Place a Boolean button **Square Button** from **Controls >> Boolean** in the front panel.

Step 2: Select the button using the Arrow Tool, and choose **Edit Control** from the **Edit** pull-down menu. This will bring up another panel similar to the regular front panel of LabVIEW.

Step 3: Click on the button with a tool image on it to initiate the *Customize Mode* (Figure 15.2).

Step 4: Go to a graphics editing program of your choice to create an image as shown in Figure 15.1. Save the image as a bitmap. Any program that can create bitmap images should work. Note that you need to create two images: one for OFF and OFF transition states, and the other for ON and ON transition states.

Step 5: Copy one of the images onto the clipboard in the graphics editing program. Usually the shortcut **Ctrl-C** on PC platforms, or **Command-C** on Macintosh platforms, does such a task.

Step 6: Go to LabVIEW, where you currently have a Boolean button in the *Customize Mode*. Right click on the button, and choose **Import Picture**. This will paste the image from the clipboard onto the button. Since there are four different modes on Boolean buttons (OFF, ON, OFF transition, ON transition), you need to copy each corresponding image onto each different mode.

Step 7: Right click on the button and select **Picture Item** from its pop-up menu. It will show you all of the four states, and you just finished the OFF state. Select the second state ON, and copy the other image from the graphics editing program to the clipboard and onto the Boolean button, as you did in Steps 5 and 6.

Figure 15.2 Edit Mode/Customize Mode switch.

Step 8: The OFF transition and ON transition states will be of concern if you select nontrivial mechanical action, such as **Latch When Released**. Depending on the action of the buttons, you may need to create four separate images. In this example, you are using only two.

Step 9: Save the control under a proper name, and close the editing window. You just finished creating a customized button. Now include it in the **Controls** palette. Go to the front panel and choose **Edit Control & Function Palette** in the **Edit** pull-down menu. This will bring up both the **Controls** palette and the **Functions** palette.

Step 10: Click and pin down the sub control palette **User Controls** (Figure 15.3).

Step 11: Right click in the **User Controls** sub control palette, and choose **Insert >> Custom Control(s)**. This will bring up a directory navigation window. Find the button you just created in the previous steps, and select it.

Step 12: Save the changes by choosing **Save Changes**. Now, the new button with the new images will appear under the **User Controls** palette.

The preceding example demonstrates both how to create a customized control and how to include it in the **Controls** palette. The same method applies in order to include your own sub VIs in the **Functions** palette. The only difference is to go to the **Functions** palette and open and pin down the sub function palette **User Libraries** instead of **User Controls** in Step 10. Once you pin down the **User Libraries** subpalette, right click to bring up its pop-up menu, insert the new sub VIs in **User Libraries**, and save it as you did in Steps 11 and 12. Therefore, those steps will be left as an exercise for you.

In fact, you can include not only individual sub VIs and customized controls but LabVIEW library files with extension **.llb** (Dot El El Be) or folders containing multiple sub VIs also. Furthermore, LabVIEW allows you to create a link to the file **dir.mnu**, which contains the menu information about its current directory. LabVIEW automatically creates such a file when you include a folder containing customized sub VIs in **User Libraries**. Therefore, if you transport your sub VIs saved in a folder to another machine, you only

Figure 15.3 **User Controls** in the **Controls** palette.

need to link to the **dir.mnu** file instead of manually including the directory in the **User Libraries** subpalette again. If you manually add the folder to **User Libraries** again, all of the information, such as the names of the sub VIs and the icon position in the palettes, will be lost.

Ask yourself the following question regarding palette customization before continuing to the next topic: "Should I use the LabVIEW library with extension **.llb** or a folder (or a directory) to save more than one sub VIs and include them in **User Libraries**?" To answer this question, you need a better understanding of the LabVIEW library file structure.

The LabVIEW library with extension **.llb** was initially implemented to have more than eight characters for a VI name when the operating system did not allow a long file name. If you use the **.llb** LabVIEW library file structure, you can have more than eight characters for any VI name regardless of the operating system. The advantage of **.llb** files is that usually the size is smaller than keeping all of the VIs individually since they are compressed. The disadvantage is that if the size of the **.llb** file becomes larger, it takes longer to save a small change in one of the embedded VIs in the **.llb** file. For example, suppose that there are 20 VIs saved in **project.llb**. If you make a small change in one of the VIs in it, the operating system will understand the change as that of **project.llb**, not a change of a single VI in it, since the operating system does not understand the LabVIEW library structure. Consequently, you will not be able to utilize all of the features of the operating system; for example, you cannot use the operating system feature *Find* to search for any individual VI saved in **project.llb**. If you save all of the 20 VIs as an individual VI in a folder or a directory, however, you can use the *Find* feature and the operating system can deal with each VI individually. Also, saving any change in the VI individually will be faster than updating the entire **.llb** file. Therefore, the answer to the question would be to use the LabVIEW library **.llb** when the size is fairly small, such as less than 2 megabytes (MBs), or when you would not be using any operating system feature pertaining to file management. Otherwise, save the VIs individually.

15.2 Occurrences

One of the unique advantages of LabVIEW over the text-based programming languages is the capability of running multiple **While Loops** simultaneously. In fact, there would be no true parallelism unless multiple tasks are run on multiple processors individually. LabVIEW will, however, make such emulation with the least amount of hassle: multiple **While Loop**s in the diagram window. Then when would running multiple **While Loop**s be necessary in G-programming? Suppose that you want to perform data acquisition while monitoring the user input, such as a password in an attempt to access a database. In such a situation, one **While Loop** will execute the data acquisition process while another **While Loop** monitors if any button in the front panel for password input has been pressed. When you use more than one **While Loop** in a VI, which will mostly be the case, you should recall one fundamental principle of the loops: Once they start, nothing else will enter or leave them. Consider a two **While Loop** scenario, which was briefly studied during the discussion of **Local Variable**s in Chapter 4 (Figure 15.4).

You have a continuous analog input process in the first loop while the second one continuously runs **User Monitor System.vi**. First, recall from Chapter 4 that the first loop will never get to run at full speed. Second, the VI will never stop once it starts. The reason why the first loop will never run at full speed is because LabVIEW will never be able to allocate the CPU time that the first loop needs to the first loop because of the second one. To make matters worse, if the second loop happens to fully occupy the CPU time, the first loop will never get a chance to start. Second, the second loop will never stop once it initiates its loop since there is no Boolean value to stop the iteration. The switch **QUIT** in the first loop initiates the second one, but the value FALSE of **QUIT** will never reach the conditional terminal of the second loop, resulting in an infinite loop. A modified version of Case 1 is shown in Figure 15.5.

The example in Figure 15.5 will now allow the data acquisition to run at much faster speed. A **Local Variable** of **QUIT** is used to control the iteration of the second loop. Note that there is no true parallelism in LabVIEW, so the

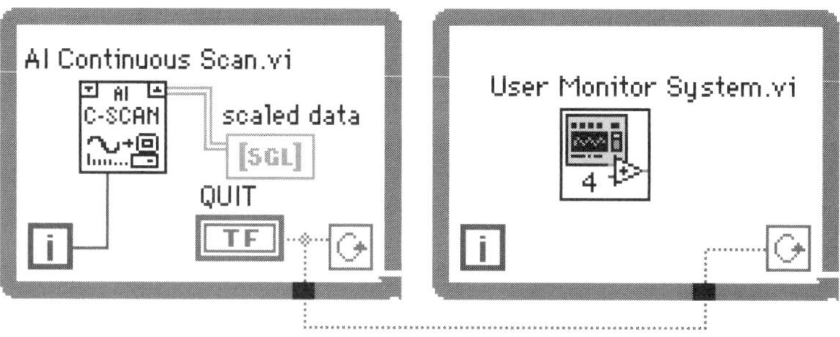

Figure 15.4
Multiple **While Loop**s (Case 1).

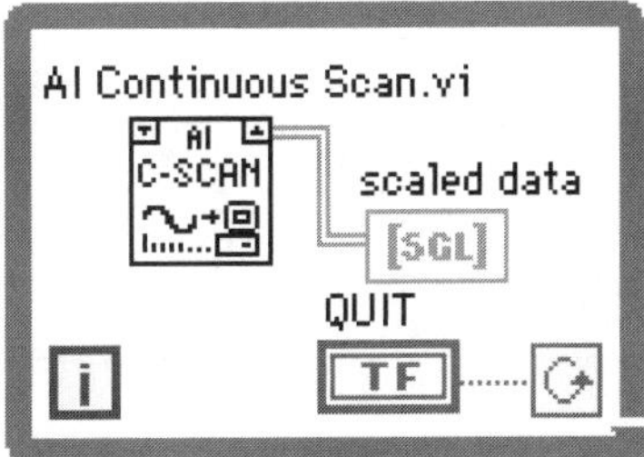

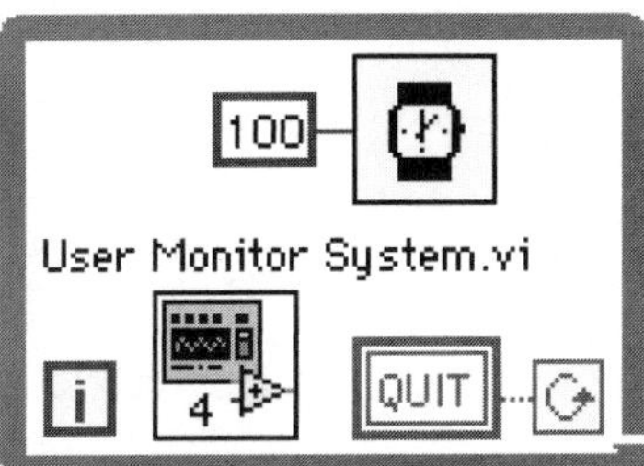

Figure 15.5
Multiple **While Loop**s (Case 2).

two loops will use the CPU time alternatively. However, due to the **Wait** function in the second loop, the first one reserves more CPU time so that the data acquisition part will not be disturbed. Therefore, place the **Wait** function in a loop with a lower-priority task. Checking the user input every 100 msec is usually frequent enough to recognize a change in the user input.

*Use the **Wait** function in lower-priority **While Loop**s to run multiple **While Loop**s simultaneously.*

The example in Case 2 is, however, still not complete. What if the time wait value was large? What if the second loop repeats once a day to check only the date? In such a case, when you click **QUIT** to stop both loops, the second one will not stop until it completes the current cycle, which will take 24 hours! In order to stop the second loop immediately when **QUIT** is pressed in the first loop, you need to introduce *occurrences*. Consider three **While Loops** running at different rates to illustrate the occurrences to stop all of them simultaneously.

Figures 15.6 and 15.7 show the two cases of three **While Loop**s with and without occurrences. In Figure 15.6, the first one has the highest priority, such as data acquisition, and the third one has the lowest priority, running at 500 seconds per iteration. Without using the occurrences, pressing **QUIT** will stop all three loops after completing one cycle of the last one, which will take 500 seconds. On the other hand, the modified version with occurrences in Figure 15.7 stops immediately after **QUIT** is pressed. Note the extra few steps that are required to achieve such a task.

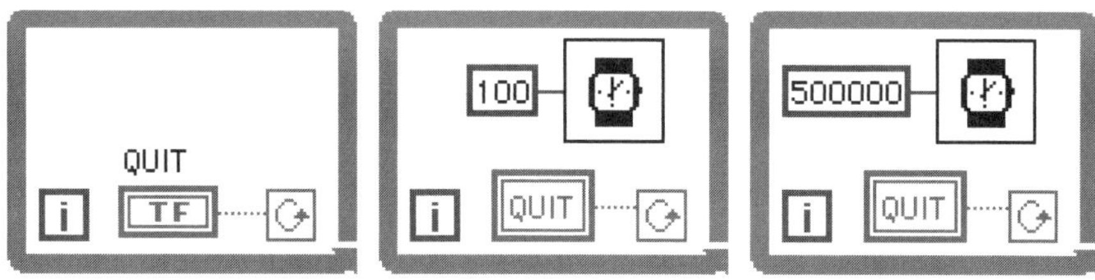

Figure 15.6
Three **While Loop**s running at different rates without occurrences.

The feature occurrences utilizes three functions: **Generate Occurrence**, **Wait on Occurrence**, and **Set Occurrence**. Without getting into too much detail about each of them, the following analogy may describe their functionality without causing any confusion. Think of a wireless communication between you and your boss. Each of you are holding a walkie-talkie, and you are waiting for the signal from your boss. In this scenario, there are three steps involved: Turn on the walkie-talkie units, wait for the signal, and execute the instruction from the boss. Each step corresponds to **Generate Occurrence**, **Wait on Occurrence**, and **Set Occurrence**, respectively (Figure 15.8), and they can be found in **Functions >> Advanced >> Synchronization >> Occurrences**.

In Figure 15.7, the first **While Loop** has the highest priority running at the fastest rate. When the VI starts running, a signal is generated by **Generate Occurrence** (turning on the walkie-talkie units). The signal then enters each **While Loop**, and **Wait on Occurrence** in each loop receives it while waiting for the instruction from **Set Occurrence** (your boss). When **QUIT** generates

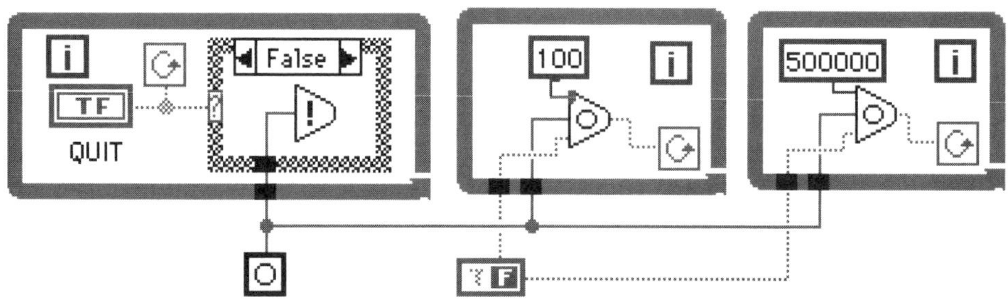

Figure 15.7
Three **While Loop**s with occurrences. The True frame of the **Case** structure in the first loop is empty.

Generate Occurrence   Wait on Occurrence   Set Occurrence

Figure 15.8
Three functions for occurrences.

FALSE to stop the first loop, **Set Occurrence** initiates the call to your walkie-talkie, delivering the stop command. Then the two of the **Wait on Occurrence**, connected to the conditional terminal, will generate FALSE to stop the second and the third **While Loops**. Note that **Wait on Occurrence** uses the top input **ms timeout** to control the iteration rate of the second and the third **While Loops**. A Boolean value FALSE is wired to the last input **ignore previous** to ignore the previous occurrence generated by **Generate Occurrence** during the last time out duration. Now, you can use the method in Figure 15.7 as a template for occurrences to control multiple **While Loops** more efficiently.

15.3 Memory Management

One core part of every application is memory management, and a good understanding of the memory usage of LabVIEW is an important aspect in writing your own application. When you open a VI, LabVIEW loads the front panel, the data stored in the VI, and the machine code of the VI into the memory. If the VI contains any sub VI, LabVIEW will also additionally load only the data stored in it and the machine code. If you make any change in the top-level VI and attempt to save it, LabVIEW will load the diagram of the top-level VI into the memory in order to recompile the code. If a sub VI controls any of its objects, such as controls and indicators, using their Attribute Nodes, LabVIEW will load the front panel of the sub VI into the memory as well. Such a behavior of LabVIEW, therefore, suggests that utilizing as many sub VIs as possible will save the memory space when you load the top-level VI. This explains why it takes a while to open a large VI because LabVIEW needs to load the proper portions of the top-level VI and of its sub VIs. In order to avoid such an inefficiency in memory usage, you can leave the sub VIs outside the top-level VI and load them into the memory only when you need them using VI Server functions. (A detailed discussion of them will take place in a later section of this chapter.)

When you write a multifunctional application in LabVIEW, the size of the VI can quickly grow, and soon it can be an order of a few megabytes. Consequently, a proper memory management will become a critical factor in the design process. LabVIEW also has overhead on the diagram editor, and to save a VI of just about 2 MB can take a few minutes to complete depending on the configuration of the computer. Therefore, as your VI becomes larger, the issue of memory management becomes more critically important.

There are some general rules that will maximize efficiency in memory usage:

1. Maximize the usage of sub VIs. Break up large VIs into small sub VIs.
2. Minimize the number of live updates on the front panel.
3. Minimize any dynamic change in data size, especially arrays and strings.
4. Minimize the usage of **Local Variable**s and **Global Variable**s.
5. Minimize the usage of Attribute Nodes, especially in sub VIs.
6. Avoid coercion dots.
7. Use simple data structure.

You already know the reason why sub VIs will take up less memory if they are kept outside the top-level VI. As for the live updates, displaying a continuous change in an indicator imposes much more overhead than displaying an intermittent change. The same rule applies to charts and graphs. Instead of displaying a single data a thousand times to display 1000 samples, display all of the samples once, reducing the number of live updates. Also, try to avoid any unnecessary displays on the front panel. As for arrays, make every possible effort to avoid the usage of the **Build Array** function to create an array, especially one of large size. This issue will play a critical role if a large number of data samples need to be acquired and analyzed in your application. By the same token, refrain from using **Concatenate Strings** to build a long string value. As for the usage of **Local Variable**s and **Global Variable**s, each of them creates a new copy of the original data. The Attribute Nodes in sub VIs will force LabVIEW to keep their front panel in the memory. The coercion dots indicate a data type conflict, and each incident will create a new copy of data of proper type. In some cases, it can be inevitable to have a data type conflict. Then try to convert the type to the desired one at the earliest stage. A simple example is shown in Figure 15.9, where the worst and the best situations are compared. Case 1 is the worst

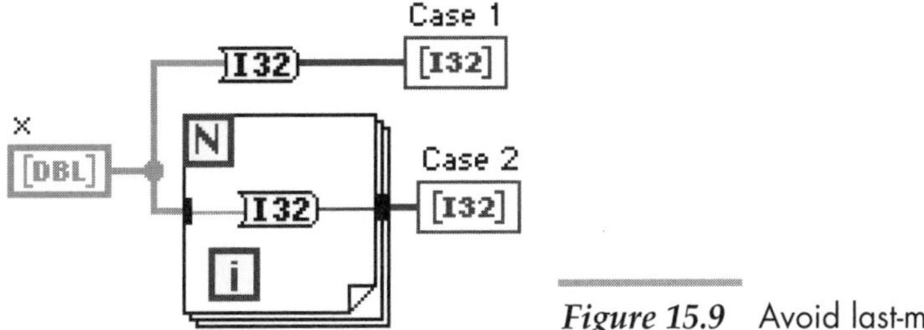

Figure 15.9 Avoid last-minute conversion.

case, whereas Case 2 is the best one. Case 1 creates another copy of the entire array **x**, whereas Case 2 simply creates a copy of a scalar. If you have a problematic source, remove it as early as possible!

Also, use simple data structure instead of a complicated hybrid version such as clusters. Unless it is absolutely necessary, use simple data type and structure. A further discussion of performance issues, such as execution speed, is presented in a later section of this chapter. Last in this section is an illustration of a situation where Mr. Stubborn is not using the memory effectively at all. Italic expressions indicate where he can improve his codes. Can you describe how you could improve where Mr. Stubborn went wrong?

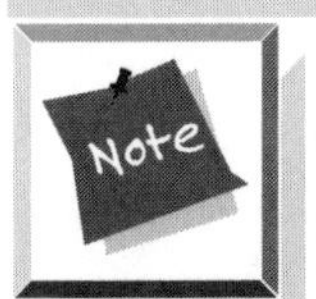

Mr. Stubborn is writing a database program in LabVIEW. He uses an array of clusters to keep the data, where each cluster element contains a numeric control, a ***Boolean array****, and* ***another cluster*** *containing three different types of data elements. Every time he needs to add a new element to the array, he* ***uses Build Array*** *while creating a new BIO record* ***using Concatenate Strings****. When he prompts the user to enter his or her salary and age, he uses* ***SGL*** *and* ***U32*** *types, but in the cluster record, he* ***uses DBL*** *for both. In order to update an item in the record, he* ***uses a Local Variable*** *to use* ***Bundle by Name****. Since he does not like to keep things in different places, he keeps* ***everything in one VI****. He also has an* ***indicator to display the time with 1/10 of a second precision****.*

15.4 Use of C/C++ Codes in LabVIEW

If you are a C/C++ programmer and like all of the advantages that LabVIEW offers and want to utilize both C/C++ and LabVIEW, there exists a solution to such a request: **Code Interface Node** (CIN) and **Call Library Function**. Those two functions can be found in **Functions >> Advanced**, and they allow you to use external libraries or C/C++ codes. This section will focus on CINs since the settings in LabVIEW for using **Call Library Function** are subject to programmers' different needs. Therefore, you are recommended to refer to the LabVIEW user manual for detailed settings about **Call Library Function**. Before proceeding to the details of CINs, consider the advantage and disadvantage of CINs and **Call Library Function.**

The advantage is that you can still utilize all of the prewritten C/C++ codes and dynamic link libraries (DLLs) in LabVIEW, saving development time. Also, execution time can be much less than realizing everything with LabVIEW sub VIs. The functions in LabVIEW with yellow color run very fast, but this may not be the case with higher-level VIs. The disadvantage is that you may encounter new problems to solve, such as memory allocation, since LabVIEW will no longer be managing such issues for you, and the programmer must handle them properly. Also, you should be very cautious in every setting in your compiler program. Any slight mistake can cause improper behavior or even a system crash. For example, you must exercise caution if your C/C++ code deals with arrays. As the array size becomes larger, the memory allocation problem can increase, and such a case usually requires a thorough understanding of the memory management in your system environment. In LabVIEW, you see that many of the VIs are already using CINs; an example is shown in Figure 15.10.

In order to create your own CIN, you need a C or C++ compiler program. Some of the compilers that can be used are (but not limited to)

Mac OS: Metrowerks Code Warrior by Metrowerks; THINK C or Symantec C++ by Symantec; Macintosh Programmer's Workshop (MPW) by Apple Computer

Microsoft Windows and NT: Win32 Microsoft Software Developer's Kit (SDK) C/C++ command line compiler, Visual C++, Watcom C/386

Solaris: Sun ANSI C Compiler

HP -UX: HP-UX C/ANSI C Compiler

You may check the latest compatibility listing with National Instruments before purchasing any compiler. A few examples of creating a CIN on Power Macintosh platforms using Metrowerks Code Warrior are included in

Figure 15.10 Diagram window of **Real FFT.vi** where a CIN is used.

Appendix A. Examples for PC platforms can be easily obtained from National Instruments, so interested readers should contact them directly for the latest examples on PC platforms.

15.5 Building an Application

This section will discuss more practical aspects along with examples in creating a stand-alone executable in LabVIEW. Based on the material that has been covered in the previous chapters, this section may play the most important role in programming using LabVIEW and may be considered as a summary of all of the topics in the previous chapters. Most of the examples shown in this section are based on actual applications. The last subsection discusses different aspects of stand-alone applications created by LabVIEW. This section will begin with error handler design.

15.5.1 Error Handler Design

You encountered error clusters when the data acquisition VIs, file I/O VIs, and instrumentation I/O VIs were studied. Error clusters contain error information to help programmers trace any error easily since such information indicates where the error was generated. This subsection covers how to create your own error control mechanism in LabVIEW. The main idea is as follows: Suppose that your application calls a series of sub VIs to perform a certain task, such as data acquisition, or a certain testing sequence. In order to ease the error tracing, you integrate a cluster that contains the error status into each sub VI. If the error status indicates that an error has occurred, all of the subsequent sub VIs will not execute their codes except to propagate the error information to the next VI. At the last stage, you display the error message, possibly with an error code.

In order to realize such a mechanism, you need to use the **Case** structure either to execute the code in each sub VI or to skip it and pass the error information to the next stage. Also, you need an error dictionary, which contains all of the error messages. Then, when an error occurs, the last stage of the process will look up the error string and display it. Consider a simple example where a customized error cluster is implemented. Figures 15.11 and 15.12 show an error handler design example where if the sum of a and b exceeds 10, error 1 message is brought up, and error 2 message if the sum is zero.

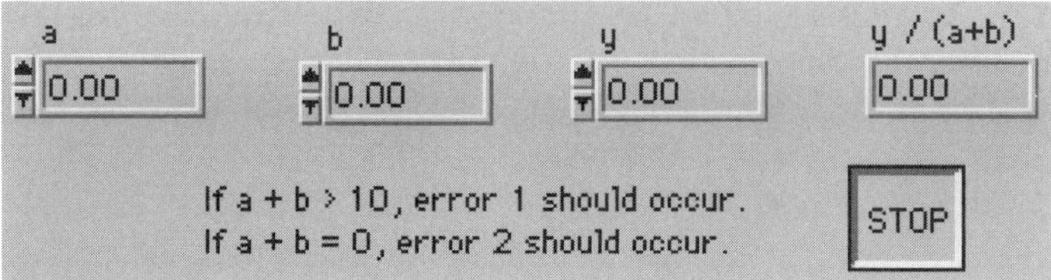

Figure 15.11
Error handler design example. Error 1 or 2 will be displayed at the end of the process if any of them has occurred.

The structure of the process is very much like the data acquisition sequence or file I/O sequence. The error cluster is wired throughout the entire chain of sub VIs, and each sub VI will execute its code only if no error has occurred from the previous sub VIs. Therefore, each sub VI has a **Case** structure to perform such a task. The last sub VI **Proc3.vi** then receives the propagated error information, looks up the error dictionary, and displays proper messages depending on the error status. The front panels and the diagram windows of **Proc1.vi**, **Proc2.vi**, and **Proc3.vi** can be found in Appendix B.

The advantage of having such an error handling mechanism in an application is the ease of error tracing, and the preceding example can be used as a template in designing error handler clusters.

15.5.2 Dynamic Sub VI Load and Unload

As you develop your VI, its size can grow quickly if many tasks are involved. As the size grows, it will take longer to load the VI into the memory as well as to save it after making any change. One of the test results shows that the operating system waited for minutes until it responded to the

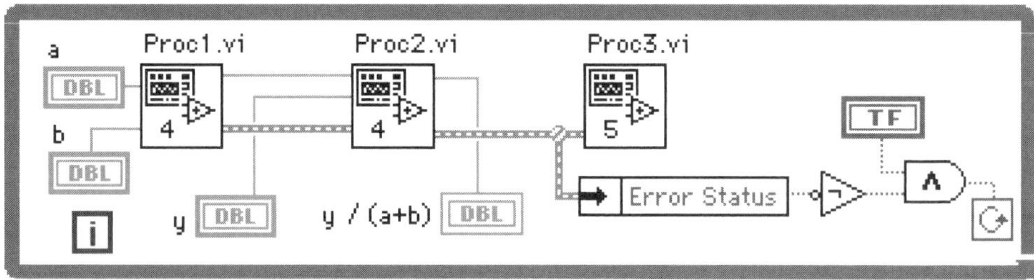

Figure 15.12
Diagram window of the error handler design example.

change that the programmer had entered, and the size of the VI was about 3 MB! (The test was performed on a PowerMac G3 400 MHz with 208 physical memory with Virtual Memory turned off. 40 MB of physical memory was allocated to LabVIEW. The same test was also administered on a 400 MHz Pentium II machine with 128 MB system RAM.) If the application involves a fair amount of special effects, the size of the VI quickly grows to be about that size or even far larger than that. One way to avoid such slow loading and response time is to break up the VI into smaller sub VIs and have them stay outside the top-level VI. Such a scheme is often called *plug-ins*. While the sub VIs remain independent of the top-level VI, they will be loaded into the memory when the top-level VI needs them. Such a task can be achieved by using VI Server functions.

The functionality of VI Server functions is not just limited to loading and unloading sub VIs dynamically. They can even control the property of VIs and external application. For example, you can change the settings in the **VI Setup** window of a VI or launch a third-party application programmatically by using VI Server functions. Another good example would be changing the window size or the location of the front panel dynamically from another VI. You can find a few examples of VI Server functions in the folder **viserver** in the LabVIEW **examples** directory.

By not having sub VIs included in the top-level VI, it will open, close, and save quickly since it does not need to load the data and the machine codes of the sub VIs. There are pros and cons of keeping sub VIs outside the top-level VI. The advantages are as follows: (1) The loading and unloading time of the top-level VI will decrease significantly. (2) In case you need to add more functionality, you don't have to recompile the top-level VI. You simply add the new sub VI to a designated folder, where you keep such plug-in sub VIs. The major disadvantage is that doing so will not improve the memory usage at all if you are building a stand-alone executable out of the top-level VI. One test result shows that it can even use up more memory space. (The test was performed on a PowerMac G3 400 MHz with 208 MB physical memory with Virtual Memory turned off.) Therefore, it should be restated that dynamic loading and unloading of sub VIs improve the performance of the top-level VI only if it is run in LabVIEW, and not as a stand-alone executable.

Figure 15.13 is a simple example where a sub VI is called dynamically. The sub VI will take a value and return the incremented version of it. The figure shows the front panel and the diagram window of the example. We present step-by-step instructions to create it.

The VI Server functions can be found in the **Application Control** subpalette in the **Functions** palette, and **Open VI Refenrece, Call By Reference**

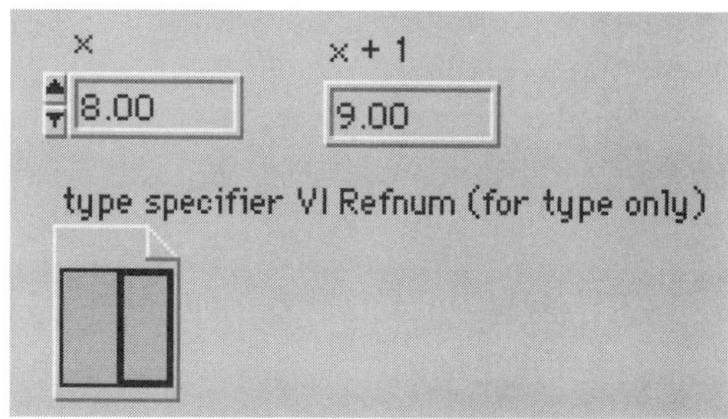

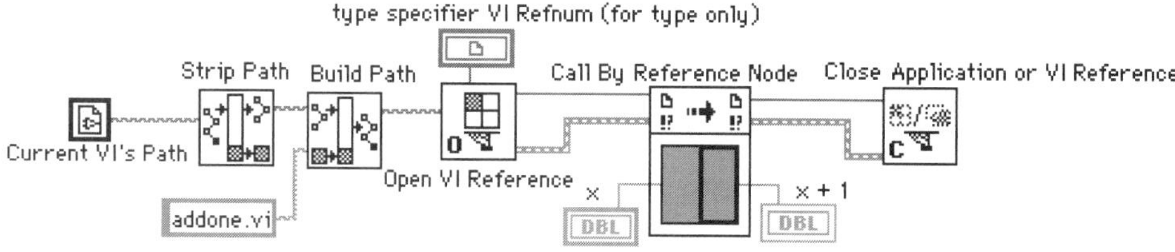

Figure 15.13
Example of dynamic loading and unloading of the sub VI **addone.vi**. It increments the input value and returns the result.

Node, and **Close Application or VI Reference** are the ones used in Figure 15.13. In this example, the sub VI **addone.vi** is called, which increments the value in **x** and returns the new result in **x + 1**. This example assumes that the sub VI **addone.vi** and the top-level VI exist in the same folder.

First, the top-level VI finds its current file path and builds a new file path for the sub VI **addone.vi**, which is then passed to **Open VI Reference**. The input control **type specifier VI Refnum** of **Open VI Reference** is an important parameter to set, and creating and setting it correctly will be described in the step-by-step instructions. Once the sub VI is open, the input value in **x** is passed to it, and the output value is displayed in **x + 1**. At the last stage, the reference to the sub VI is terminated by **Close Application or VI Reference**. To help create such an example, the following steps are provided:

Step 1: Open a new VI, and save it as **top.vi**. Place one numeric control and one numeric indicator, and label them **x** and **x + 1**, respectively.

Step 2: Create a simple sub VI that takes one input and returns the incremented input by 1. Make sure to create terminals for the input and the output in the sub VI. Save it as **addone.vi** in the same directory as **top.vi** is saved.

Figure 15.14 File I/O function **List Directory**.

Step 3: Finish **top.vi**, as shown in Figure 15.13. As for the input **type specifier VI Refnum**, right click on the top input of **Open VI Reference** and choose **Create Control**.

Step 4: Go to the front panel of **top.vi** and right click on the control input **type specifier VI Refnum** that you just created in Step 3. Choose **Select VI Server Class >> Browse . . .** from the pop-up menu. This will allow you to navigate through directories. Find **addone.vi**, and select it. This tells **Call By Reference Node** that you will be loading **addone.vi** dynamically into the memory.

Step 5: Once **top.vi** is completed, as shown in Figure 15.13, save and run it to test if **x + 1** displays the incremented value of **x**.

The five steps are for creating the example in Figure 15.13, which can be used as a template for dynamic sub VI calls from the top-level VI. If you keep the sub VIs in a designated folder, the only change to be made in Figure 15.13 is to provide the correct file path. If you would like to display all of the VIs available in the file path, the function **List Directory** in **Functions >> File I/O >> Advanced File Functions** can be useful (Figure 15.14). It lists all of the files with extensions specified in the input **pattern** in the file path provided. One more interesting case of using VI Server functions is presented in the next subsection.

15.5.3 Initial Startup Screen

When you launch any program, it always starts with its initial screen, which stays for a few seconds, then enters the main window. If you are writing an application in LabVIEW and will eventually be creating a stand-alone executable, you may want to have such a professional effect in your application. One quick thought might be using the settings in **VI Setup . . .** of a sub VI, which displays the initial screen. In other words, if you create a sub VI with the initial message and graphics on its front panel, and choose **Show Front Panel When Called** and **Close Afterwards if Originally Closed** in its **VI Setup . . .** window, it will display the front panel for a certain specified time duration and disappear. Even though it is a valid idea, one problem is that the front panel of the top-level VI must remain open on the screen. However, you want only the initial message and the graphics to stay on the screen for a while and disappear, followed by the main window showing up. Of course, there could be some workarounds, such as changing the location of the front panel of the top-level VI to a coordinate off the screen, then bringing it back to the center when the sub VI completes, or minimizing the size of the front panel of the top-level VI and putting it behind the sub VI front panel, while the sub VI is showing its initial messages and the graphics. However, none of the workarounds seems to create such a professional startup screen after all. Also, note that the same effect should occur continuously as you call any subsequent sub VI windows: The caller disappears, and the callee appears.

Such a dilemma can be resolved by using the VI Server functions, and Figure 15.15 illustrates the order of callers and callees. The first VI **window_init.vi** will display the initial message and graphics at the time of launch. This usually stays on the screen for a few seconds and then disappears. When it disappears, the next window of **window_main.vi** comes up on the screen. In this window, you can choose different subwindows, such as **window_sub1.vi, window_sub2.vi,** and so forth. A similar structure can continue below those windows again.

Note that each sub VI must initiate immediately after closing the front panel of its callee. The strategy, therefore, is that each caller contains its sub VIs in its diagram window, and the callees will close the front panel of their callers when they get called. The LabVIEW functions that are used in the following example are as follows:

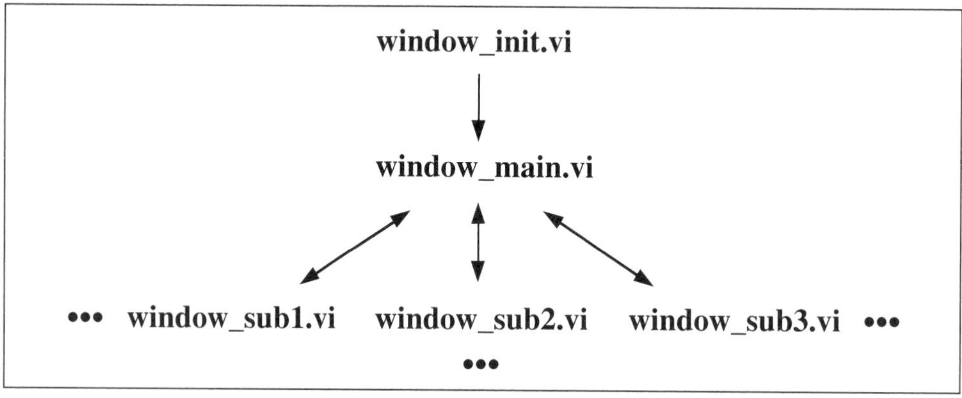

Figure 15.15
General hierarchy of stand-alone applications.

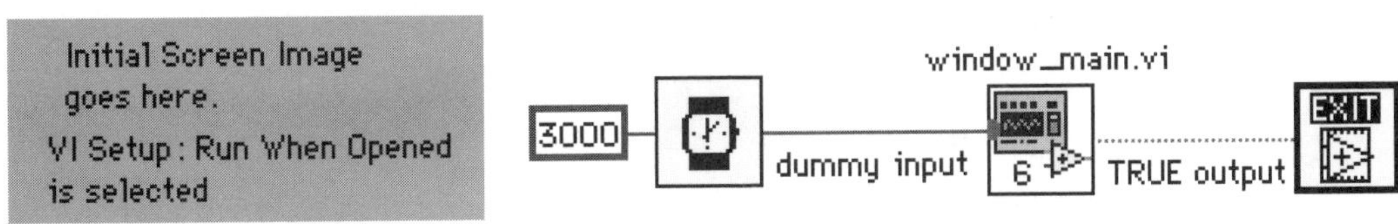

Figure 15.16
The front panel and the diagram window of **window_init.vi**, which displays the initial message and graphics.

Functions >> Application Control >> Open VI Reference
Functions >> Application Control >> Property Node
Functions >> Application Control >> Close Application or VI Reference
Functions >> Application Control >> Quit LabVIEW
Functions >> File I/O >> Build Path, and **Strip Path**
Functions >> File I/O >> File Constants >> Current VI's Path
Functions >> Time & Dialog >> Wait (ms)

Figures 15.16 through 15.20 are the front panels and the diagram windows of the sub VIs used in this example. Figure 15.16 shows that the initial screen stays active on the screen for 3 seconds. Then the next window **window_main.vi** is called. The input **dummy input** is used in order to specify the order of execution without using the **Sequence** structure. Since the use of **Sequence** structure can slow down the execution of VIs, try to use natural execution flow without using it. When **window_main.vi** terminates, it returns the TRUE constant to the function **Quit LabVIEW**, which will terminate the application. Since this is the first VI to run, the **Run When Opened** option is selected in the **VI Setup . . .** window of **window_init.vi**. In order to bring images onto the front panel, bitmap format is recommended rather

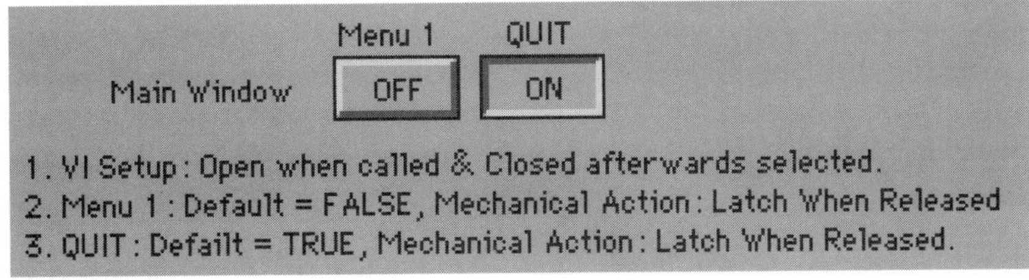

Figure 15.17
Front panel of **window_main.vi**. It will have a link to all of the submenus.

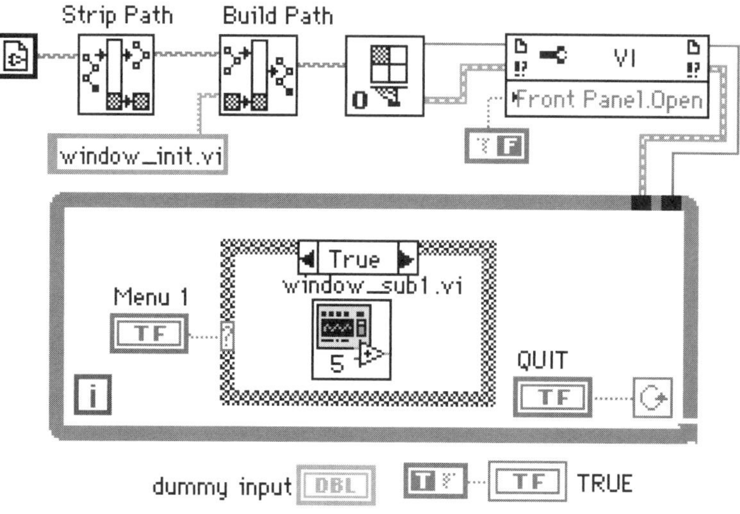

Figure 15.18
Diagram window of **window_main.vi**. When this VI is called, it closes the front panel of the callee, **window_init.vi**. The False frame of the **Case** structure is empty.

than Joint Photographic Experts Group (JPEG), Graphics Interchange Format (GIF), or any other formats.

The next window will be similar to the front panel in Figure 15.17. The button **Menu 1** allows you to enter the next lower level, and **QUIT** will terminate the entire program. Note that **Show Front Panel When Called** and **Close Afterwards if Originally Closed** options are selected in its **VI Setup . . .** window. Also, the mechanical action **Latch When Released** is chosen for both **Menu 1** and **QUIT**. The default value of each is set to FALSE and

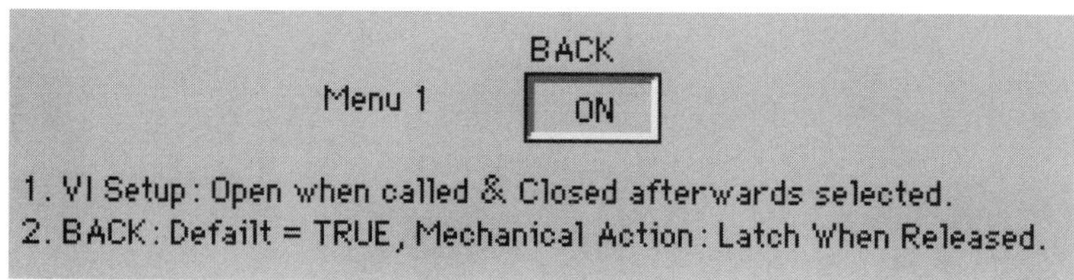

Figure 15.19
Front panel of **window_sub1.vi**. The button **BACK** allows you to go back to the main menu window.

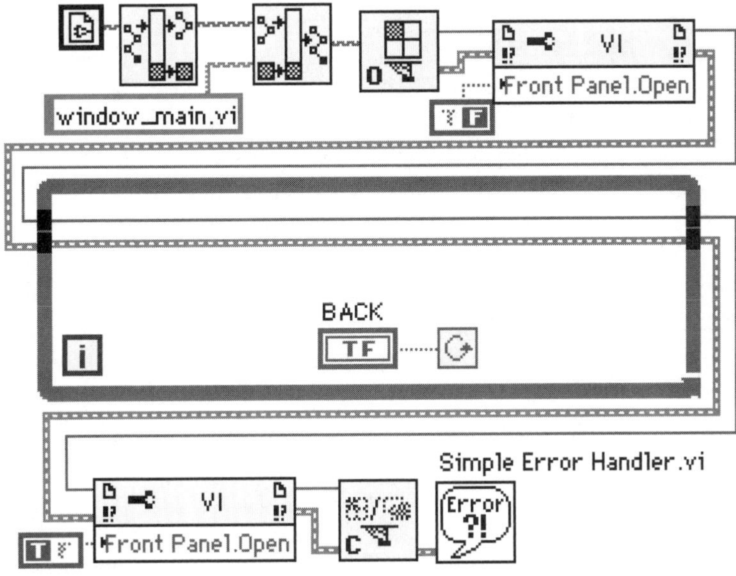

Figure 15.20
Diagram window of **window_sub1.vi**. When this VI is called, it closes the front panel of its callee **window_main.vi**.

TRUE, respectively. The default value of any object in the front panel can be set by right clicking on the object and selecting **Data Operations >> Make Current Value Default**. If you want to set all of the current values displayed in the front panel to all of the corresponding objects as their default values all at once, go to the pull-down menu **Operate** and choose **Make Current Values Default**.

The diagram window of **window_main.vi** is shown in Figure 15.18. The function **Current VI's Path** returns the current path, and the next two file I/O functions build a path of the caller VI **window_init.vi** based on the VI path of **window_main.vi**. Note that this example assumes that all of the VIs (**window_init.vi**, **window_main.vi**, and **window_sub1.vi**) are stored in the same directory. (It is a convention to keep all of the source files in the same project folder in any compiler program.) Using the new VI path, **Open VI Reference** generates a reference number to **window_init.vi**, and the next function **Property Node** closes its front panel by writing FALSE to the property **Front Panel Open**. Note that the function **Close Application or VI Reference** is not used to terminate the reference to **window_init.vi**. If you terminate it, it will also terminate the highest-level VI, which will stop everything below. Therefore, it is important not to close the reference to **window_init.vi**, which is the highest-level VI.

Once the front panel of the caller **window_init.vi** is closed, the **While Loop** initiates and you can call different submenus in the **Case** structures. When you press **QUIT**, the front panel of **window_main.vi** will close because of its **VI Setup . . .** window settings, and returns to **window_init.vi** while returning the Boolean constant TRUE, which is wired to the LabVIEW function **Quit LabVIEW** as shown in Figure 15.16. Then the entire application terminates. If you press **Menu 1** in Figure 15.17, it will initiate **window_sub1.vi** in the True frame of the **Case** structure in Figure 15.18. The state of **Menu 1** returns to its default state FALSE due to its mechanical action setting (**Latch When Released**) so that **window_sub1.vi** does not run repeatedly.

The front panel of **window_sub1.vi** is shown in Figure 15.19. It can also contain images and many buttons for lower-level processes. The button **BACK** will bring the control back to its caller **window_main.vi**. The two options **Show Front Panel When Called** and **Close Afterwards if Originally Closed** in its **VI Setup . . .** window are again selected, and the mechanical action **Latch When Released** is chosen for **BACK**.

The diagram window of **window_sub1.vi** is shown in Figure 15.20. It repeats procedures similar to Figure 15.18: Compute the VI path of the caller VI **window_main.vi** and close its front panel. When you terminate the process in **window_sub1.vi** and go back to the main window, the front panel of the caller **window_main.vi** needs to be open again. This step is shown after the **While Loop**. Such execution flow is defined by passing the two outputs from the **Property Node** through the **While Loop**, and into the two input terminals of the **Property Node** after the **While Loop**. Note the difference from Figure 15.18 such that you do terminate the reference to **window_main.vi** in **window_sub1.vi**, whereas you leave the reference to the highest-level VI **window_init.vi** open in **window_main.vi**. If you do not terminate the reference to the VI you are returning to (**window_main.vi** in this example), it will cause a problem since you will be in a VI, which is being referenced by another VI at the same time. Therefore, you must terminate the reference to **window_main.vi** before returning to it.

This example can be used as a template in your application to create a professional startup screen when you launch the application. The same routine can be repeated for any other lower-level sub VI calls: Close the front panel of the callers using the VI Server VIs, and open the front panel of the callee using **VI Setup . . .** options. Make sure to terminate the references to the caller sub VIs before returning to them except for the highest-level VI, which shows its front panel only once at the beginning of the application.

15.5.4 Status Checking Mechanism

You will most likely be using **While Loops** to run your application continuously until you decide to terminate it. In order to keep running the VI as fast as possible, you should make every possible effort to avoid any redundant calculation or status checking. For example, your top-level VI takes an input from the user and performs one of many functions. Each function will be chosen based on the user input, implying the use of **Case** structures. If you have about 20 **Case** structures that execute only if the user input is of a certain kind, it would be a waste of time to keep checking the validity of all of them every time the **While Loop** repeats. Instead, it will be much more efficient and faster to put them in another **Case** structure (it should be big enough to contain all of the 20 **Case** structures), which will skip them unless the user input tells it to run one of them.

This subsection will focus on a technique that allows you to monitor the change in the user input status and notifies LabVIEW to take action based on the change. The following three examples, in Figures 15.21 through 15.23, show different ways of detecting when a Boolean button is pressed.

To detect every time a Boolean button is pressed, the example in Figure 15.21 can be used. When the state of **Option 1** is changed—either from TRUE to FALSE, or from FALSE to TRUE—the content in the True frame of the **Case** structure will execute only once. A modified version of such a case is shown in Figure 15.22, where only the change from FALSE to TRUE is detected. If you set the mechanical action of **Option 1** to **Latch When Released**, it will automatically reset to the false state (with the default of **Option 1** set to FALSE), and ready to execute the True frame of the **Case** structure for the next state change. The example in Figure 15.23 is similar to Figure 15.21 but detects the change in both directions individually. In this case, the mechanical action of **Option 1** is set back to its default setting **Switch When Pressed**. When a change in the status of **Option 1** is detected, the True frame of either **Case**

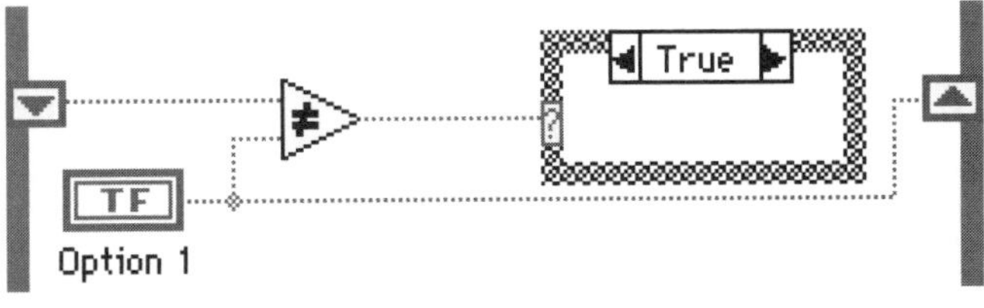

Figure 15.21
Status change checking mechanism using a shift register. This detects the status change either from TRUE to FALSE, or from FALSE to TRUE in the state of **Option 1**.

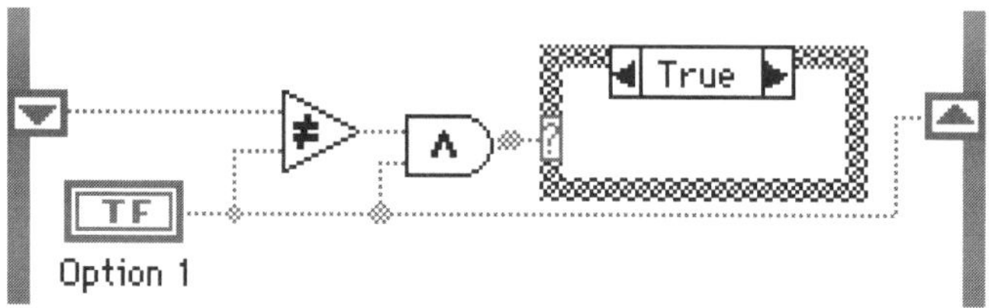

Figure 15.22
Modified status change checking method. This detects the status change only from FALSE to TRUE in the state of **Option 1**.

structures will be executed once. Utilizing those three examples, you can run a certain process once only when there is a request by the user.

15.5.5 Toggle Switch Design

In many applications, you have seen toggle switches that allow you to make an alternative selection. This section will present such an example created in LabVIEW. Consider the example in Figure 15.24.

The goal is to realize a mechanism that allows you to make an alternative decision among **Choice 1** and **Choice 2**. Those buttons are **Round Radio Button** in **Controls >> Boolean**. The pitfall is that you cannot just write the opposite state of one button to the other because it will end up with infinite flip-flop states. (Recall that you will be in a **While Loop** to run your application continuously.) Therefore, you need to write the opposite state to the other only once when the state changes. Using the technique that was presented in the previous subsection, you can easily write a code that allows you to detect the state change, as shown in Figure 15.24. When the state of

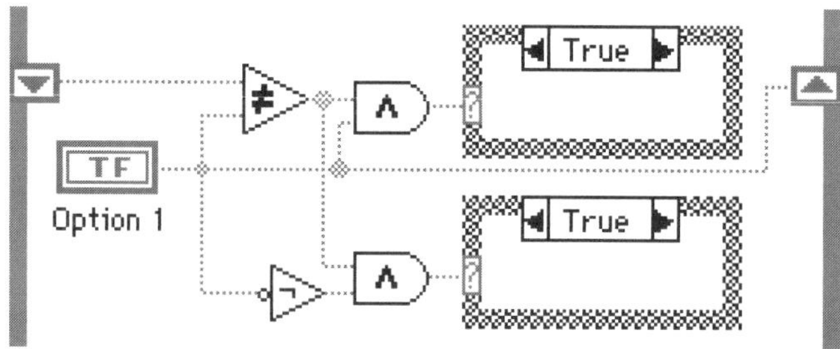

Figure 15.23
Modified status change checking method. This is similar to Figure 15.21 but detects both ways separately.

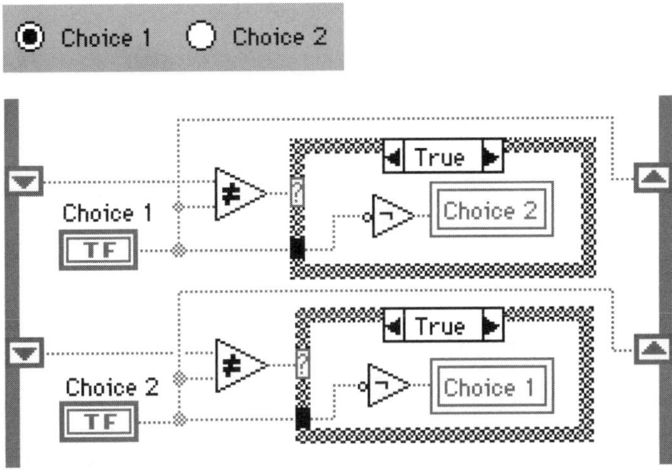

Figure 15.24
Example of toggle switch design.

either switches changes, you write the opposite Boolean value of itself to the **Local Variable** of the other one. Once one of the switches is selected, the selected state remains the same. (If a button is selected, it remains at TRUE; otherwise, it remains at FALSE.) Therefore, you can execute proper procedures such as calling proper sub VI(s), based on the Boolean values of **Choice 1** and **Choice 2**.

The design of toggle switches consisting of more than two Boolean buttons is somewhat trickier and will be left as an exercise for you. The basic idea remains the same except for the contents in the **Case** structures. In Figure 15.24, you write the opposite Boolean value to the other Boolean control. If the number of switches is more than two, you will need to conditionally write the opposite Boolean value to the rest of the switches. For example, if you have three switches **Choice 1**, **Choice 2**, and **Choice 3**, and the user selects **Choice 1** while the previous selection was **Choice 3**, you should write FALSE (the opposite state of the new selection **Choice 1**) only to **Choice 3**, and **Choice 2** must remain the same. Therefore, in this three-switch user exercise, you will need to know what the previous selection was, which can be found by monitoring the previous values of all three switches. Therefore, you would need an extra decision-making routine in each **Case** structure, but again the basic concept remains the same as the case in Figure 15.24. A general example of more than two selections is treated in the hands-on exercise manual, and you are recommended to refer to it.

15.5.6 Save As and Save A Copy As

The title of this section may raise some doubts due to its trivial content; however, it will point out some important issues in saving VIs, which could cause a confusing and frustrating situation, especially when the size of the application is large. This section will begin by considering the following scenario:

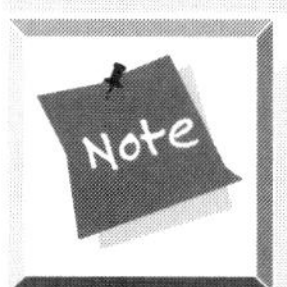

Mrs. Careful, who is the wife of Mr. Stubborn, has been writing a VI ***main.vi****, which contains many sub VIs. At one point, she realizes that a copy of* ***prompt.vi*** *with some changes needs to be used in* ***main.vi****. So she double clicks on* ***prompt.vi****, which is already being used in numerous locations—not only in* ***main.vi*** *but in many other sub VIs also. When the front panel of the sub VI* ***prompt.vi*** *comes up on the screen, she makes some changes in its diagram window and chooses* ***Save As . . .*** *under the pull-down menu* ***File****. After saving it as* ***prompt_new.vi****, she promptly saves* ***main.vi*** *to save the changes just made before any strange thing happens such as power failure. (She is a very cautious programmer so it is a routine for her to save the current VI every now and then, and immediately after any change is made.) Then she realizes that the original* ***prompt.vi*** *now has the label* ***prompt_new.vi*** *in the diagram window of* ***main.vi****. So she right clicks on it and selects* ***Replace*** *to replace it by the original* ***prompt.vi****. But she sees that it does not replace all of the copies of* ***prompt_new.vi*** *by its original one, and she does not remember where all of the copies of the original* ***prompt.vi*** *are used since it has been being used everywhere. The only way to correct this problem is to find every copy of* ***prompt_new.vi*** *and replace it by* ***prompt.vi*** *one by one. Since she already saved the change, the* ***Undo*** *feature will not remember the previous setting.*

Can you see where Mrs. Careful went wrong in the scenario? Such a situation can easily happen to any G-programmer. Therefore, it is important to be aware of the difference between **Save**, **Save As . . .** , and **Save A Copy As . . .** and when to use each one.

The mistake that Mrs. Careful made was using **Save As . . .** instead of **Save A Copy As . . .** . She should have saved **prompt.vi** as **prompt_new.vi** using **Save A Copy As . . .** and loaded it into the diagram window of **main.vi** manually. The command **Save** is to save any change in the current VI. **Save As . . .** is to save the current VI under a different label. Then all of the copies of the original VI will have the new label in every place where the copies are used. The command **Save A Copy As . . .** does not alter the content nor the label of the original VI but merely creates a copy of the original

VI and saves it under a new name in the directory you specify. The new copy would have nothing to do with the current VI until you loaded it into the diagram of the current VI. These differences may seem to be too trivial to be aware of, but it is better to be safe than sorry.

15.5.7 Replacing Local Variables by Shift Registers

It was mentioned earlier that in order to maximize the memory usage, you should minimize the use of **Local Variable**s and **Global Variable**s. This can indeed be the most important rule to be kept if your data siz becomes large. Suppose that your application needs to load from a file, an array of size 10,000 by 35, which is the acquired data from 35 channels and 10,000 samples from each channel. Once you bring in the data, you need to process them when the user makes a request. While monitoring the user selection using the technique presented in the discussion of status checking, you may need to pass the data to a sub VI for processing. However, you do not want to load or process the data continuously, but to do it only once when there is a request by the user. Then you can accomplish such a task by using the **Case** structure. Consider the example in Figure 15.25.

The example in Figure 15.25 is not complete because the two **Case** structures need an output in their False frames. The simplest way to correct the problem is to create an indicator of each output of the two **Case** structures and use their **Local Variable**s in the False frames of each **Case** structure, as shown in Figure 15.26.

However, the solution in Figure 15.26 would not be the optimal one, especially with a large data size. It creates a total of four extra copies of the origi-

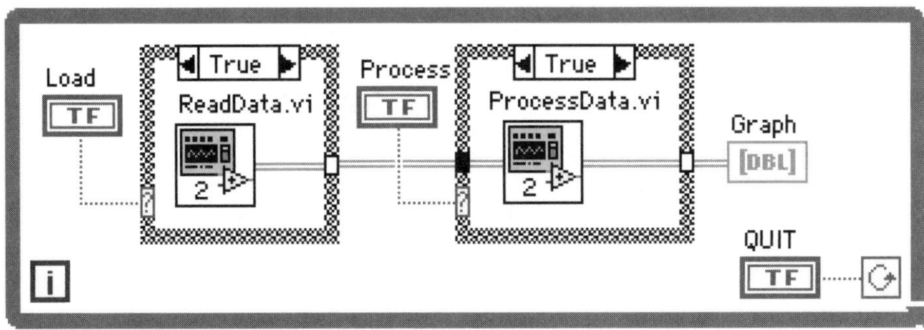

Figure 15.25
Selective data load and process.

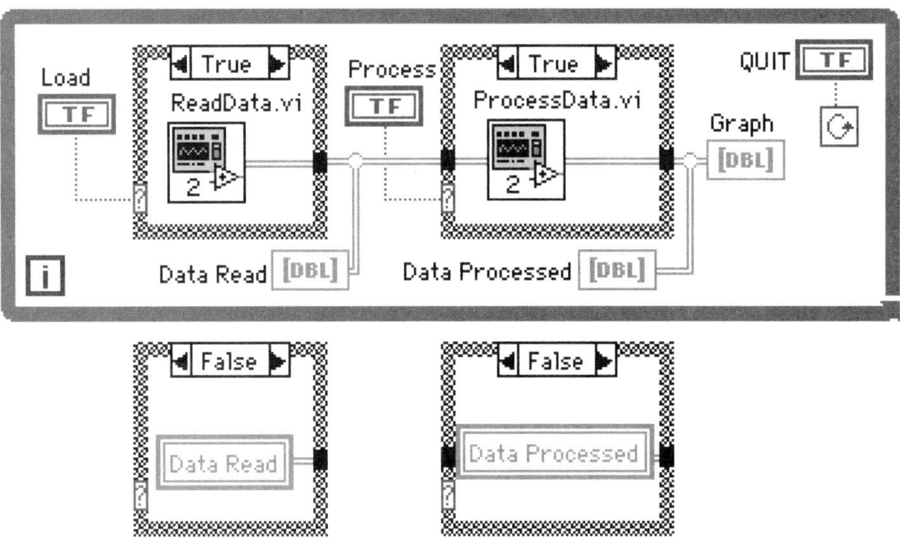

Figure 15.26
Corrected version of the example in Figure 15.25.

nal data by creating two indicators **Data Read** and **Data Processed**, and two **Local Variable**s. Therefore, it is obvious that Figure 15.26 is not the solution to the problem in Figure 15.25. Then what could be done to avoid creating extra copies of data such as indicators and their **Local Variable**s while being able to pass the data selectively and process them?

Consider the solution given in Figure 15.27, where **ReadData.vi** and **ProcessData.vi** have been modified so that they only read or process the data when **Load** or **Process** is TRUE. The diagram windows of **ReadData.vi** and **ProcessData.vi** are shown in Figure 15.28 and 15.29, respectively. How-

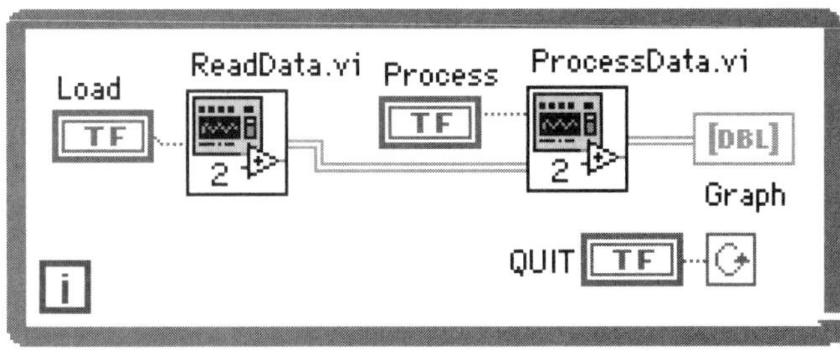

Figure 15.27
Another possible solution to Figure 15.25.

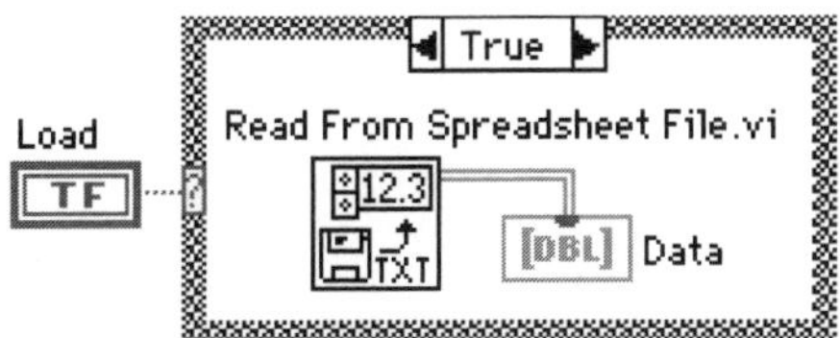

Figure 15.28 Diagram window of **Read Data.vi**. The False frame is empty.

ever, Figure 15.27 still poses a problem. What about the data that both **ReadData.vi** and **ProcessData.vi** return when the Boolean inputs are not TRUE? They will both return an *empty* array if they do not execute the contents in the True frame. Therefore, Figure 15.27 would still not work. Now, consider another solution in Figure 15.30.

The solution given in Figure 15.30 uses neither extra indicators nor **Local Variables** and has the data available in the loop without repeatedly reading or processing the data. The two VIs **ReadData.vi** and **ProcessData.vi** still use the same structure as shown in Figure 15.28 and Figure 15.29. Both **ReadData.vi** and **ProcessData.vi** perform their task—data read or data process—only if their Boolean inputs are TRUE, and return a valid array data. When the Boolean inputs are not TRUE, each VI will return an empty array. However, the empty array has no effect on the data that will be available in the **While Loop** thanks to the extra two **While Loops**, which repeat only once, and two **Case** structures inside them. When **Load** is TRUE, **ReadData.vi** returns the valid data, which are passed to the shift register on the right side. Assuming that the mechanical action of **Load** is **Latch When Released**, it will automatically go back to the default state FALSE, which will make **ReadData.vi** return an empty array. However, when **Load** is FALSE, the data leaving the right shift register will be the valid data that were passed to the shift register when **Load** was TRUE. By the same token, the data entering the **Graph** will always be the valid processed data that were returned by **ProcessData.vi** when **Process** was TRUE. The purpose of the two **While Loops** is only to use the shift registers, not to iterate! Therefore, you have avoided the use of neither indicators nor **Local Variables** and still managed to keep the valid data.

This example, indeed, delivers an important message in programming with LabVIEW: Do not repeat any redundant process. This issue will be re-

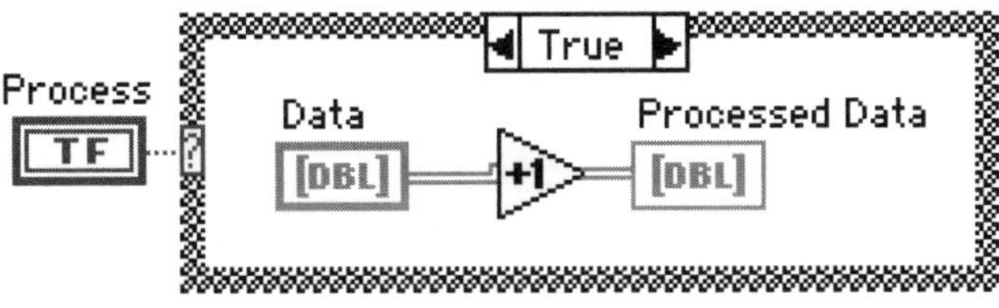

Figure 15.29 Diagram window of **ProcessData.vi**. The False frame is empty.

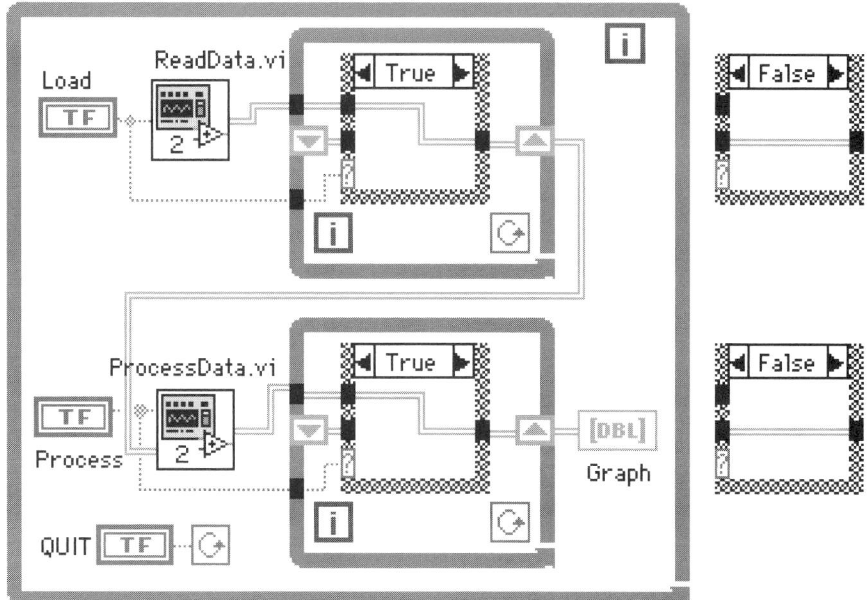

Figure 15.30
Selective data load and process without using **Local Variable**s.

visited in a later subsection since it is often forgotten by programmers even though it sounds natural to them.

15.5.8 Menu Customization

The only way to make a selection had always been by pressing a Boolean button in the front panel until LabVIEW introduced the menu customization capability. This subsection will first present the front panel and the diagram window of an example and then go through the steps to create it.

The example in Figure 15.31 increments or decrements the value in **Current No** whenever **Add One** or **Subtract One** is pressed, respectively, and pressing **QUIT** will stop the VI. You will achieve those three actions not only with the three Boolean buttons but also through the pull-down menu that you are about to customize. The following are the steps to create the example.

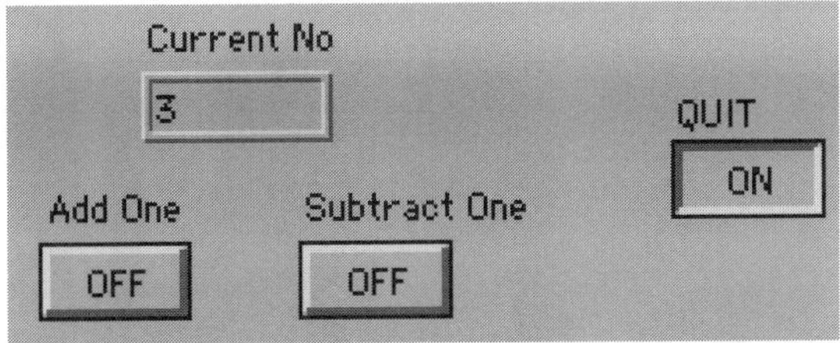

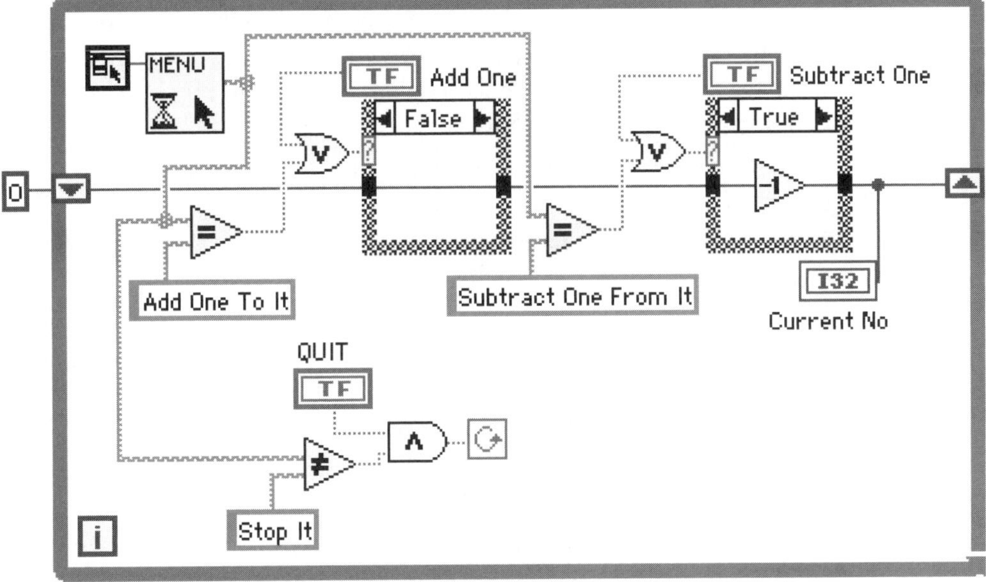

Figure 15.31
Customized menu control in LabVIEW.

Step 1: Open a new VI, and save it as **custom_menu.vi**.

Step 2: Complete the front panel and the diagram window as shown in Figure 15.31. Note that the mechanical action of all three Boolean buttons is set to **Latch When Released**. The default value of **QUIT** is changed to TRUE. (Do you remember how to change the default value individually?)

Step 3: The two core LabVIEW functions used for menu control in this example are **Current VI's Menu** and **Get Menu Selection**, which can be found in **Functions >> Application Control >> Menu**. Note that the output from **Current VI's Menu** is wired to the top input **menubar** of **Get Menu Selection**.

Figure 15.32 Insert and Delete buttons in the **Menu Editor** window.

Step 4: Select **Edit Menu ...** from the pull-down menu **Edit**. It will bring up the **Menu Editor** window.

Step 5: Choose **Custom** from the menu ring in the center right below the title bar of the window. The previous selection is **Default**.

Step 6: When you change the item in the menu ring to **Custom**, the **Single Selection Listbox** (that can be found in **Controls** >> **List & Ring**) will come up with a **???** sign. (Yes. This feature is written by LabVIEW VIs. Can you duplicate such a window by yourself?)

Step 7: Enter `Action List` in **Item Name**. Notice that it also replicates in **Item Tag**.

Step 8: Select **Item Tag**, and enter `List of Options` instead of **Action List**, which was automatically copied from **Item Name**. Click on the button with the + sign to add it to the listbox. (Refer to Figure 15.32.)

Step 9: Add `Increment`, `Decrement`, and `EXIT` to the listbox using Steps 7 and 8. Enter `Add One To It`, `Subtract One From It`, and `Stop It` as their tags in **Item Tag**, respectively.

Step 10: Highlight **Increment** in the listbox, and click the **Right** button to make it as a sublevel to **Action List**. (Refer to Figure 15.33.)

Step 11: Highlight **Decrement** and **EXIT** to lower its level by using the button **Right** using Step 10. When you complete this step, your listbox should look like Figure 15.34.

Step 12: Choose **Save As ...** from the pull-down menu **File**. Save the menu setting as **custom_menu.rtm**. The extension **.rtm** stands for *run-time menu*.

Step 13: Run the VI. You will see that the runtime menu will have only two pull-down menus: **Action List** and **Help**. Under **Action List**, you should have the three items listed: **Increment**, **Decrement**, and **EXIT**. Those are item names, and their item tags are used in the diagram window, as shown in Figure 15.31. This completes this example. The value in **Current No** should increment or decrement either by pressing the buttons or selecting the options from the pull-down menu **Action List**.

Figure 15.33 The button **Right** in the **Menu Editor** window.

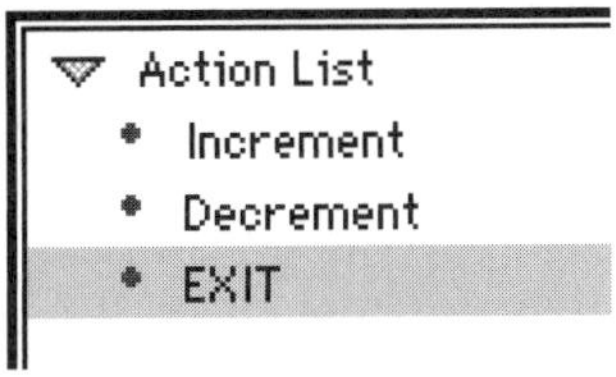

Action List
• Increment
• Decrement
• EXIT

Figure 15.34 Partial view of the listbox after completing the first 11 steps.

15.5.9 Creating Efficient Executables

This subsection will discuss how to maximize the efficiency of your application not just in terms of memory usage but in the execution speed and development strategy as well. As you become a more experienced LabVIEW programmer, the size of the VI will tend to become larger since the application you will be developing can include more than trivial tasks and effects. As a result, you will likely be using many Attribute Nodes to have various types of customized effects on the screen. Some common purposes of Attribute Nodes are (1) setting the range or initial value of x- and y-axes of multiple graphs, (2) displaying mode of x- and y-axes (linear or logarithmic), (3) changing the text in captions, (4) enabling or disabling Boolean or numeric controls, (5) showing or hiding controls and indicators, (6) changing the legend text, (7) changing the plot color and plot style, and (8) controlling the cursor display of graphs, and many more. As those lists show, the size of the VI can quickly grow, and soon be an order of a few megabytes!

The VIs of large size bring two major disadvantages: (1) They will slow down the editing process. (2) The execution speed will also slow down. The execution speed, however, will speed up once the VI becomes a stand-alone executable, which is the topic of the next subsection. What about the editing process? Should you carry such overhead throughout the entire developing process? The answer is yes and no. If you can break up the large top-level VI into small pieces, you should definitely do that. Even though such a process should be done before building a large VI, you can still find incidents where further downsizing of sub VIs is possible by looking back at the top-level VI. Also, use the VI Server functions to call sub VIs to reduce the overhead upon the diagram editor in LabVIEW. As was mentioned previously, one test result shows that it can take up to a couple of minutes until the operating system responds to the user change such as **Save**, if the size of the VI is on the order of a few megabytes. (Refer to Section 15.5.2.) Therefore, utilizing the VI Server functions can greatly reduce the over-head on the diagram editor. In those cases where you can utilize VI Server functions or break up large pieces of codes into smaller sub VIs, the answer is no. On the other hand, there could be some cases where you must keep a lot of Attribute Nodes and **Local Variable**s in the top-level VI since you cannot control the objects in higher-level VIs from lower-level sub VIs through their Attribute Nodes. In other words, you may share the *values* of the Attribute Nodes between the caller and the callee VIs but cannot keep the Attribute Nodes of the objects in the caller VIs in the callee VIs. Since Attribute Nodes, not their values, are the ones that take up a lot of memory and impose overhead on the diagram editor, carrying such overhead can sometimes be unavoidable. Then the answer is yes, and the only hope you can have in that case is a faster processor and a lot of physical (not virtual) memory on the computer. (In fact, a faster processor seems to be more important than the amount of physical memory based on numerous test results.)

How about the reliability of LabVIEW in dealing with large-size VIs? The experiences of and the comments from many LabVIEW programmers, including the author, indicate that LabVIEW stays stable in terms of editing, compiling, and executing VIs. LabVIEW also seems to complete the process of creating a stand-alone executable out of a large VI smoothly. Therefore, it can be concluded that LabVIEW is quite a stable compiler and functions smoothly with VIs of fairly large size. However, such an issue about the overhead differs between PC platforms and Macintosh platforms. That is because each platform manages the memory in different ways. On Macintosh platforms, a fixed amount of memory is allocated to an application by the user, and the application cannot use more than what is preassigned to itself. However, the Windows operating system handles the memory dynamically, and the allocated amount of memory varies and is controlled by the operating system. Therefore, the degree of deficiency in the performance of VIs due to the overhead can differ between different platforms.

As was mentioned earlier, the VI Server functions do help in loading, editing, and saving VIs but have no effect on the executables. Since the plug-in feature originated from the VI Server functions, the same result applies to the implementation of plug-in feature.

As for **Local Variable**s and **Global Variable**s, their usage must be minimized to increase the efficiency in both the stand-alone executables and the individual VIs to be executed in the LabVIEW environment. Such minimization should not be understood in terms of the *number* of **Local Variable**s or **Global Variable**s but in terms of the data type and the size of their original objects. For example, having one **Local Variable** of an array of size 30 by 40,000 will not benefit over having two **Local Variable**s of simple numeric controls. The **Local Variable**s or **Global Variable**s of large-size data type or complicated data structure must be prevented or minimized. The technique covered to avoid the usage of **Local Variable**s by a **While Loop** and a shift register can be helpful.

Try to avoid any redundant condition checking or execution. If a big portion of your diagram contains nothing but **Case** structures, which do not execute unless the user makes a request, keeping that portion in a big **Case** structure, which only depends on the change in the user request, will speed up the VI in a great deal. That is because the entire **Case** structures will be skipped by a single big **Case** structure, and this completes one iteration of the outermost **While Loop** much quicker.

Minimize the usage of **Sequence** structure. Try to use the natural execution flow if the use of **Sequence** structure can be avoided. Sometimes creating a dummy input and a dummy output in sub VIs will help create a transparent execution flow.

Additionally, all of the suggestions made in the discussion of memory management are applicable in this subsection, too. They are listed again for your convenience:

1. Maximize the usage of sub VIs. Break up large VIs into small sub VIs.
2. Minimize the number of live updates on the front panel.
3. Minimize any dynamic change in data size, especially arrays and strings.
4. Minimize the usage of **Local Variable**s and **Global Variable**s.
5. Minimize the usage of Attribute Nodes, especially in sub VIs.
6. Avoid coercion dots. To convert data type, do it at the simplest data type stage.
7. Use simple data structure.

15.5.10 Application Builder

Having covered all of the important techniques and tips in writing VIs with LabVIEW, naturally the last topic is how to create a stand-alone executable (or application) out of VIs. In order to create one, you must have an add-on toolkit "Application Builder," which allows you not only to create an executable but to create distribution disks also. (The distribution disk capability applies only to PC platforms.)

This section is indeed the most difficult section to present because LabVIEW keeps being updated, and as newer versions appear, what is currently valid could quickly become obsolete (and that is contrary to the intent of this book; this book is intended to cover the fundamentals in G-programming, not simple mechanics of a specific feature). Therefore, this section will not cover *how* to build an executable in LabVIEW except for the discussion of its features. As for how to create a stand-alone application, it is recommend that you refer to the instructions about Application Builder, which are usually only a couple of pages long.

The basic concept in creating a stand-alone executable with Application Builder is very similar to that of other programming language compilers. Almost all of the text-based programming language compilers keep all of the source files in the same folder, which is often referred to as a *project* folder. When they compile codes, they create an executable by linking all of the source codes in the project folder. LabVIEW creates an executable in a similar manner except that the entire process seems to be much easier than conventional compilers. What corresponds to the project folder of the conventional compilers is the LabVIEW directory with extension **.llb** (Dot El El Be). When LabVIEW creates an executable, it first collects all of the user-made sub VIs, LabVIEW functions, and sub VIs, and compresses them into a **.llb** file by removing the diagram window of each sub VI. (This step can be done manually or automatically using **Save With Options . . .** under the pull-down menu **File.**) Then it creates an executable based on the **.llb** file. One important point to be addressed is that if your executable performs either data acquisition or serial port I/O, you *must* include either **daqdrv** or **serpdrv**, respectively, in the same directory where the executable is saved. This applies only to PC platform executables, and those files can be found in the LabVIEW directory. Additionally, you must include **lvrt.dll** in the same directory where the executable is located, if you create an executable in LabVIEW 5.1 or higher. Versions prior to 5.1 will include in the executables the LabVIEW runtime engine, which allows the executables to run without the full version of LabVIEW. However, versions higher than 5.1 will not install the runtime engine in order to reduce the size of the executables. Therefore, you must include **lvrt.dll** in the same directory where the executable is saved. The DLL **lvrt.dll** can be found in the **shared** directory of LabVIEW 5.1 or higher.

As for Macintosh platform executables, if the executables use any of the analysis VIs in LabVIEW, you *must* include the folder **Shared Libraries**, which contains four files—**LVAnalysis**, **LVPPCPortMgr**, **lvsound.lib**, and **vximodul**—in the same folder where the executable is saved. (Actually, you need only LV Analysis if you are using the analysis VIs.) The **Shared Libraries** folder can be found in the LabVIEW folder, too. Forgetting those files and folders is the most common mistake made during the process of creating executables.

The advantage of executables is definitely their portability. You do not need LabVIEW to run the application. Also, the execution speed is much faster than running the VI in LabVIEW since any overhead that could have been incurred in LabVIEW would have no effect on the executables. Therefore, if the performance of your VI suffers from such overhead in LabVIEW, you may want to create an executable and run the stand-alone version instead. The disadvantages are as follows: (1) Of course, you will not be able to retrieve the VIs from the executables, and (2) you need an extra add-on toolkit to be able to build one since LabVIEW basic package does not come with Application Builder.

15.6 Suggestions for Further Reading

G-Programming Reference Manual, National Instruments, 1998.

LabVIEW Code Interface Reference Manual, National Instruments, 1996.

LabVIEW Function and VI Reference Manual, National Instruments, 1998.

R. A. DeCarlo, *Linear Systems,* Prentice Hall, 1989.

G. H. Golub, and C. F. Van Loan, *Matrix Computations,* 2nd ed., Johns Hopkins, 1989.

L. B. Jackson, *Digital Filters and Signal Processing,* 2nd ed., Kluwer Academic Publishers, 1989.

A. V. Oppenheim, and A. S. Willsky, *Signals and Systems,* Prentice Hall, 1983.

A. Papoulis, *Probability, Random Variables, and Stochastic Processes,* 3rd ed., McGraw-Hill, 1991.

P. Z. Peebles, Jr., *Probability, Random Variables, and Random Signal Principles,* 2nd ed., McGraw-Hill, 1987.

J. G. Proakis, C. M. Rader, F. Ling, and C. L. Nikias, *Advanced Digital Signal Processing,* Macmillan Publishing Co., 1992.

R. A. Roberts, and C. T. Mullis, *Digital Signal Processing,* Addison-Wesley, 1987.

L. L. Scharf, *Statistical Signal Processing: Detection, Estimation, and Time Series Analysis,* Addison-Wesley, 1991.

C. W. Therrien, *Discrete Random Signals and Statistical Signal Processing,* Prentice Hall, 1992.

H. L. Van Trees, *Detection, Estimation, and Modulation Theory,* John Wiley & Sons, 1968.

Appendix A
CIN and Code Warrior for Power Macintosh

A.1 Creating a CIN Using Metrowerks Code Warrior

The following are step-by-step instructions to create a CIN using Code Warrior Professional Release with LabVIEW 4.0 or higher. These steps will allow you to create a simple CIN that performs **y** = 2**x**, where **x** is a numeric control and **y** is a numeric indicator in LabVIEW. Examples for PC platforms can easily be obtained by contacting National Instruments directly at 1-800-433-3488, or http://www.ni.com.

Step 1: Place the Code Warrior CD in the CD tray of your computer, and install it. Installing the basic package would be just fine to create CINs.

Step 2: Go to LabVIEW, create a new VI, and save it as **test.vi**.

Step 3: In the front panel, place one digital control and one digital indicator, and label them **x** and **y**, respectively.

Step 4: Go to **Advanced** in the **Functions** palette, and place a **Code Interface Node** in the diagram window.

Step 5: Expand the **Code Interface Node** to have two inputs and two outputs. Wire the digital control **x** to the top left input. Wire the digital indicator **y** to the bottom right output.

Step 6: **Command**-click on either the bottom input or the bottom output and select **Output Only**.

If you have followed the steps correctly so far, you should have one digital control and one digital indicator in the front panel, and one CIN node with its top input wired to **x** and the bottom output to **y** with its input grayed out. If so, please continue; otherwise, go back and make proper corrections.

Step 7: **Command**-click on the CIN node, and select **Create .c File ...**. This will create **test.c** in the folder where **test.vi** resides.

Step 8: Quit LabVIEW to save the memory.

Step 9: Start Code Warrior by double clicking on CodeWarrior IDE 2.0 in the Metrowerks Code Warrior folder.

Step 10: Select **File >> New Project ...**. This will bring up a window **New Project**. Click on the arrow in front of **MacOS** to expand it. Expand **C/C++** again by clicking on its arrow, and select **Basic Toolbox PPC** at the bottom.

Step 11: Enter the new project name as **test**, and save it.

Step 12: When Code Warrior opens the new project window **test**, you will see four folders: **Sources, Resources, Mac Libraries**, and **ANSI Libraries** under **File.** Select **ANSI Libraries**, and press **Option-Delete** to delete it since you don't need it. Delete the folder **Resources** by the same fashion. Also, delete the file **SillyBalls.c**, which was created by Code Warrior under **Sources** folder by selecting it and hitting **Option-Delete**.

If you have followed the steps correctly so far, you should have two folders: **Sources** and **Mac Libraries**. You should have no file under the folder **Sources**, and three ones under **Mac Libraries**: **InterfaceLib**, **MathLib**, and **MSL RuntimePPC.Lib**. If so, please continue; otherwise, go back and make proper adjustments.

Step 13: You need to add two more files to **Mac Libraries**. Select **Mac Libraries**, and choose **Project >> Add Files . . . .** Navigate to LabVIEW directory, and find **CINLib.ppc.mwerks** in **cintools** >> **Metrowerks Files** >> **PPC Libraries**. Also, add **LabVIEW.xcoff** from **cintools** >> **PowerPC Libraries.** Now, you have five libraries under **Mac Libraries**: **InterfaceLib**, **MathLib**, **MSL RuntimePPC.Lib**, **CINLib.ppc.mwerks**, and **LabVIEW.xcoff**.

Step 14: Add the source file **test.c** by selecting **Sources** and **Project** >> **Add Files . . . . test.c** is the source file that you created in Step 2.

Step 15: Double click on **test.c** to modify it. Enter the following line of code where it says **/*ENTER YOUR CODE HERE */**.

```
*y = *x * 2;
```

Then save it. You should now have **test.c** listed under **Sources** and five libraries under **Mac Libraries** at this point.

Step 16: Make a copy of **projectName.exp** from LabVIEW directory >> **cintools** >> **Metrowerks Files** >> **PPC Libraries**, and paste it in the directory where you saved your Code Warrior project file **test**. This is a good time to make sure you have the following files in the same directory: **test** (Code Warrior project file), **test.c**, **test Data** (Code Warrior creates this folder automatically.), and **projectName.exp**.

Step 17: Change the name **projectName.exp** to **test.exp**. *This step is important.* Use the same name as the project **test** before the extension **.exp**.

If you saved your project under a different name, use that name for the **.exp** file.

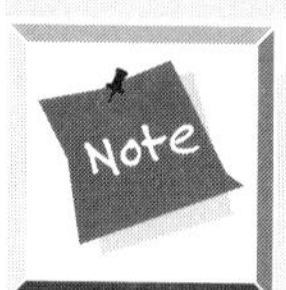

CHECK POINT: DID YOU USE THE SAME NAME FOR THE ***.exp*** *FILE?*

Step 18: Go to **Edit** and select **Basic Toolbox PPC Settings....** The window should display **Target**, **Language Settings**, **Code Generation**, **Linker**, and **Editor** with their sub items.

Step 19: Go to **Access Paths** under **Target**. Under **User Paths**, add the following three directories from LabVIEW directory:

```
cintools >> Metrowerks Files >> PPC Libraries
cintools
cintools >> PowerPC Libraries
```

Step 20: Go to **PPC Target** under **Target**. If it asks you to save the changes when moving from **Access Paths** to **PPC Target**, click **Yes** or **OK**. Under **PPC Target**, make the following settings:

```
Project Type: Shared Library
File Name: test.tmp
Creator: LVsb
Type: .tmp
```

The Type is "period tmp." Make sure not to forget the period in front of tmp. Also, when you are told to make certain settings, you need to leave the other settings unchanged.

Step 21: Go to **C/C++ Language** under **Language Settings**. Again, if it asks you to save the change, click **Yes** or **OK**. Under this section, make the following two settings:

```
Source Model: Apple C
Prefix File:
```

There is none specified for Prefix File; therefore, you need to *Empty* the Prefix File entry.

Step 22: Go to **PPC Processor** under **Code Generation**. Here, just make one setting:

```
Struct Alignment: 68K
```

Step 23. Go to **PPC Linker** under **Linker**, and empty the entries of **Entry Points** (i.e., make the following settings):

```
Initialization:
Main:
termination:
```

Step 24. Go to **PPC PEF** under **Linker**, and make the following setting:

```
Export Symbols: Use ".exp" file
```

Step 25: Close the **Basic Toolbox PPC Settings** window. If it prompts you to save the changes, click **Yes** or **OK**.

Step 26: Go to **Project**, and select **Make**. (Shortcut: **Command-M**) This will create **test.tmp** in the project folder. Now, quit Code Warrior.

Step 27: Go to the LabVIEW directory and find **lvsbutil.app**. It is located in **cintools** directory.

Step 28: Launch **lvsbutil.app** by double clicking on it.

Step 29: Select **File** >> **Convert .tmp File**. (Shortcut: **Command-K**)

Step 30: Select **test.tmp** that you just created in Code Warrior. This will create **test.lsb** in the same folder where you have **test.tmp**.

Step 31: You are done with **lvsbutil.app**. Quit the application.

Step 32: Launch LabVIEW again, open **test.vi**, and go to the diagram window.

Step 33: **Command**-click on the **Code Interface Node** and select **Load Code Resources . . .** from its pop-up menu.

Step 34: Select the **test.lsb** that was created in Step 30.

This completes the example. Save the VI, and test it. Once you have linked the **.lsb** file to the **Code Interface Node**, you don't need the source code anymore. The VI **test.vi** will behave just like LabVIEW functions.

A.2 Sample C++ Code for 1-D Array Data Type

This section will show a sample C++ code **add1D.c**, which adds all of the elements in the 1-D input array. In the previous example, the input and the output were a scalar. The steps to create, compile and link to a VI are identical; therefore, only the source code will be listed. The diagram window of the VI **add1D.vi** is shown in Figure A.1. Note that the input type is DBL, and LabVIEW converts it to the type float as seen in the source code.

```
#include "extcode.h"
/*
 * typedefs
 */

typedef struct {
       int32 dimSize;
       float64 Numeric[1];
       } TD1;
typedef TD1 **TD1Hdl;

CIN MgErr CINRun(TD1Hdl X, float64 *sum);

CIN MgErr CINRun(TD1Hdl X, float64 *sum) {

      /* ENTER YOUR CODE HERE */

      int32  k;
      float64*x_value_p;

      x_value_p = (*X)->Numeric;

      *sum = 0;
      for (k=0; k<(*X)->dimSize; k++){
              *sum += x_value_p[k];
```

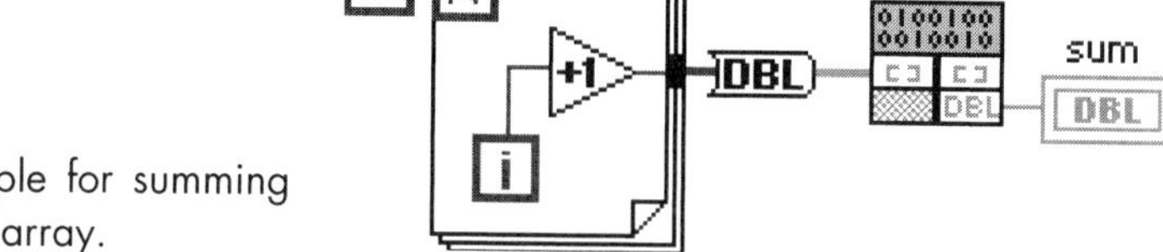

Figure A.1 CIN example for summing the elements in the input array.

```
            // *sum += (*X)->Numeric[k]; //This command also works.
            cout << *sum;
      }
      return noErr;
}
```

A.3 Sample C++ Code for 2-D Array Data Type

This section shows a C++ source code to extract the first two columns of the input 2-D array. Note that the 2-D array is treated as a 1-D array when it gets loaded in the memory, and the order is the first row first, the second row second, and so on. Therefore, if you have a 2-D array of size 3 by 4 (3 rows and 4 columns), it is treated as a 1-D array with the following order. Note that the index starts at zero in LabVIEW, and the first coordinate indicates the row index, and the second coordinate indicates the column index.

```
(1,1) (1,2) (1,3) (1,4) (2,1) (2,2) (2,3) (2,4) (3,1) (3,2) (3,3) (3,4)
```

Therefore, in order to extract the first column, you should index every fourth element starting with the first one since the first column consists of (1,1), (2,1), and (3,1). The purpose of the sample code is to show where you can get the array size information and how to index each element of the input 2-D array.

```
/*
 * CIN source file
 */

#include "extcode.h"

/*
 * typedefs
 */

typedef struct {
       int32 dimSizes[2];
       float64 Numeric[1];
       } TD1;
typedef TD1 **TD1Hdl;
```

```
typedef struct {
       int32 dimSize;
       int32 Numeric[1];
       } TD2;
typedef TD2 **TD2Hdl;

typedef struct {
       int32 dimSize;
       float64 Numeric[1];
       } TD3;
typedef TD3 **TD3Hdl;

CIN MgErr CINRun(TD1Hdl X, TD2Hdl array_size, TD3Hdl col1, TD3Hdl
col2);

CIN MgErr CINRun(TD1Hdl X, TD2Hdl array_size, TD3Hdl col1, TD3Hdl col2)
{

       int32 k, rsize, csize;

       rsize = (*X)->dimSizes[0]; // Size of X (row)
       csize = (*X)->dimSizes[1]; // Size of X (column)

       // Array Size calculation
       (*array_size)->dimSize = 2; // Length of array_size
       (*array_size)->Numeric[0] = rsize;
       (*array_size)->Numeric[1] = csize;
       (*col1)->dimSize = rsize; // Length of col1
       (*col2)->dimSize = rsize; // Length of col2

       for (k=0; k<rsize; k++){
               (*col1)->Numeric[k] = (*X)->Numeric[3*k];
               (*col2)->Numeric[k] = (*X)->Numeric[3*k+1];
       }

       return noErr;
       }
```

Appendix B
Error Handler Design

Front Panel and Diagram Window of the Example VIs

Figures B.1 through B.4 show the front panels and the diagram windows of the sub VIs that are used in the example during the error handler design discussion in Chapter 15.

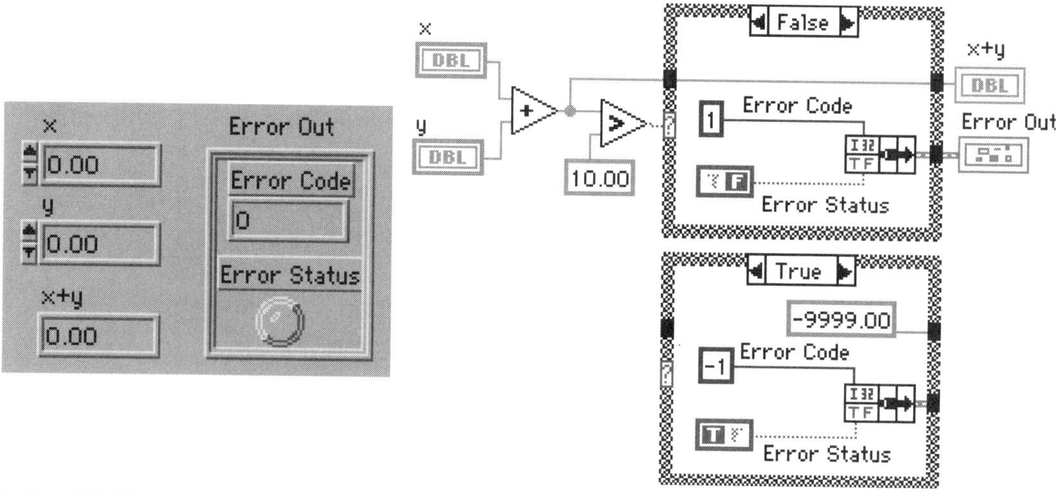

Figure B.1
The front panel and the diagram window of **Proc1.vi**. If there is no error, the error code is –1, and the Boolean **Error Status** becomes TRUE. Otherwise, the error code is 1. This checks if the sum of **x** and **y** exceeds 10.

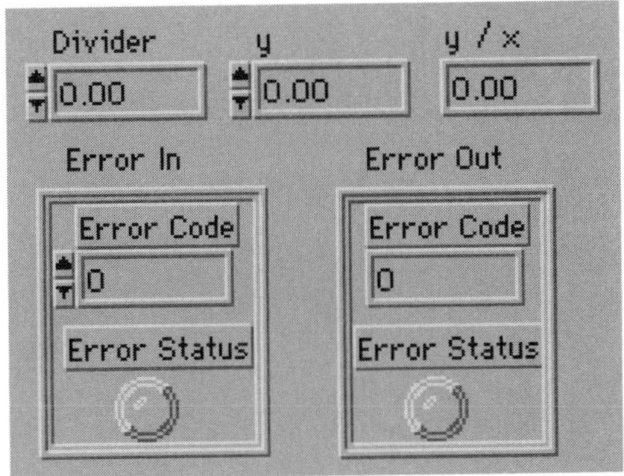

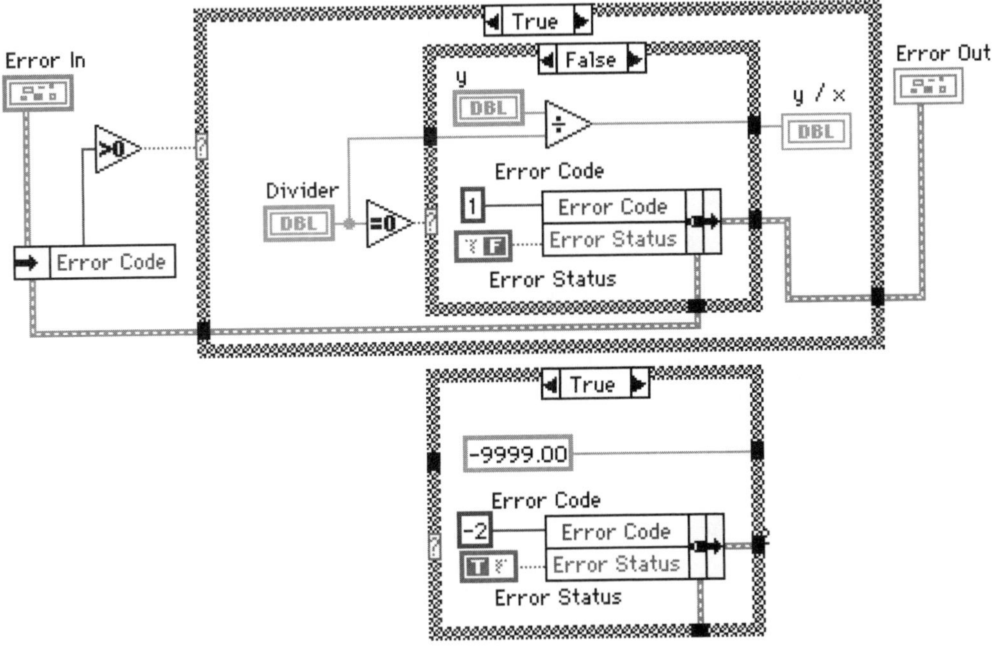

Figure B.2
The front panel and the diagram window of **Proc2.vi**. This checks if the sum of **x** and **y** is zero. If so, the error code becomes –2, and the Boolean **Error Status** becomes TRUE. The False frame of the outer **Case** structure is empty while passing the **Error In** to the output **Error Out**.

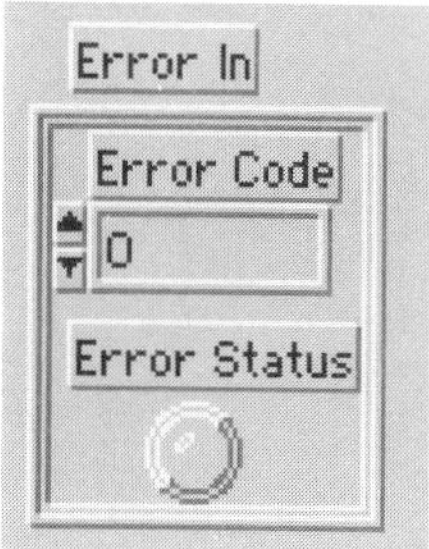

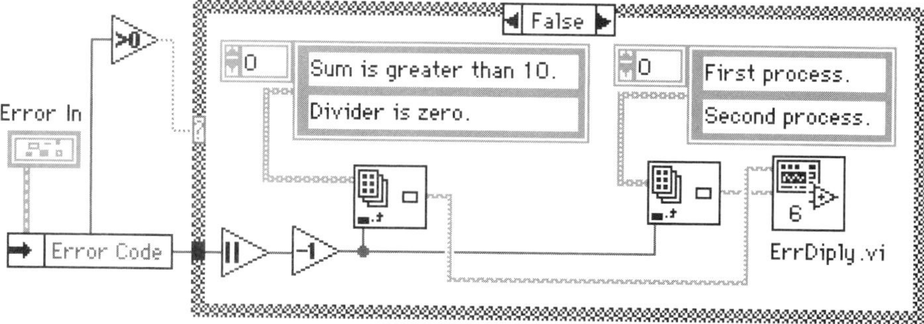

Figure B.3
The front panel and the diagram window of **Proc3.vi**. Based on the error code—either −1 or −2 if there has been an error in the previous stage, or 1 if there is no error—this VI passes proper messages to another sub VI **Err-Diply.vi**, which is solely responsible for displaying the messages. The True frame is empty.

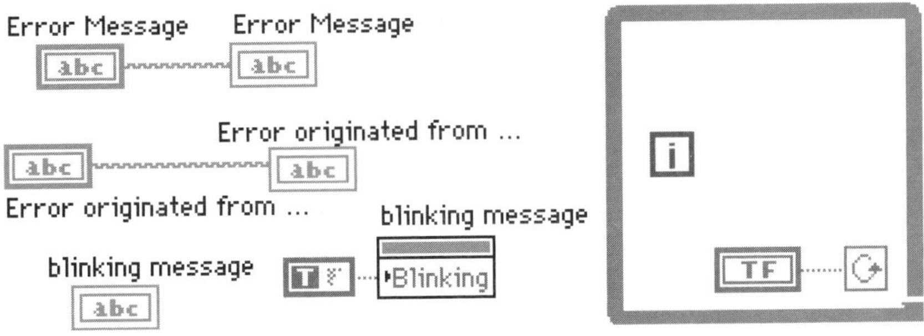

Figure B.4
The front panel and the diagram window of **ErrDiply.vi**. This VI receives the error messages from the previous stages and displays them properly.

Index

A

B

C

D

E

F

G

H

I

L

M

N

O

P

R

S

T

U

V

W

X

Z

LICENSE AGREEMENT AND LIMITED WARRANTY

READ THE FOLLOWING TERMS AND CONDITIONS CAREFULLY BEFORE OPENING THIS DISK PACKAGE. THIS LEGAL DOCUMENT IS AN AGREEMENT BETWEEN YOU AND PRENTICE-HALL, INC. (THE "COMPANY"). BY OPENING THIS SEALED DISK PACKAGE, YOU ARE AGREEING TO BE BOUND BY THESE TERMS AND CONDITIONS. IF YOU DO NOT AGREE WITH THESE TERMS AND CONDITIONS, DO NOT OPEN THE DISK PACKAGE. PROMPTLY RETURN THE UNOPENED DISK PACKAGE AND ALL ACCOMPANYING ITEMS TO THE PLACE YOU OBTAINED THEM FOR A FULL REFUND OF ANY SUMS YOU HAVE PAID.

1. **GRANT OF LICENSE:** In consideration of your payment of the license fee, which is part of the price you paid for this product, and your agreement to abide by the terms and conditions of this Agreement, the Company grants to you a nonexclusive right to use and display the copy of the enclosed software program (hereinafter the "SOFTWARE") on a single computer (i.e., with a single CPU) at a single location so long as you comply with the terms of this Agreement. The Company reserves all rights not expressly granted to you under this Agreement.

2. **OWNERSHIP OF SOFTWARE:** You own only the magnetic or physical media (the enclosed disks) on which the SOFTWARE is recorded or fixed, but the Company retains all the rights, title, and ownership to the SOFTWARE recorded on the original disk copy(ies) and all subsequent copies of the SOFTWARE, regardless of the form or media on which the original or other copies may exist. This license is not a sale of the original SOFTWARE or any copy to you.

3. **COPY RESTRICTIONS:** This SOFTWARE and the accompanying printed materials and user manual (the "Documentation") are the subject of copyright. You may not copy the Documentation or the SOFTWARE, except that you may make a single copy of the SOFTWARE for backup or archival purposes only. You may be held legally responsible for any copying or copyright infringement which is caused or encouraged by your failure to abide by the terms of this restriction.

4. **USE RESTRICTIONS:** You may not network the SOFTWARE or otherwise use it on more than one computer or computer terminal at the same time. You may physically transfer the SOFTWARE from one computer to another provided that the SOFTWARE is used on only one computer at a time. You may not distribute copies of the SOFTWARE or Documentation to others. You may not reverse engineer, disassemble, decompile, modify, adapt, translate, or create derivative works based on the SOFTWARE or the Documentation without the prior written consent of the Company.

5. **TRANSFER RESTRICTIONS:** The enclosed SOFTWARE is licensed only to you and may not be transferred to any one else without the prior written consent of the Company. Any unauthorized transfer of the SOFTWARE shall result in the immediate termination of this Agreement.

6. **TERMINATION:** This license is effective until terminated. This license will terminate automatically without notice from the Company and become null and void if you fail to comply with any provisions or limitations of this license. Upon termination, you shall destroy the Documentation and all copies of the SOFTWARE. All provisions of this Agreement as to warranties, limitation of liability, remedies or damages, and our ownership rights shall survive termination.

7. **MISCELLANEOUS:** This Agreement shall be construed in accordance with the laws of the United States of America and the State of New York and shall benefit the Company, its affiliates, and assignees.

8. **LIMITED WARRANTY AND DISCLAIMER OF WARRANTY:** The Company warrants that the SOFTWARE, when properly used in accordance with the Documentation, will operate in substantial conformity with the description of the SOFTWARE set forth in the Documentation. The Company does not warrant that the SOFTWARE will meet your requirements or that the operation of the SOFTWARE will be uninterrupted or error-free. The Company warrants that the media on which the SOFTWARE is

delivered shall be free from defects in materials and workmanship under normal use for a period of thirty (30) days from the date of your purchase. Your only remedy and the Company's only obligation under these limited warranties is, at the Company's option, return of the warranted item for a refund of any amounts paid by you or replacement of the item. Any replacement of SOFTWARE or media under the warranties shall not extend the original warranty period. The limited warranty set forth above shall not apply to any SOFTWARE which the Company determines in good faith has been subject to misuse, neglect, improper installation, repair, alteration, or damage by you. EXCEPT FOR THE EXPRESSED WARRANTIES SET FORTH ABOVE, THE COMPANY DISCLAIMS ALL WARRANTIES, EXPRESS OR IMPLIED, INCLUDING WITHOUT LIMITATION, THE IMPLIED WARRANTIES OF MERCHANTABILITY AND FITNESS FOR A PARTICULAR PURPOSE. EXCEPT FOR THE EXPRESS WARRANTY SET FORTH ABOVE, THE COMPANY DOES NOT WARRANT, GUARANTEE, OR MAKE ANY REPRESENTATION REGARDING THE USE OR THE RESULTS OF THE USE OF THE SOFTWARE IN TERMS OF ITS CORRECTNESS, ACCURACY, RELIABILITY, CURRENTNESS, OR OTHERWISE.

IN NO EVENT, SHALL THE COMPANY OR ITS EMPLOYEES, AGENTS, SUPPLIERS, OR CONTRACTORS BE LIABLE FOR ANY INCIDENTAL, INDIRECT, SPECIAL, OR CONSEQUENTIAL DAMAGES ARISING OUT OF OR IN CONNECTION WITH THE LICENSE GRANTED UNDER THIS AGREEMENT, OR FOR LOSS OF USE, LOSS OF DATA, LOSS OF INCOME OR PROFIT, OR OTHER LOSSES, SUSTAINED AS A RESULT OF INJURY TO ANY PERSON, OR LOSS OF OR DAMAGE TO PROPERTY, OR CLAIMS OF THIRD PARTIES, EVEN IF THE COMPANY OR AN AUTHORIZED REPRESENTATIVE OF THE COMPANY HAS BEEN ADVISED OF THE POSSIBILITY OF SUCH DAMAGES. IN NO EVENT SHALL LIABILITY OF THE COMPANY FOR DAMAGES WITH RESPECT TO THE SOFTWARE EXCEED THE AMOUNTS ACTUALLY PAID BY YOU, IF ANY, FOR THE SOFTWARE.

SOME JURISDICTIONS DO NOT ALLOW THE LIMITATION OF IMPLIED WARRANTIES OR LIABILITY FOR INCIDENTAL, INDIRECT, SPECIAL, OR CONSEQUENTIAL DAMAGES, SO THE ABOVE LIMITATIONS MAY NOT ALWAYS APPLY. THE WARRANTIES IN THIS AGREEMENT GIVE YOU SPECIFIC LEGAL RIGHTS AND YOU MAY ALSO HAVE OTHER RIGHTS WHICH VARY IN ACCORDANCE WITH LOCAL LAW.

ACKNOWLEDGMENT

YOU ACKNOWLEDGE THAT YOU HAVE READ THIS AGREEMENT, UNDERSTAND IT, AND AGREE TO BE BOUND BY ITS TERMS AND CONDITIONS. YOU ALSO AGREE THAT THIS AGREEMENT IS THE COMPLETE AND EXCLUSIVE STATEMENT OF THE AGREEMENT BETWEEN YOU AND THE COMPANY AND SUPERSEDES ALL PROPOSALS OR PRIOR AGREEMENTS, ORAL, OR WRITTEN, AND ANY OTHER COMMUNICATIONS BETWEEN YOU AND THE COMPANY OR ANY REPRESENTATIVE OF THE COMPANY RELATING TO THE SUBJECT MATTER OF THIS AGREEMENT.

Should you have any questions concerning this Agreement or if you wish to contact the Company for any reason, please contact in writing at the address below.

Robin Short
Prentice Hall PTR
One Lake Street
Upper Saddle River, New Jersey 07458